SPECIAL RELATIVITY, ELECTRODYNAMICS, AND GENERAL RELATIVITY

SPECIAL RELATIVITY, ELECTRODYNAMICS, AND GENERAL RELATIVITY

From Newton to Einstein

SECOND EDITION

JOHN B. KOGUT
University of Maryland, College Park, MD, USA

ELSEVIER

ACADEMIC PRESS
An imprint of Elsevier

Academic Press is an imprint of Elsevier
125 London Wall, London EC2Y 5AS, United Kingdom
525 B Street, Suite 1800, San Diego, CA 92101-4495, United States
50 Hampshire Street, 5th Floor, Cambridge, MA 02139, United States
The Boulevard, Langford Lane, Kidlington, Oxford OX5 1GB, United Kingdom

Library of Congress Cataloging-in-Publication Data
A catalog record for this book is available from the Library of Congress

British Library Cataloguing-in-Publication Data
A catalogue record for this book is available from the British Library

ISBN: 978-0-12-813720-8

Publisher: John Fedor
Acquisition Editor: Anita A. Koch
Editorial Project Manager: Amy M. Clark
Production Project Manager: Bharatwaj Varatharajan
Designer: Christian J. Bilbow

Typeset by TNQ Books and Journals

CONTENTS

ABOUT THE BOOK COVER

The book cover presents pictures of five of the most famous physicist–mathematicians of history. On the left-hand side we find Carl Friedrich Gauss (1777–1855) and above him Bernhard Riemann (1826–66). On the right-hand side we have Isaac Newton (1642–1727) and above him James Clerk Maxwell (1831–79). And at the pinnacle, Albert Einstein (1879–1955). The figure depicts two parallel developments, mathematics on one side and physics on the other, that culminated with the founding of modern physics: Gauss and Riemann developed the mathematics critical for the expressions of the physical laws discovered by Newton and Maxwell, all of which led to the grand synthesis: Einstein's theories of special and general relativity. This book focuses on the concepts these great scientists developed. It shows which ones have survived the tests of time (locality—interactions occur at space–time points, causality—information travels at the speed of light, and covariance—physical laws should hold in all frames of reference) and which have fallen over time (action at a distance, nonlocal conservation laws). With a minimum of technical trickery but a large dose of fundamentals, the book develops special relativity, electromagnetism, and general relativity from these basic principles.

PREFACE

1. SPECIAL RELATIVITY WITH A MISSION

This book is a revision and extension of "Introduction to Relativity." That book presents an approach to relativity in which the theory is developed through "thought experiments" that illustrate the concepts in simple fashions. For example, to derive time dilation and Lorentz contraction, a clock is constructed out of mirrors and a light beam, and we learn how it works in its rest frame and in a frame in which it moves with velocity **v**.

This analysis is simple and direct because light beams and rods are governed by the postulates of special relativity:

1. The laws of physics are the same in all inertial frames of reference.
2. There is a common finite speed limit c in all inertial frames.

Throughout the book the rules of nonrelativistic physics (Newton's world) are contrasted with relativistic physics (Einstein's world). It is the second postulate that distinguishes Newtonian rules from Einstein's. In Newton's world, velocities are unbounded but in Einstein's world, the speed of light c is an upper bound. The existence of a speed limit combined with Postulate 1, that all inertial frames are equivalent, produces relativistic space—time.

The book shows that time dilation, Lorentz contraction, and the relativity of simultaneity, that clocks which are separated in space and synchronized in one frame are *not* synchronized in a frame in relative motion, are intimately related.

The book builds the subject from the ground up. It uses Minkowski space—time diagrams, which show the spatial and temporal coordinates of two frames in relative motion v. Time dilation, Lorenz contraction, and the relativity of simultaneity can be understood through simple pictures (cartoons, really), both qualitatively and quantitatively. From this perspective, we study the twin paradox and see that there is nothing really paradoxical about it—when one twin leaves the other and takes a round trip, she returns younger than her sibling. It is fun to see how it works out!

Once we have mastered kinematics—the measurement of space and time intervals—we move to dynamics, the study of energy, momentum, interactions and equations of motion. In Newtonian mechanics there is momentum conservation, on the one hand, and mass conservation, on the other. In relativity, however, once we have figured out how spatial and temporal measurements are related in inertial frames, the properties of momentum, energy, and mass are determined. We present another of Einstein's famous thought experiments that show that energy and inertia (mass) are two aspects of one concept, relativistic energy, and find $E = mc^2$. The properties of relativistic momentum then follow from the properties of spatial and temporal measurements and we learn that momentum and energy conservation must go hand in hand.

Since Newton's second law, force equals mass time acceleration, is not consistent with the existence of a speed limit, we must modify it and invent a new force law that is relativistic. We do this aided, again, by the original thought experiments of the masters, Einstein and Max Born. In this case, a close look at an inelastic collision in several frames of reference leads us to relativistic momenta and energy. From there we obtain the relativistic version of Newton's second law. We then consider high-energy collisions between particles and illustrate how energy can be converted into mass and how mass can be converted into energy. These illustrations show that Newton's concept of the integrity of a particle is violated in relativity. There is no way in relativity to restrict one's attention to the "single particle sector." The fundamentals of relativity simply do not allow it. This fact has consequences beyond the scope of this book and finds its fruition in relativistic field theories of particle physics.

In the first edition of this book we were satisfied with introductory work on relativistic dynamics. In this extended book we have greater ambitions. We will start with electrostatics, that there is a static force which falls off as the square of the distance between charges. We will use this fact coupled with special relativity to deduce:

1. The existence of magnetism, a field that is generated by electric currents,
2. The Lorentz force law, which states that particles are deflected by electric fields and also by magnetic fields in which case the force is proportional to the particle's velocity,
3. Maxwell's equations: the Ampere—Maxwell equation predicting how currents produce magnetic fields and Faraday's law, which predicts how changing magnetic fields produce electric forces (electromotive forces),
4. The wave equation for electromagnetism and the demonstration that light travels at the speed limit c of special relativity in all inertial reference frames.

The book emphasizes that magnetism is a consequence of relativity—that Newton's world, a world without a speed limit, has no place for this phenomenon. In addition, magnetism produces velocity-dependent forces on charged particles that do not satisfy Newton's third law, that action equals reaction. Once we understand that Newton's third law fails, we must look elsewhere for an understanding of momentum conservation in dynamical interactions.

Once we have derived the Lorentz force law, we are able to deduce how electric and magnetic fields transform under boosts. We learn from carrying out this exercise that electric and magnetic fields mix under transformations between frames of reference. These effects are consequences of the fact that space and time mix under such transformations. These results illustrate one of Einstein's famous remarks that "electric forces in one frame can be interpreted as magnetic effects in another." The problem sets in this part of the book consider more traditional ways of obtaining the same results, considering capacitors and solenoids in various reference frames.

These arguments elucidate the nature of magnetism, but they do not quite get to the heart of the matter. Electrostatics states that charges are the sources of electric fields. This is stated quantitatively through Gauss' law. On the other hand, we learn that electric currents are the sources of magnetic fields. Since electric currents are produced from charge densities by Lorentz transformations ("boosts"), it must be that the equation that predicts that magnetic fields are produced from electric currents can be obtained by boosting Gauss' law. We carry out this exercise and discover Ampere—Maxwell's equation. This is Ampere's law corrected for nonstationary currents by Maxwell's displacement current. Similarly, we obtain Faraday's law from the divergence-free character of magnetic fields.

This book shows the student that a scientist can *build* the theory of electromagnetism from electrostatics and special relativity. Unlike conventional textbooks, we learn to manipulate differential equations and accomplish our goals in general fundamental terms rather than working through myriad illustrations of devices. Once we have Maxwell's equations, it is an easy task to show that relativity elevates electromagnetic fields to independent dynamical degrees of freedom, which propagate at the speed limit c.

In this journey we find it necessary to forsake Newtonian ideas and replace them with modern concepts: forces lead to fields, which become dynamical objects that propagate nonlocal conservation laws such as

Newton's law of action—reaction are thrown over and are replaced by local conservation laws, which are compatible with the relativistic notion of causality and the finiteness of the speed of propagation of information. In addition, the integrity of particle number, so crucial in Newton's world, is replaced by particle creation and destruction, which is forced upon us by the energy—mass relationship, causality and locality of relativistic interactions of relativity, etc.

Thus our introduction to electrodynamics is an alternative to the traditional approach popularized by the textbooks of J. D. Jackson [1] and his many followers. Our book starts with postulates 1 and 2 of special relativity and the primitive notion that charges are the local sources of electric fields, and develops the fundamentals of electromagnetism: unified magnetic and electric fields, the Lorentz force law, Maxwell's equations, and the wave equation. After mastering these fundamentals, the student is in a position to do more complex applications of relativity and electrodynamics. The approach in this book is simple and straightforward. The derivation of Maxwell's equations, having come from such humble beginnings in Chapter 1, should be an epiphany for the student.

2. GENERAL RELATIVITY

Our last topic in this book is the general theory of relativity, where we consider accelerated reference frames in Einstein's world. The key insight here is Einstein's version of the equivalence principle: there is no local, physical means to distinguish a uniform gravitational field from an accelerated reference frame.

This principle has many interesting forms and applications. Suppose you are on the surface of Earth and want to understand the influence of the gravitational field on your measuring sticks and clocks compared to those of your assistant who is at a greater height in the Empire State Building. Einstein suggests that the assistant jump out the window, because in a freely falling frame all effects of gravity are eliminated and we have a perfectly inertial environment where special relativity holds to arbitrary precision! During his descent, your assistant can make measurements of clocks and meter sticks fixed at various heights along the building and measure how their operation depends on their gravitational potential. We pursue ideas like this one in the book to derive the gravitational redshift, the fact that clocks close to stars run more slowly than those far away from stars; the resolution of the twin paradox as a problem in accelerating reference frames; and the bending of light by gravitational fields.

A fascinating aspect of the equivalence principle is its universality, which becomes particularly clear when we calculate the bending of light as a ray glances by the Sun. Why does the light ray feel the presence of the mass of the Sun? The equivalence principle states that an environment with a uniform gravitational field is equivalent to an environment in an accelerating reference frame—explicit acceleration clearly affects the trajectory of light and any other physical phenomena. So gravity becomes a problem in accelerated reference frames, which is just a problem in coordinate transformations, which is an aspect of geometry! We show that this problem in geometry must be done in the context of four-dimensional space—time, our world of Minkowski diagrams. Einstein's theory of gravity brought modern geometry, the study of curved spaces, into physics forever.

As long as we concentrate on uniform gravitational fields of ordinary strength, we are able to use the equivalence principle to make reliable, accurate predictions. The equivalence principle reduces gravity to an apparent force, much like the centrifugal or Coriolis forces that we feel when riding on a carousel. In fact, we use relativistic turntables and rotating reference frames as an aide to studying and deriving relativistic gravitational effects.

However, we are also interested in strong non-uniform gravity, situations where the equivalence principle is not sufficient to describe the physics. We discuss tidal forces, which are at the heart of the dynamics of general relativity. This is done both within Newton's world and general relativity where we learn that tidal forces and the curvature of space—time are two sides of the same coin. Since curvature is the essence of the dynamics of general relativity, we review classical differential geometry and learn about the curvature of curves and the curvature of two-dimensional surfaces. Gaussian curvature is introduced and the fundamental theorems of classical differential geometry are derived and discussed.

These developments lead to a discussion of modern general relativity. The critical symmetries of special relativity were Lorentz transformations (boosts) and in general relativity these symmetries are elevated to local symmetries, general coordinate transformations. The Riemann tensor is introduced as a measure of the local, intrinsic curvature of space—time. These ideas lead to the central idea of general relativity, that local energy momentum produce local space—time curvature. This idea is expressed quantitatively through Einstein's field equation, which is introduced as a grand extension of Newton's equation for the gravitational potential. We solve these equations in the case of a static, symmetric mass distribution, which gives rise to the Schwarzschild metric. This metric contains a black hole, which we study in considerable detail. We find that there is an "event horizon," the Schwarzschild radius, inside of which there is no escape. We also consider the production and propagation of gravitational waves. We see that gravity waves are described by fluctuations of the space—time metric that travel at the speed limit of special relativity and are generated by oscillating quadrupole moments of mass distributions. We review the "Advanced LIGO" experiment that discovered gravity waves generated from the violent merger of two black holes over one billion light-years away from earth.

We end our discussion of general relativity by comparing its fundamentals with modern theories of elementary particles, which are also based on local symmetry groups, as well as a look at current puzzles and unsolved problems, such as dark energy.

3. BACKGROUND READING AND RECOMMENDATIONS FOR FUTURE READING

Although the perspective of this book is its own, it owes much to other presentations. The influence of A. P. French's 1968 book [2] *Special Relativity* is considerable, and references throughout the text indicate where my discussions follow his. To my knowledge, this is the finest textbook written on the subject because it balances theory and experiment perfectly. The reader will find discussions of the Michelson—Morley experiment and early tests of relativity there. French's discussions of energy and momentum are reflected in my later chapters, and his problem sets are a significant influence on those included here. The present book goes beyond French's when we derive the Lorentz transformation laws for the electric and magnetic field and then deduce all of Maxwell's equations by boosting Gauss' law. We include a shortened version of French's demonstration that the magnetic attraction between parallel current carrying wires is due to a minuscule Lorentz contraction effect. We show that the force between such wires is strong enough to bend typical copper wires because the huge number of electrons in the wire multiplies the tiny Lorentz contraction they experience due to their tiny velocities relative to the wire's stationary Cu+ ions. This observation explains why magnetic forces can be substantial under ordinary lab conditions even though magnetism is a pure relativistic effect. These numbers underlie how relativity was discovered at the turn of the 20th century, long before the modern era of particle accelerators where particle velocities close to the speed limit were possible.

Another influential book is *The Classical Theory of Fields* by L. D. Landau and E. M. Lifshitz [3]. Unlike most books on electromagnetism, which follow the historical development of the field and introduce relativity toward the end, Landau and Lifshitz start with relativity and build the theory of relativistic particles and light. The emphasis is on fundamental principles rather than applications. Unfortunately, the book is aimed at advanced students and most editions are flush with typographical errors. Nonetheless, the book is very inspiring and has unique insights into the subject. Several of the more challenging problems in chapters 7–10 in this volume were inspired by Landau and Lifshitz. This book also introduces general relativity and includes derivations of the Schwarzschild metric, gravitational waves, the bending of light rays in a gravitational field, etc. The discussions are brief and powerful, the essence of Landau's brilliant approach to physics.

The exposition by N. D. Mermin [4] influenced several discussions of the paradoxes of special relativity. This book is also recommended to the student because it shows a condensed matter physicist learning the subject and finding a comfort level in it through thought-provoking analyses that avoid lengthy algebraic developments. The huge book by J. A. Wheeler and E. F. Taylor [5] titled *Spacetime Physics* inspired several of our discussions and problem sets. This book, a work that only the unique, creative soul of John Archibald Wheeler could produce, is recommended for its leisurely, interactive, thought-provoking character. Finally, the books by W. Rindler [6,7], a pioneer in modern general relativity, are also recommended. His book *Essential Relativity* [7] is a solid introduction to general relativity rooted in the era of Einstein and the pioneers. After the student has mastered electricity and magnetism and Lagrangian mechanics, he or she could tackle *The Classical Theory of Fields* by L. D. Landau and E. M. Lifshitz [3] and *Essential Relativity* [7].

The future of research in relativity and field theory is bright. Hopefully this book will spark some interest in its future practitioners.

"Time travels in diverse paces with diverse persons," Rosalind, from Act 3, Scene 2, "As You Like It" by William Shakespeare.

"Mathematics allows you to exceed your imagination," Lev Landau, Moscow, 1952.

"Magic!", student, anonymous, 2018.

REFERENCES

[1] J.D. Jackson, Classical Electrodynamics, John Wiley & Sons, New York, 1962.
[2] A.P. French, Special Relativity, W. W. Norton, New York, 1968.
[3] L.D. Landau, E.M. Lifshitz, The Classical Theory of Fields, Pergamon Press, Oxford, 1962.
[4] N.D. Mermin, Space Time in Special Relativity, Waveland Press, Prospect Heights, IL, 1968.
[5] E.F. Taylor, J.A. Wheeler, Spacetime Physics, W. H. Freeman, New York, 1992.
[6] W. Rindler, Introduction to Special Relativity, Oxford University Press, Oxford, 1991.
[7] W. Rindler, Essential Relativity, Springer-Verlag, Berlin, 1971.

CHAPTER 1

Physics According to Newton—A World With No Speed Limit

Contents

1.1 NEWTON'S WORLD: LAWS AND MEASUREMENTS

When you set up a problem in Newtonian mechanics, you choose a reference frame. This means that you set up a three-dimensional coordinate system, for example, so any point $\boldsymbol{r}$ can be labeled with an x measurement of length, a y measurement, and a z measurement, $\boldsymbol{r} = (x, y, z)$. In addition, you place clocks at convenient points in the coordinate system so you can make measurements and record when and where they occurred.

Newton imagined carrying out experiments on a mass point m in this coordinate system. He imagined that the coordinate system was far from any external influences, and under those conditions he claimed, on the basis of the experiments of Galileo and others, that the mass point would move in a straight line at a constant velocity. Newton labeled such a frame of reference "inertial." This idea is codified in Newton's first law:

Law 1. A body in an inertial reference frame remains at rest or in uniform motion unless acted on by a force.

Newton and others realized that there must be a wide class of inertial frames. If we discovered one frame that was inertial, then Newton argued that other inertial reference frames could be generated by

1. Translation—move the coordinate system to a new origin and use that system.
2. Rotation—rotate the coordinate system about some axis to a fixed, new orientation.
3. Boosts—consider a frame moving at velocity v with respect to the first.

Properties 1 and 2 are referred to as the uniformity and isotropy of space. Property 3 is referred to as Galilean invariance: the Galilean boost

Special Relativity, Electrodynamics, and General Relativity
ISBN 978-0-12-813720-8
https://doi.org/10.1016/B978-0-12-813720-8.00001-5

is a symmetry—it maps the physical system to another inertial reference frame, and this process has no physical impact on the system.

Newton's concept of relativity can now be stated: All inertial frames are physically equivalent. This means that the laws of physics are the same in all of them.

The reader should be aware that Newton and his colleagues argued constantly about these points. What is the origin of inertia? Why are inertial frames so special? Can't we generalize the Properties 1–3 to a wider class of reference frames? We will not discuss these historical issues here, but the reader might want to pursue them for a greater interesting perspective. Our approach is complementary to the traditional one and is forward looking rather than historical.

Underlying Newtonian mechanics are his era's concepts of space and time. To measure the distance between points, we imagine a measuring rod. Starting from an origin (0, 0, 0), we lay down markers in the x and y and z directions so we can measure a particle's position $\boldsymbol{r} = (x, y, z)$ in this inertial frame. Next we need a clock. A simple device such as a simple harmonic oscillator will do. Take a mass point m on the end of a spring, and let it execute periodic motion back and forth. Make a convention that one unit of time passes when the mass point goes through one cycle of motion. In this way we construct a clock and measure speeds of other masses by noting how far they move in several units of time—in other words, we compare the motion of our "standard" simple harmonic oscillator with the motion of experimental particles. Note that this is just what we mean by time in day-to-day situations. To make an accurate clock, we need one with a sufficiently short unit of time, or period. Clocks based on the inner workings of the atom can be used in demanding, modern circumstances.

Just as it was convenient to place distance markers along the three spatial axes $\boldsymbol{r} = (x, y, z)$, it is convenient to place clocks on the spatial gridwork. This will make it easy to measure velocities of moving particles—we just record the positions and times of the moving particle on our grid of measuring rods and clocks. When the moving particle is at the position $\boldsymbol{r}$ and the clock there reads a time t, we record the event. Measuring two such events allows us to calculate the particle's velocity in this frame at and near that spatial position. In setting up the grid of clocks, we must synchronize them so we can obtain meaningful time differences. This is easy to do. Place a clock at the origin and one at $\boldsymbol{r} = (1, 0, 0)$. Then, at the halfway point between them, place a beacon that sends out a signal in all directions. Because space is homogeneous and isotropic, the signal travels at the same speed toward both clocks. Set the clocks to zero, say, when they both receive the signal. The two clocks are synchronized, and we did not even need to know the speed of the signal emitted from the beacon. Clearly, we could use this method to synchronize all the clocks on the grid.

Now we are ready to do experiments involving space–time measurements in this frame of reference. Call this frame S. It consists of the gridwork of measuring rods and clocks, all at rest, with respect to each other. Now, suppose we want to compare our experimental results with those obtained by a friend of ours at rest in another frame that moves at constant velocity $\boldsymbol{v} = (v_x, 0, 0)$ with respect to us. According to our postulates, his or her measurements are as good as ours and all our physical laws can be written in his or her frame of reference S' without any change (Fig. 1.1).

The measuring rods in S' are identical to those in S; the clocks in S' are also identical to those in S. Suppose that the origin of S and S' coincide when the clock at the origin in S reads time $t = 0$ and the clock at the origin in S' reads $t' = 0$. Now for the crucial question: Do the other grid markings and clocks at those markings also agree in the two frames S and S'? There are two distinct physics issues to consider here. The first is the operation of the clocks and rods in each frame. Following Newton's principle of relativity, that all inertial reference frames are equivalent, all the clocks and rods in S work exactly the same as those at rest in S'. We need only know that both frames are inertial—the relative velocity between the frames is physically irrelevant to the dynamics within each. The second issue is the physical mechanism by which we can transmit the information in one grid of rods and clocks to the other at relative velocity $\boldsymbol{v}$. In Newton's world, objects can move with unbounded relative velocities relative to one another, as we shall discuss further in Eq. (1.2). Accordingly, signals and information can be transmitted at unbounded velocities, essentially

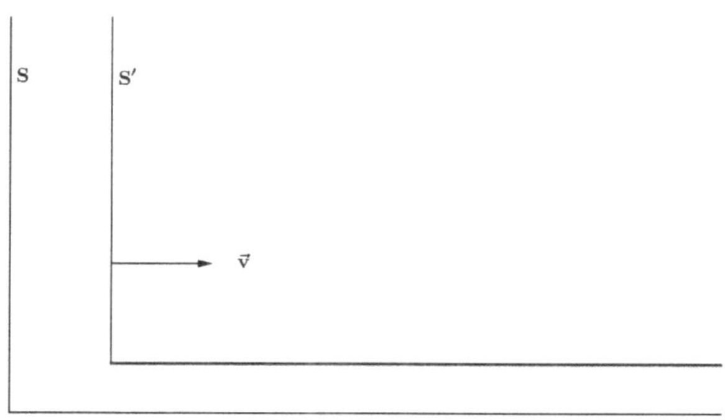

Figure 1.1 Two inertial frames of reference in relative motion.

instantaneously. Therefore, the spatial gridwork and the times on each clock in S' can be instantaneously broadcast to the corresponding rods and clocks in the frame S. Therefore, the gridwork of coordinates and times in both frames must be identical, even though they are in relative motion. So, the lengths of measuring rods and the rates of clocks are independent of their relative velocities. We therefore need only one measuring rod and one clock at one point in one inertial frame, and we know the positions and times of all measuring rods and clocks in any other inertial frame. For the purposes of this book, this serves as the meaning of "absolute space" and "absolute time" in Newtonian mechanics. We need not hypothesize about these notions from a philosophical basis, as was done historically, but can deduce them from the fact that in Newtonian mechanics there is no speed limit—information can be transmitted instantaneously.

Now consider the rule by which times and position measurements are compared between the two frames, S and S', in relative motion in Newton's world,

$$\begin{aligned} x &= x' + vt \\ y &= y' \\ z &= z' \\ t &= t', \end{aligned} \tag{1.1}$$

where x, y, z, and t are measurements in S and x', y', z', and t' are the corresponding measurements in S'. For example, if we measure the position of a particle in S' to be x', y', and z', at time t', then the coordinates of this event (measurement) in frame S are given by Eq. (1.1). These relations, which are so familiar and "obvious," are called Galilean transformations. Note that the first of them, $x = x' + vt$, states that the position of the particle at time t consists of two pieces: (1) the distance vt between the origins of the two frames at time t and (2) the distance x' from the origin to the particle in the frame S'. This rule contains the notion "absolute" space—it uses the fact that in a Newtonian world the distance x' in the frame S' is also measured as x' in the frame S. The last equation in Eq. (1.1) is the statement of "absolute" time. Note also that if the particle has a velocity v'_p with respect to the frame S' in the x' direction, then $x' = v'_p t' = v'_p t$, and its position in frame S is $x = \left(v'_p + v\right)t$, so its velocity relative to the origin of frame S is v_p,

$$v_p = v'_p + v, \tag{1.2}$$

which is the familiar rule called "addition of velocities." It rests on Newton's ideas of absolute space and time. It will be interesting indeed to see how the results from Eqs. (1.1) and (1.2) are different in Einstein's world!

We can go slightly further than Eq. (1.2) by considering a sequence of frames. Let S_1 move at velocity v_{10} along the x-axis with respect to S, let S_2 move at velocity v_{21} along the x-axis with respect to S_1. Then the velocity of a particle in S, $v_{p,0}$, is related to its velocity measured in frame S_2 by adding up the relative velocities,

$$v_{p,0} = v_{p,2} + v_{2,1} + v_{1,0} \tag{1.3}$$

There is nothing stopping us from generalizing this process again and again to S_n frames until $v_{p,0}$ is a velocity as large as one wishes. In other words, velocities in Newton's world are unbounded because of the addition law for velocities.

Perhaps, it has not occurred to the reader before that information could be transmitted instantaneously in a Newtonian world. Actually, if the reader recalls some elementary problems in dynamics, the assumption was lurking just under the surface.

Suppose you consider the collision of a point particle and a rod, as shown in Fig. 1.2.

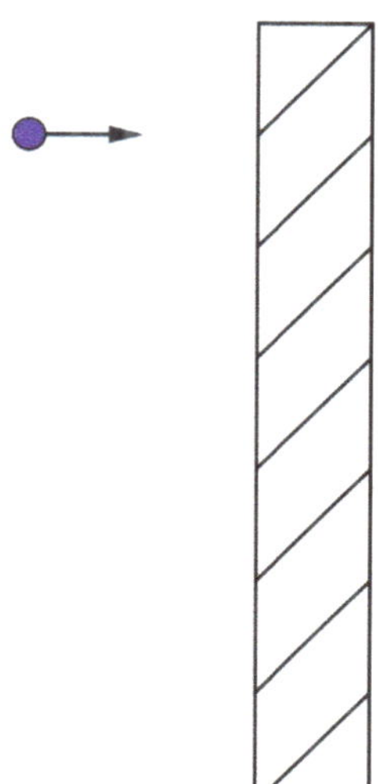

Figure 1.2 Collision between a particle and the upper edge of a rod.

When you solve this problem using Newtonian mechanics using conservation of linear and angular momentum, you implicitly assume that the impact at the upper edge of the rigid rod instantaneously accelerates the center of mass of the rod and begins rotating the bottom end of the rod to the left. If information was transmitted at a finite velocity, call it c, the lower edge of the rod could not know about the collision until a later time, $t = \ell/c$ where ℓ is the length of the rod. It is clear that the very notion of rigid bodies requires a theoretical basis where information is transmitted instantaneously. Rigid bodies cannot exist in Einstein's world. Of course, in practical problems where ℓ is a few meters, say, and c is enormous, $c \approx 3.0 \cdot 10^8$ m/s, the time delay encountered here is usually considered negligible. (But it certainly would be significant if ℓ were the diameter of a star.) Nonetheless, the matter of principle is our major concern in these discussions.

Another place where instantaneous transmission of information occurs in Newtonian mechanics problems is in the use of potentials. When we have two bodies interacting through a mutual force that is described through a potential that depends on the distance between the particles, we are assuming that the force on each particle is determined by the relative position of the other particle at that exact moment. No account is given of the transmission of that information between the separate particles. This Newtonian notion is called "action at a distance," and it underlies some of the greatest successes of nonrelativistic mechanics. We will see that it is absolutely impossible in Einstein's world.

Now let us move on to discussions of Newton's second and third laws. The second law introduces dynamics:

Law 2. If a particle of mass m is subject to a force $\boldsymbol{f}$, then it experiences an acceleration $\boldsymbol{a}$ given by,

$$\boldsymbol{f} = m\boldsymbol{a}. \tag{1.4}$$

In slightly more generality,

$$\boldsymbol{f} = \frac{d}{dt}\boldsymbol{p} \tag{1.5}$$

where $\boldsymbol{p}$ is the particle's momentum, $\boldsymbol{p} = m\boldsymbol{v}$. Before discussing the second law further, it helps to introduce Newton's third law:

Law 3. When one particle exerts a force on a second particle, the second one exerts an equal and opposite force on the first.

This law is the basis of a conservation law: the conservation of momentum. Consider two interacting particles. Denote the force of particle 1 on particle 2 $\boldsymbol{F}_{12}$ and the force 2 on 1 $\boldsymbol{F}_{21}$. The third law states that,

$$\boldsymbol{F}_{12} = -\boldsymbol{F}_{21} \tag{1.6}$$

Notice that this is a vector equation: the forces are equal in magnitude but opposite in direction. We can see that the total momentum of the system, $\boldsymbol{p}_1 + \boldsymbol{p}_2$ is constant in time,

$$\frac{d}{dt}(\boldsymbol{p}_1 + \boldsymbol{p}_2) = \frac{d}{dt}\boldsymbol{p}_1 + \frac{d}{dt}\boldsymbol{p}_2 = \boldsymbol{F}_{21} + \boldsymbol{F}_{12} = \boldsymbol{F}_{21} - \boldsymbol{F}_{21} = 0 \tag{1.7}$$

Conservation laws are fundamental to all subfields of physics. Newton's third law is an essential ingredient in mechanics. But it is important to note something very perplexing about the third law: it is a *nonlocal* conservation law. The forces of action and reaction cancel as stated in Eq. (1.7), but the action and reaction forces act on *different* particles, which could be very far from each other if the force acts over long distances such as electrostatics or gravity. Eq. (1.7) makes the remarkable statement that at every instant of time t, particle 1 responds to the motion of particle 2 just so the sum of their momenta is a constant even if the particles are thousands of meters apart. This is possible in a causal world only if information travels instantaneously. Clearly the third law will not survive in this form in Einstein's world where information cannot travel faster than c, the speed of light. The implications of the failure of the third law will be more profound than we can imagine at this point in our studies.

The mass m in Eq. (1.4) is clarified in part by the third law, the law of action–reaction, which implies in the case of two interacting particles that

$$m_1\boldsymbol{a}_1 = -m_2\boldsymbol{a}_2. \tag{1.8}$$

So, if the first body is chosen to set the scale for inertia, in other words, if we define $m_1 \equiv 1$, then m_2 is determined from Eq. (1.8). The third law provides an operational meaning to the "mass" or "inertia" of a particle.

Are the laws of dynamics, Newton's second and third laws, compatible with Newton's principle of relativity, that all inertial reference frames are physically equivalent? The key observation is that acceleration is a "Galilean invariant," which means that an acceleration $\boldsymbol{a}$ is the same in all inertial reference frames. This follows from Eq. (1.2). Differentiate it with respect to time, use the fact that v is a constant and learn that,

$$\boldsymbol{a}_p = \boldsymbol{a}'_p \tag{1.9}$$

We learn that forces are Galilean invariants. We are using the fact that masses are Galilean invariant in Newton's world to make this observation. This property was assumed in Newton's era and was verified experimentally to reasonable precision at modest velocities. We shall see that masses are not invariant in Einstein's world, and the implications of this fact will prove to be very significant.

We have dealt with these issues very explicitly because they will help us appreciate special relativity, where there is a speed limit, the speed of light. In all the rules of Newtonian mechanics, the rules of how things work, the second and third laws, do not distinguish between inertial frames. The dynamics do satisfy a principle of relativity. The difference between the two theories comes from the fact that one has a speed limit and the other does not. This affects how information is shared between frames. It also affects the dynamics within each frame—Einstein's form of Newton's "force equals mass times acceleration" is different because it must not permit velocities greater than the speed limit. But each theory is consistent within its own rules.

1.2 NEWTON'S WORLD: NO PLACE FOR MAGNETISM

It is instructive to return to our example of electrostatics and consider Newtonian dynamics in more detail. The goal to this discussion is to demonstrate that if you accept Newton's laws 1–3 and you know Coulomb's law, then you can rule out the very existence of magnetism!

What do we mean by magnetism in this context? We accept the experimental fact that electric charges produce electrostatic forces. We will write down Coulomb's law and will manipulate it and study it. Magnetic forces are, by definition, those forces produced by electric currents. For example, if a thin wire carries a current I, this means that electrons are flowing in the wire at a rate of I Coulomb's per second passing a fixed point along the wire. The wire is uncharged but electrons flow along it. We know from our experiences in the lab, perhaps making an electromagnet and bending charged particle trajectories with it, that there is a magnetic field in the lab and its strength is proportional to the current I. In addition, if a single charge such as a free electron, whose charge we label q and measure in units of coulombs, has a fixed position $\boldsymbol{r}$ in a reference frame S then it produces a Coulomb field emanating from $\boldsymbol{r}$. If the electron is viewed from the perspective of a frame S', which moves with velocity $\boldsymbol{v}$ along the x axis, then the electron has a velocity $-\boldsymbol{v}$ in this frame so it also has a current, $-q\boldsymbol{v}$, in S'. From our experiences in the lab, we know that there will be a magnetic field in S' whose strength is proportional to $-q\boldsymbol{v}$.

Let us see where Newton gets into trouble with these simple observations. Let us return to our example of electrostatics. Let particle 1 be at rest at the origin $\boldsymbol{r} = (0,0,0)$ in a frame S. The electric field generated by the charge q_1 is,

$$\boldsymbol{E}(\boldsymbol{r}) = k\frac{q_1}{\boldsymbol{r}^2}\widehat{\boldsymbol{r}} \tag{1.10}$$

where k is constant determined experimentally. The electric field is introduced so the force on charge q_2 at position $\boldsymbol{r} = \boldsymbol{r}_2$ can be written as,

$$\boldsymbol{F} = q_2\boldsymbol{E}(\boldsymbol{r}_2) \tag{1.11}$$

As we will motivate in later chapters, these equations mean that the charge q_1 at the origin generates a static electric field $\boldsymbol{E}(\boldsymbol{r})$ around it that produces a force on other charged particles. Now suppose that the charge q_2 has a velocity $\boldsymbol{v}_2(t)$ in frame S. At time t q_2 is at position $\boldsymbol{r}_2(t)$ and experiences the electric field there. So, the force it experiences becomes time dependent,

$$\boldsymbol{F}(t) = q_2\boldsymbol{E}(\boldsymbol{r}_2(t)) \tag{1.12}$$

According to Newton's third law, the charge q_1 experiences a force $-\boldsymbol{F}(t)$. We can view the force in the following way: the moving charge q_2 generates an electric field of magnitude $kq_2/|\boldsymbol{r} - \boldsymbol{r}_2(t)|^2$ in the direction $\boldsymbol{r}-\boldsymbol{r}_2(t)$. Multiplying this field by q_1 gives the force on particle 1.

We learn from all this that the force between particle 1 and particle 2 is independent of the velocities of either particle and depends only on the instantaneous positions of each particle.

$$\boldsymbol{F}_{12}(t) = k\frac{q_1q_2}{|\boldsymbol{r}_1(t) - \boldsymbol{r}_2(t)|^2}\widehat{\boldsymbol{r}}_{12}(t) \tag{1.13}$$

So, in Newton's world, the force between charge q_1 and q_2 depends only on the particles' instantaneous positions and not on their velocities. This is a very perplexing result because, as promised, it rules out the existence of magnetism in Newton's world. This vital flaw will be cured by special relativity. Special relativity and Coulomb's law will be shown to imply magnetic forces of just the sort observed in the lab and will, in fact, produce all of Maxwell's equations, and the wave equation for electromagnetic waves that travel at the speed limit of relativity, c.

To emphasize the failure of Newton's laws to accommodate magnetism, consider an even simpler example that does not use the third law. Imagine two charges q_1 and q_2 moving parallel to one another with a common velocity $\boldsymbol{v}$ as shown in Fig. 1.3. We want to calculate the force between them. To do this, we view the charges from the perspective of their rest frame. In this frame, we know the answer immediately: Coulomb's law applies and the particles experience a force that acts on the line between them and that falls off as the square of the distance between them. This produces a formula for each particle's acceleration. But in Newton's world, accelerations are independent of the inertial frame in which they are calculated. Therefore, the particles experience the same acceleration $\boldsymbol{a}$ in the frame S where they are moving and are producing currents. This demonstrates that these currents do not effect the acceleration of the charges, and magnetic forces are ruled out yet again.

In Chapter 9, we will see that the origin of magnetic forces and the magnetic field $\mathbf{B}$ is the space–time of special relativity. We will go further and derive all of electromagnetism (Maxwell's laws, including Ampere's law of magnetic fields and the displacement current and Faraday's law of electromagnetic induction) from special relativity and Coulomb's law of electrostatics. From there we shall see that electromagnetic waves exist and travel with a universal speed limit, c. But before we head off in that direction let us learn the basics of the physics of space–time measurements according to Einstein.

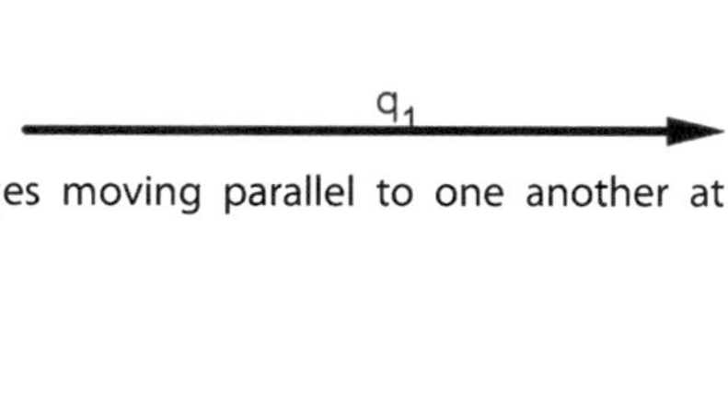

Figure 1.3 Two charges moving parallel to one another at a common velocity in the lab.

CHAPTER 2

Space–Time Measurements According to Einstein

Contents

2.1 A WORLD WITH A SPEED LIMIT

According to Newtonian mechanics, you could place a charged particle (charge q, mass m) into a constant electric field E and accelerate it to an arbitrary velocity, $v(t) = (qE/m)t$. In addition, you could consider inertial frames in relative motion and the relative velocity could be arbitrarily large.

Unfortunately, nature does not allow such freedom. Modern accelerator experiments indicate that particles cannot be accelerated beyond a universal speed limit, which is the speed of light, $c = 2.9979\ldots\cdot 10^8$ m/s. Light is nature's fastest runner. Although historically the fact that the speed of light is the speed limit was extraordinarily important, it is not important for the logical development of the subject. Just the existence of a speed limit is all that really matters. We do not need to know anything about electromagnetism to derive time dilation and Lorentz contraction and thereby overthrow Newton's vision of absolute space and time. But it is crucial to understand that the existence of a speed limit must be compatible with Postulate 1 of relativity, that the laws of physics are the same in all inertial frames. This means that all inertial frames must find the same universal value for the speed limit through their own experiments.

In particular, suppose that an experimenter at rest in frame S' shown in Fig. 2.1 turns on a flashlight and points it to the right. Those light rays can

Special Relativity, Electrodynamics, and General Relativity
ISBN 978-0-12-813720-8
https://doi.org/10.1016/B978-0-12-813720-8.00002-7

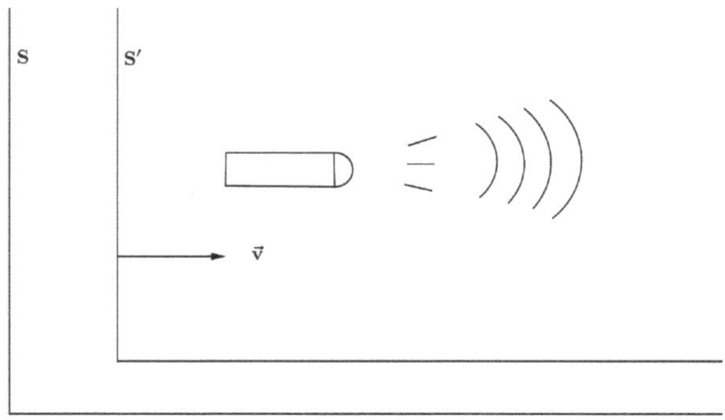

Figure 2.1 A flashlight at rest in a moving frame beams its light in its direction of motion.

be measured as traveling at the speed limit c with respect to any observer at rest in frame S'. But an experimenter in frame S must also measure the speed of the light ray to be the speed limit c, independent of v! Just for fun, take an extreme example. Choose $v/c = 0.99999999$ and point the flashlight to the left, in the direction opposing the velocity v. An observer at rest in frame S' measures the velocity of the light rays emanating from the flashlight and finds $c \approx 3 \cdot 10^8$ m/s to the left, as usual. But an observer at rest in frame S measures the velocity of light rays, too, and finds $c \approx 3 \cdot 10^8$ m/s to the left as well, instead of the value 3 m/s to the left that we would predict in Newton's world! The speed of the source of light, the flashlight, is utterly irrelevant! Both observers must measure the same speed limit because the laws of physics are identical in both of their frames. The Galilean result of Eq. (1.3), known as addition of velocities, which we have held as obvious, is just plain wrong. It will take us some time and effort to feel comfortable with the physics in Einstein's world!

In summary, the two postulates of Einstein's special relativity are as follows:

1. The laws of physics are the same in all inertial frames.
2. There is a speed limit.

The rest of this book will work out the implications of these two postulates. We begin by studying how measurements of rods and clocks are related between different inertial frames. Then we turn back to Postulate 1 and see how Newton's laws must be modified, how momentum and energy should be defined and related, to be compatible with the existence of a universal speed limit.

2.2 MAKING A CLOCK WITH MIRRORS AND LIGHT

Just as in our discussion of Newtonian measurements of the positions and times of events, we set up a gridwork of coordinates and clocks in two frames, S and S', which are both inertial and have a relative velocity $\boldsymbol{v}$ in the x direction, as depicted Fig. 2.1. We synchronize clocks in frames S and S' just as we described in the Newtonian world of Chapter 1. Our next task is to compare the positions and times on the clocks in S with those in S'.

We need to construct a clock so that we can understand its operation using just Postulates 1 and 2. Einstein thought up a simple clock whose two ingredients, a basic length and a basic velocity, are easily understood in terms of Postulates 1 and 2. Take two mirrors at rest in S' and let them be separated by a distance ℓ_o in the y' direction. Both mirrors move relative to S in the x direction at velocity $\boldsymbol{v}$ and maintain their separation ℓ_o as measured in either frame. Now let a beam of light bounce between the mirrors. Because light travels at the speed limit c, it takes a time $\Delta t' = 2\ell_o/c$ for light to travel from one mirror, call it A, to the second mirror, call it B, and then back to mirror A (see Fig. 2.2). Because the clock is at rest in frame S', it is conventional to call the interval $\Delta t'$ "proper time" and denote it $\Delta\tau = 2\ell_o/c$. These simple clocks could be the ones we place along the gridwork in the frame S' to measure the position and time of events.

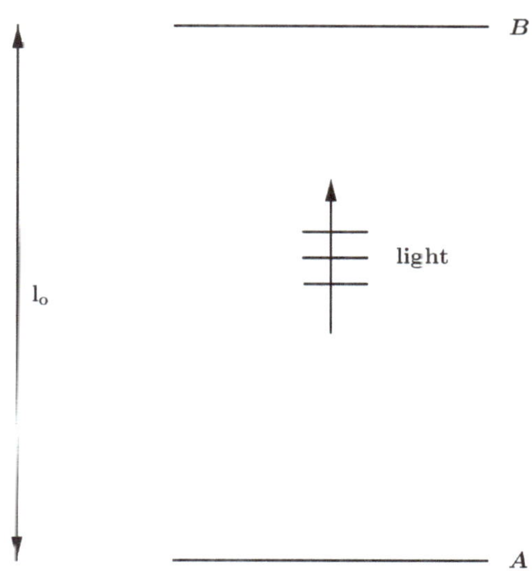

Figure 2.2 A beam of light travels from mirror A to mirror B at rest in frame S'.

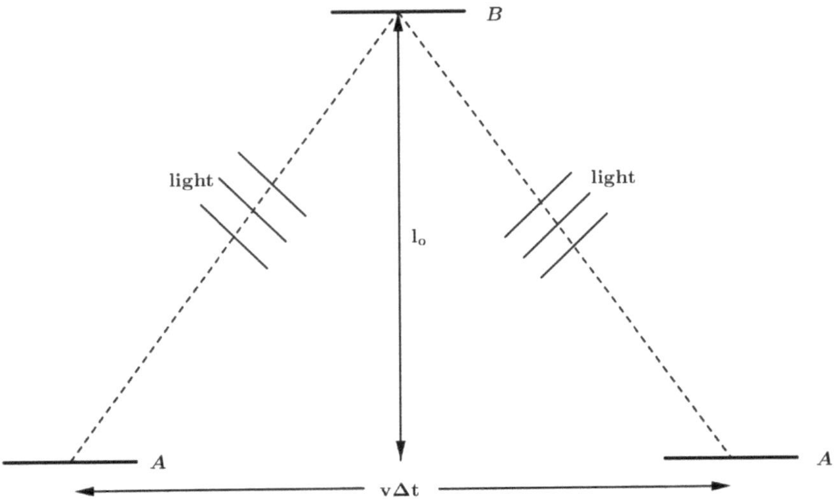

Figure 2.3 Fig. 2.2 viewed from frame S'.

Now view the clock's operation from the perspective of an observer at rest in S. Now the clock moves to the right at velocity $\boldsymbol{v}$, and the light ray takes the path in frame S as shown in Fig. 2.3. In the frame S, the light ray travels along the line segment $\overline{\mathrm{AB}}$ and then along $\overline{\mathrm{BA}}$ back to mirror A. The distance mirror A in Fig. 2.3 travels between sending and receiving the light ray is $v\Delta t$. From Fig. 2.3, the distance the light ray travels is

$$\overline{\mathrm{AB}} + \overline{\mathrm{BA}} = 2\sqrt{\ell_o^2 + (v\Delta t/2)^2}. \tag{2.1}$$

But the light ray also travels at the speed limit c in frame S according to Postulate 2, so

$$\overline{\mathrm{AB}} + \overline{\mathrm{BA}} = c\Delta t. \tag{2.2}$$

This is the essential point in this argument—the universal speed limit enters here, and we have a clear qualitative distinction from the rules of Newtonian mechanics. Combining Eqs. (2.1) and (2.2), we find

$$c\Delta t = 2\sqrt{\ell_o^2 + (v\Delta t/2)^2}, \tag{2.3}$$

which allows us to solve for Δt,

$$\begin{aligned} \Delta t &= \frac{2\ell_o}{\sqrt{c^2 - v^2}} = \frac{2\ell_o/c}{\sqrt{1 - v^2/c^2}} \\ \Delta t &= \Delta\tau\Big/\sqrt{1 - v^2/c^2}. \end{aligned} \tag{2.4}$$

This result is called "time dilation"—the time interval in frame S, Δt, is dilated by a factor of

$$\gamma \equiv 1\Big/\sqrt{1 - v^2/c^2} \tag{2.5}$$

compared with the proper time $\Delta\tau$ measured in frame S'.

A look at the figure reveals why Δt is greater than $\Delta\tau$, the proper time interval measured on the clock in its rest frame. When the light travels from A to B and back in the frame S, it does so at the speed limit, following Postulate 2. In a Newtonian world where all velocities satisfy the addition of velocities theorem of Galilean transformations, the velocity would be $v_x = v$ and $v_y = c$, producing a speed in frame S of $\sqrt{c^2 + v^2}$. Substituting this into Eq. (2.3) leads to the Newtonian result, $\Delta t = 2\ell_o/c = \Delta\tau$ (Newton). But in the real world of Einstein where Eq. (2.2) rules, the speed of light is c in all frames, and the time it takes to transverse the path $\overline{AB} + \overline{BA}$ is longer than Newton would calculate. The universal nature of the speed limit produces time dilation.

A skeptic might complain that this curious result is particular to the type of clock we made and would not be true of a clock made of springs and masses. We can rule out this objection by appealing to Postulate 1. Consider the clock made of mirrors and light at rest in frame S', and place next to it a mechanical clock made of springs and masses. Synchronize the two clocks. Now do the same in the frame S, which is moving at a relative velocity v. This is guaranteed to work in frame S as well as in the original frame S' because of Postulate 1. The two clocks made of mirrors and light experience time dilation, and the mechanical clocks must also because each is synchronized to a clock of mirrors and light at rest with respect to it. The clocks of mirrors and light can be dispensed with at this point, and we arrive at the conclusion that the mechanical clocks experience time dilation, too. (Of course, it remains to be seen if we can lay down laws of mechanics that satisfy Postulates 1 and 2. When we discuss the relativistic versions of energy, momentum, and Newton's second law, we shall carry out this promise.)

As exotic as Eq. (2.4) for time dilation might seem, the reader should understand that for relative velocities that are small compared with the speed limit, $c = 3.0\cdot10^8$ m/s, the effect is very tiny. Take $v = 1000$ mph $\approx$ 490 m/s, so $v/c \approx 1.6\cdot10^{-6}$, and using the expansion and approximations reviewed in Appendix C,

$$\Delta t \approx \Delta\tau\left(1 + \frac{1}{2}\frac{v^2}{c^2}\right) \approx \Delta\tau(1 + 1.33\cdot10^{-12}). \tag{2.6}$$

The relativistic correction to the Newtonian result of the equality of Δt and $\Delta\tau$ is suppressed by a factor of the square of the speed limit and is tiny in present everyday situations.

If you live and work at a high-energy accelerator center where elementary particles achieve speeds within a fraction of a percent of c, relativistic effects are commonplace. For example, consider the muon, a heavy relative of the electron, which is a common ingredient in cosmic rays and is produced copiously in high-energy collisions. It is an unstable particle with a half-life $\Delta\tau \approx 2.2\cdot 10^{-6}$ s, so a population of muons, at rest in the frame S', decays according to $N(t') = N_o \exp(-0.693t'/\Delta\tau)$. (The funny factor of 0.693 is just the natural logarithm of 2.) At an accelerator center, this population of muons might be moving in a beam at velocity v comparable to c and the exponential decay law in the frame S, the lab frame, would be $N(t) = N_o \exp(-0.693\, t/\gamma\Delta\tau)$. This expression for $N(t)$ has incorporated the quantity $\gamma\Delta\tau$ in its exponential, so the average expected lifetime of a moving muon is dilated to $\gamma\Delta\tau$. Modern accelerators produce beams with $v/c \approx 0.99c$ for which $\gamma \approx 7.1$. This is a huge effect that is easily detected.

For example, if you have a bunch of muons with $N_o = 1000$, they travel across a detector of 1000 m at $v/c = 0.99$ in the time interval $1000/(0.99\cdot 3\cdot 10^8) \approx 3.4\cdot 10^{-6}$ s. Therefore, Newton predicts that $1000\cdot\exp(-0.693\cdot 3.4\cdot 10^{-6}/2.2\cdot 10^{-6}) \approx 1000\exp(-1.06) \approx 346$ would survive, because in his world the muon lifetime is $\Delta\tau \approx 2.2\cdot 10^{-6}$ s in *all* inertial frames. On the other hand, Einstein predicts that $1000\exp(-1.06/7.1) \approx 1000\exp(-0.149) \approx 862$ would survive. The Einstein result (time dilation) is easily confirmed at high-energy accelerator centers.

We see that by boosting the muons to a velocity $v/c \approx 1$, we can slow down their internal workings relative to the lab frame. Accelerator centers also make beams of other particles such as protons, which are composed of quarks and gluons, according to modern theories of strongly interacting subnucleon particles. By boosting them to velocities close to the speed of light, we can slow down their internal dynamics so we can probe inside them and learn how quarks interact. We can scatter high-energy quanta of light off the accelerated protons and effectively take a snapshot of their internal state when doing high-energy scattering experiments. This is how quarks were discovered as the internal constituents of protons in the classic experiments at the Stanford Linear Accelerator Center in the 1960s and 1970s. In that case the protons were at rest in the lab and a high-energy beam of electrons provided the (virtual) light quanta that probed their structure. It was from the perspective of the electrons that the internal dynamics of the proton were slowed down and their constituents were revealed.

But there is a puzzle here. If we view the muon decay process in the frame S' where the muons are at rest, how can we understand that 862 of them survive the trip to the end of the detector and not 346? Obviously, observers in both frames must agree on how many muons get through the detector—both observers can just count them! If relativity is consistent, it must be true that $t'/\Delta\tau = t/\gamma\Delta\tau$, so the two exponential decay laws give identical predictions. We computed $t = d_o/v$, where d_o is the length of the detector in the frame S where it is at rest. Substituting into $t'/\Delta\tau = t/\gamma\Delta\tau$, we learn that $t' = d_o/\gamma v$. How can we understand this? In the frame S' the detector is racing toward the muons at velocity v, so t' should be the length of the *moving* detector divided by the velocity v. It must be that the length of the detector measured in the frame where it is moving is contracted by a factor of γ, $t' = d'/v$ with $d' = d_o/\gamma$! This is indeed the case. The effect is called Lorentz contraction, named after the physicist who first deduced the effect in the context of the theory of electricity and magnetism at the very beginning of the 20th century. We look more closely at this important effect in the next section.

2.3 LORENTZ CONTRACTION

If the constancy of the speed limit implies time dilation, it must also affect the measurements of lengths of moving objects. This effect, called Lorentz contraction, is also easily obtained from Einstein's clock.

To begin, take the clock at rest in frame S' and rotate it by 90 degrees so that its length ℓ_o is parallel to v, the relative velocity of the second frame S. It is conventional to call ℓ_o the proper length of the clock because this is its length in the frame in which it is at rest. Because space is homogeneous in an inertial frame, the rotated clock works as it did before. In frame S', we had $\Delta\tau = 2\ell_o/c$. As before, this proper time interval $\Delta\tau$ must be dilated when measured in frame S because the rotation of the clock cannot affect its internal workings. But now view the path of the light ray in the frame S, as shown in Fig. 2.4.

We label the length of the clock, from A to B, as measured in frame S as ℓ. In the first image of Fig. 2.4, light leaves the mirror and heads toward mirror B. In the second image, the light ray reaches mirror B after a time Δt_1. The mirror A has moved a distance $v\Delta t_1$ to the right, so

$$c\Delta t_1 = \ell + v\Delta t_1. \tag{2.7}$$

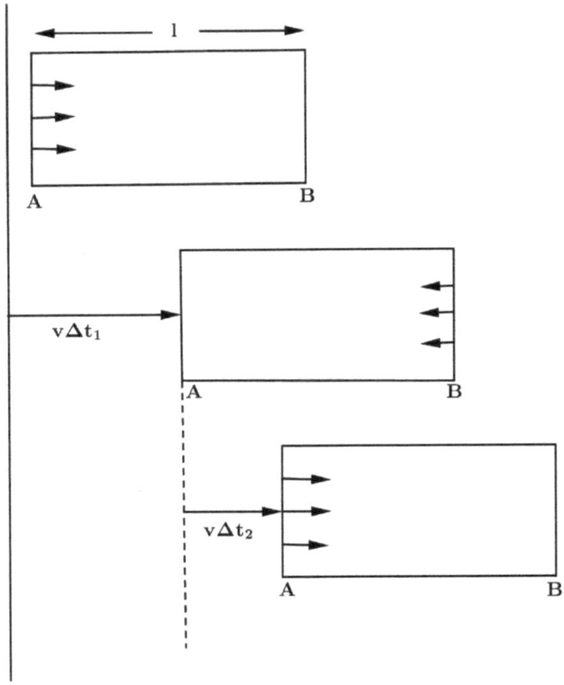

Figure 2.4 The rotated clock viewed from frame S.

In the third image of Fig. 2.4, the light that left mirror B reaches mirror A and is reflected. Between images 2 and 3, a time Δt_2 has passed in frame S, so the mirror A moves an additional distance $v\Delta t_2$,

$$c\Delta t_2 = \ell - v\Delta t_2, \tag{2.8}$$

where the minus sign occurs because the light ray is moving from mirror B to A while mirror A is moving to the right. Finally, the sum of Δt_1 and Δt_2 is the time that passes in frame S corresponding to one period of the clock. We know that this time is Δt, and we computed it to be $\gamma(2\ell_o/c)$, where γ denotes, as usual, $1\big/\sqrt{1 - v^2/c^2}$. So, using Eqs. (2.7) and (2.8),

$$\begin{aligned}\Delta t &= \Delta t_1 + \Delta t_2\\ \Delta t &= \frac{\ell}{c-v} + \frac{\ell}{c+v}\\ \Delta t &= \frac{2\ell/c}{1 - v^2/c^2}.\end{aligned} \tag{2.9}$$

But $\Delta t = \gamma(2\ell_o/c)$, so we learn that

$$\ell = \ell_o/\gamma = \sqrt{1 - v^2/c^2}\ell_o. \tag{2.10}$$

This is the result we wanted—it is called Lorentz contraction because ℓ is less than ℓ_o. Although H. A. Lorentz discovered this result, it was not until Einstein's derivation that the universality of the result was understood.

We learn that when an observer measures the length of a rod moving in the direction of its length, the observer will obtain a result contracted by a factor of γ. This derivation shows that this effect follows from the same considerations that led to time dilation—the universal character of the speed limit. But we need to understand this contraction in a more explicit fashion. There are no forces at play here, so what does it mean that the rod is contracted? The effect clearly has nothing to do with how rods are put together—the effect just states how measurements of lengths transform between inertial frames. We see in the next section that Lorentz contraction is only really clarified by a discussion of the synchronization of spatially separated moving clocks.

2.4 THE RELATIVITY OF SIMULTANEITY

It is clear from our first look at time dilation and Lorentz contraction that the universal nature of the speed of light produces puzzling results. The simplest space–time measurements we can imagine are those that started this chapter—the setting up of a grid of coordinates and clocks—so we will return to it here to understand it better. In particular, let a grid of coordinates and clocks be set up in frame S', which is moving with respect to a frame S at a relative velocity of v to the right. The clocks along the x', y', and z' axes are synchronized in frame S' in the usual way. We want to consider the synchronization procedure from the perspective of observers at rest in frame S. For example, an observer in S' synchronizes two clocks separated by a distance ℓ' in the y' direction by placing a source of light halfway between them, sending a signal out in all directions and insisting that the two clocks read the same time when the signals arrive (Fig. 2.5). From the perspective of the observer in frame S, the source and both clocks are moving to the right at velocity v. Both the light rays travel at velocity c with respect to frame S, so the light rays arrive at the two clocks simultaneously in frame S. Clearly the synchronization procedure devised in frame S' is acceptable to frame S, and all is well. Also, because each clock is

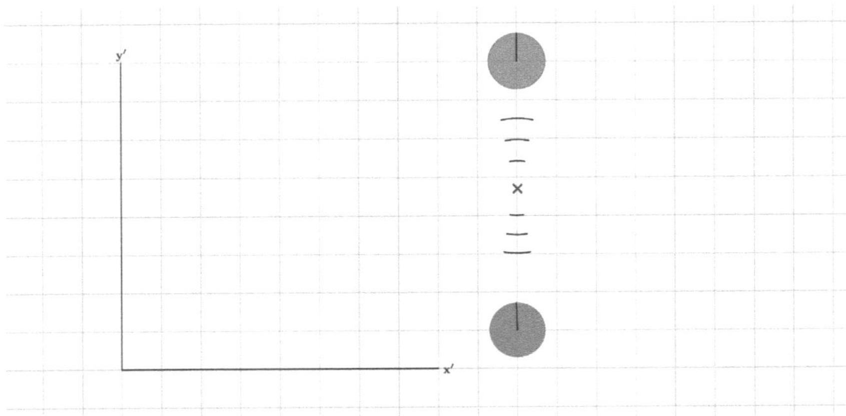

Figure 2.5 Synchronizing two clocks at rest in frame S'.

moving with velocity v transverse to the separation between them of ℓ', their positions are $x_1 = vt$, $y_1 = 0$, $z_1 = 0$ for the lower clock and $x_2 = vt$, $y_2 = \ell$, $z_2 = 0$ for the upper clock (no surprises here). These preliminary considerations underlie our discussion of time dilation of the clock with mirrors in Section 2.2.

Things get more interesting when we consider the synchronization procedure for clocks at rest in S' but separated in the x' direction. Let the distance between the clocks in their rest frame S' be $\Delta x' = \ell' = \ell_o$. A signal generator is placed halfway between them, a pulse is sent out in all directions, and the clocks are synchronized. If we say that $t' = 0$ when the light is emitted, then both clocks register $t' = \ell_o/2c$ when the light rays arrive. All observers in all inertial frames will agree that the faces of the two clocks read $t' = \ell_o/2c$ when the light rays are received (Fig. 2.6).

Now view the procedure from frame S. An observer in frame S will measure that light reaches clock 1 before it reaches clock 2 because clock 1 is racing toward the rays and clock 2 is racing away from them! The individual times are given by the same sort of considerations we saw in Section 2.3,

$$\begin{aligned} ct_2 &= \ell/2 + vt_2 \\ ct_1 &= \ell/2 - vt_1, \end{aligned} \tag{2.11}$$

where t_1 is the time for light to reach clock 1, and t_2 is the time for light to reach clock 2. Note that Eq. (2.11) incorporates the universal nature of the speed of light—regardless of its direction of motion, light travels at speed c

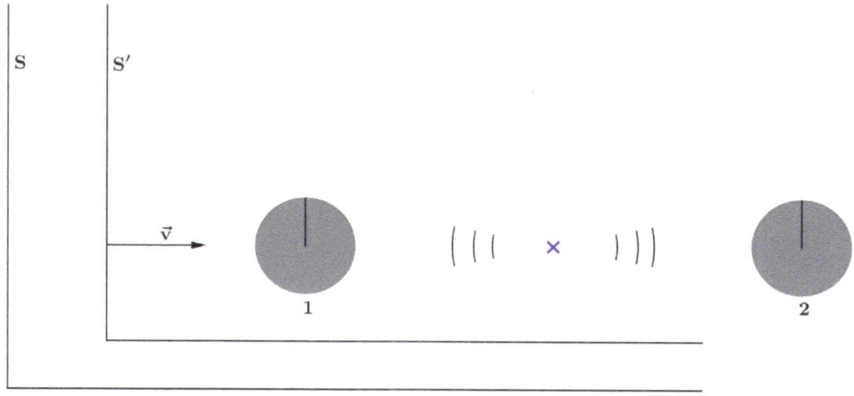

Figure 2.6 Synchronizing two clocks at rest in frame S' separated in the x' direction.

with respect to frame S. So we should appreciate, before extracting quantitative details from Eq. (2.11), that an observer in frame S concludes that the two clocks are *not* synchronized in his or her rest frame—"clocks that are synchronized in one inertial frame are *not* synchronized in another inertial frame if there is a nonzero relative velocity along the line of sight between the clocks." This startling result, which lies at the core of Lorentz contraction, follows simply from the fact that the speed limit is the same in all frames and violates Galileo's law of addition of velocities.

Now let us work out the time difference $t_2 - t_1$ as measured in frame S. The time difference is

$$\begin{aligned}\Delta t &= t_2 - t_1\\ \Delta t &= \frac{\ell/2}{c-v} - \frac{\ell/2}{c+v}\\ \Delta t &= \frac{\ell v/c^2}{1-v^2/c^2}.\end{aligned} \tag{2.12}$$

But ℓ is the length measured in the frame S moving with velocity v with respect to the frame S' where the clocks are at rest. So, ℓ is related to ℓ_o, the proper distance between them, by the Lorentz contraction effect $\ell = \ell_o\sqrt{1 - v^2/c^2}$. Thus Eq. (2.12) can be written $\Delta t = \gamma \ell_o v/c^2$.

Next we need to calculate the time that an observer in frame S measures as passing in frame S'. Recall from time dilation, that, as measured in frame S, the clocks in frame S' run slowly. Therefore, if a time $\Delta t'$ passes on a clock at rest in frame S', then $\Delta t = \gamma \Delta t'$ gives the time interval that passes on a clock at rest in frame S. So, an observer in frame S states that there was a time difference of $\Delta t' = \ell_o v/c^2$ in frame S' between the time clock 2 received the light ray and the time clock 1 received its ray. But all observers in all frames have noted that clocks 1 and 2 have identical readings when the light rays reach them. Therefore, an observer in frame S concludes that clock 1 was set *ahead* of clock 2 by an amount $\ell_o v/c^2$. In other words, when an observer at time t in frame S inquires what the time t' is in frame S', the observer finds that t' depends on x': t' varies as $x'v/c^2$.

The lesson we learn here—that clocks that are synchronized in one frame are *not* synchronized when measured in a frame at relative motion along the line between the clocks—is crucial to understanding time dilation and Lorentz contraction. It is called the "relativity of simultaneity." These three effects must be taken all together (more on this later).

2.5 TIME DILATION REVISITED

The purpose of this discussion is to understand how time dilation, the relativity of simultaneity, and Lorentz contraction conspire together to produce a consistent picture of time measurements of moving clocks [1]. We have already seen that, if two frames S and S' are in relative motion, an observer in each says that the other's clock runs slowly. This sounds paradoxical. Let us take a closer look from *both* observers' perspectives and see that each agrees with the observations of the other.

Suppose that there are two synchronized clocks at rest in the frame S and separated by a proper distance ℓ_o in the x direction. As usual, let frame S' be moving to the right at velocity v, as shown in Fig. 2.7. Let the clock in frame S' be synchronized with clock 1, which is at rest in frame S. They pass each other at $x = x' = 0$, $t = t' = 0$, as shown in Fig. 2.7A. We wish to know what S′'s clock will read when it passes S's clock 2, as shown in Fig. 2.7B. Will observers at rest in both frames make the same prediction even though they see one another's clocks running slowly?

First, consider the measurement from the perspective of an observer at rest in frame S. The clock in frame S' is moving at velocity v with respect to that observer, so clock 2 will read $t = \ell_o/v$ when the clocks pass. Because the observer in frame S sees frame S''s clocks running slowly, the observer predicts that S′'s clock will read $t' = \frac{\ell_0}{v}\sqrt{1 - v^2/c^2}$ when the clocks pass each other. So, the observer in S notes that although clock 1 and S′'s clock were synchronized, clock 2 is *not* synchronized with the moving clock!

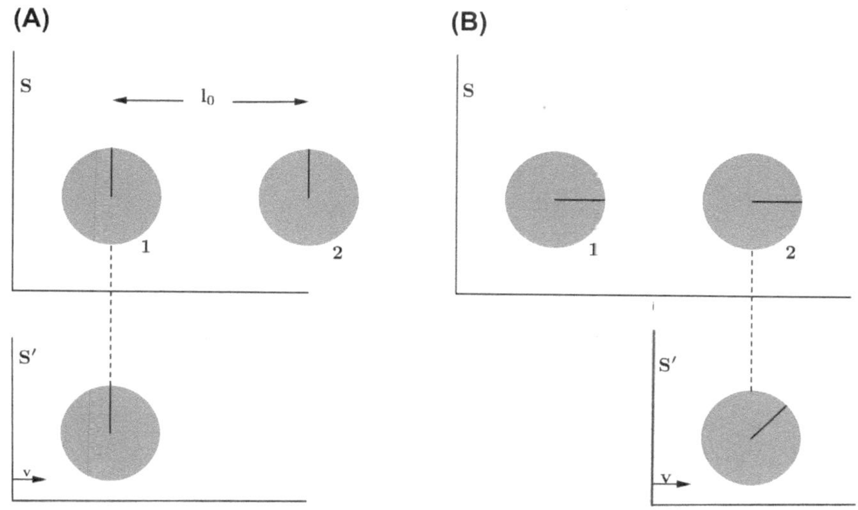

Figure 2.7 In part (A) the clock at rest in frame S' is synchronized with a clock at rest in frame S. In part (B) the time on the clock at rest in frame S' s recorded when it passes a second c ock at rest in frame S.

Now consider the measurements again by treating frame S' as the one at rest and frame S as moving to the left at velocity v, as shown in Fig. 2.8. From the perspective of the observer in frame S', the distance between clocks 1 and 2 is contracted to $\ell' = \ell_o\sqrt{1 - v^2/c^2}$, so the observer's clock

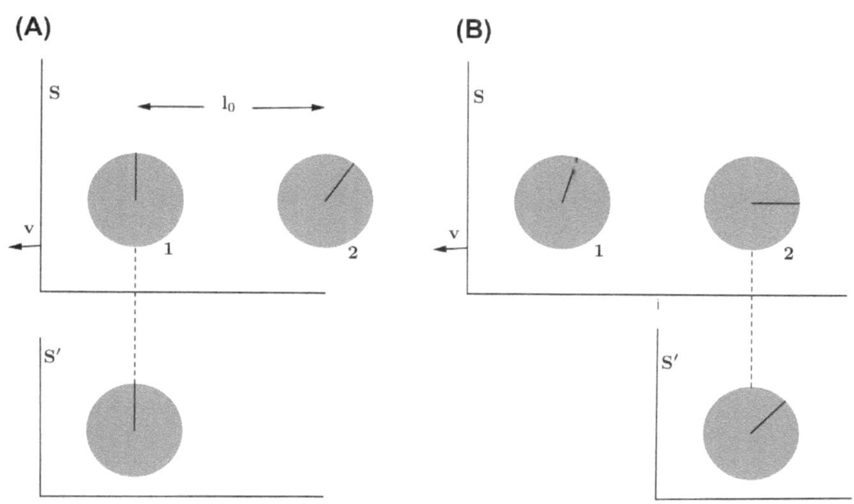

Figure 2.8 The events in Fig. 2.7A and B from the perspective of frame S'.

will read $t' = \ell_o\sqrt{1 - v^2/c^2}/v$ when it reaches clock 2. So we learn that S and S' do agree on this point, as they must.

Next consider the observation of clock 2 by the observer in S'. This is the tricky part! The left-hand clocks coincide at $t = t' = 0$, as shown in Fig. 2.8A. The clock 2 is measured by the observer in S' to be $\ell_o\sqrt{1 - v^2/c^2}$ to the right and to be $\ell_o v/c^2$ ahead of clock 1! In Fig. 2.8b, a time interval $\ell_o\sqrt{1 - v^2/c^2}//v$ has passed in S', so the observer in S' states that a time interval $\left(\ell_o\sqrt{1 - v^2/c^2}/v\right)\sqrt{1 - v^2/c^2}$ has passed in frame S (this time must dilate to $\ell_o\sqrt{1 - v^2/c^2}/v$ when measured in frame S'). So, clock 2 must read $\ell_o v/c^2 + \ell_o(1 - v^2/c^2)/v$ when the clocks coincide, the observer in frame S' reasons. And this agrees with the observer in frame S, who simply calculates the time by dividing distance by velocity, ℓ_o/v.

In summary, each observer sees the other clocks running slower than his or her own, but, because of the relativity of simultaneity and Lorentz contraction, they both agree that $t = \ell_o/v$ and $t' = \ell_o\sqrt{1 - v^2/c^2}/v$ when the right-hand clock in S coincides with the clock in S'.

2.6 LORENTZ CONTRACTION REVISITED

Now we want to consider Lorentz contraction again, but this time keep track of the times of the measurements of the positions of the ends of a moving rod [1]. We shall see in this way that the crux of the matter here is the relativity of simultaneity—that spatially separate clocks that are synchronized in one frame are not synchronized in a moving frame. So, when one observer notes the positions of the ends of a rod simultaneously in his or her frame, an observer moving by notes that those measurement events are *not* simultaneous in his or her frame. This effect, then, explains why observers in relative motion measure different lengths for a given physical rod.

For example, let there be a rod of length ℓ_o at rest in a frame S, which moves to the left at velocity v with respect to frame S' as shown in Fig. 2.9. When an observer at rest in frame S' measures the length of the rod, the observer gets an answer ℓ', which must be related to ℓ_o in a linear fashion with, perhaps, a prefactor f that can depend on the dimensionless ratio v/c, $\ell' = f(v/c)\ell_o$. The structure of this formula is a consequence of dimensional analysis. Our task is to find (*and* understand) the prefactor $f(v/c)$. When the observer in S' makes the measurement, the observer notes the positions of

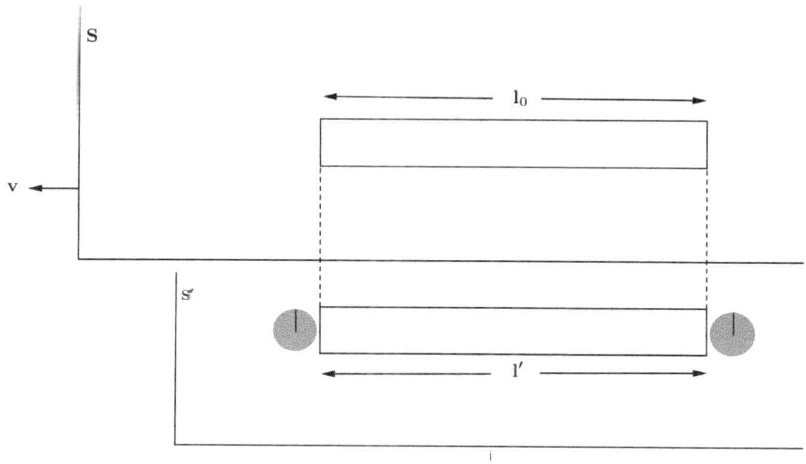

Figure 2.9 Measuring the length of a moving rod of proper length l_0 in frame S'.

the end of the rod at *one instant* in his or her frame, as shown in Fig. 2.9, because the observer does not want the rod to move during the measurement.

Consider the perspective of the observer in frame S on the measurements. The observer's rod of proper length ℓ_o is aligned along the direction of S''s relative velocity v and there is a rod of length ℓ', as measured by the observer in S', moving to the right at velocity v. At $t = t' = 0$, let the left ends of the rods coincide as shown in Fig. 2.10A. At this moment the

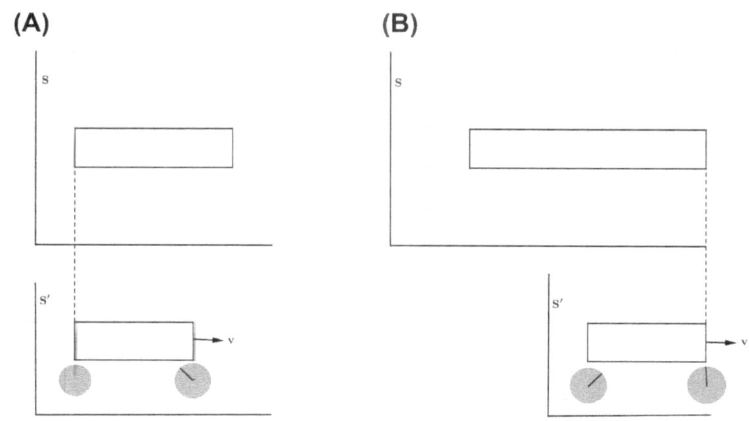

Figure 2.10 The measurements of the ends of the moving rod from the perspective of frame S: the left end in (A) and the right end in (B).

observer in S notes that the clock on the right end of S''s rod reads $t' = -\ell' v/c^2$ by the relativity of simultaneity formula. Later, the right ends of the rods are seen to coincide in S, and the clock on that end of the moving rod must read $t' = 0$, as shown in Fig. 2.10B, because the observer in S' is making the measurement at one time in his or her frame. Since $t' = -\ell' v/c^2$ is the time on the right-hand clock in Fig. 2.10A and is $t' = 0$ in Fig. 2.10B, the time elapsed is $\ell' v/c^2$, which is dilated to $\gamma\ell' v/c^2$ in frame S. When the observer in S measures the length of the moving rod, he or she measures a length $f(v/c)\ell' = f^2(v/c)\ell_o$. So, the observer in S can measure the length of the rod ℓ_o at rest in his or her frame by following Fig. 2.10: add the length of the moving rod, $f^2(v/c)\ell_o$, to the distance its right end moves, $(\gamma\ell' v/c^2)\cdot v = \gamma f\ell_o v^2/c^2$. So,

$$f^2\ell_o + f\gamma\ell_o v^2/c^2 = \ell_o, \tag{2.13}$$

which determines f through a quadratic equation,

$$f^2 + \gamma v^2/c^2 f - 1 = 0, \tag{2.14}$$

which has solutions

$$f = 1/2\left(-v^2\gamma/c^2 \pm \sqrt{v^4\gamma^2/c^4 + 4}\right) \tag{2.15}$$

and which can be simplified using $\gamma = 1\Big/\sqrt{1 - v^2/c^2}$ to $f = 1/\gamma = \sqrt{1 - v^2/c^2}$ or $f = -\gamma$. This last solution is negative and unphysical. So, we have one physical solution,

$$f = \sqrt{1 - v^2/c^2}, \tag{2.16}$$

and we have successfully rederived the Lorentz contraction formula,

$$\ell' = f(v/c)\ell_o = \ell_o\sqrt{1 - v^2/c^2}. \tag{2.17}$$

In summary, we see the consistency of the Lorentz contraction formula, $\ell' = \ell_o\sqrt{1 - v^2/c^2}$, with time dilation and the relativity of simultaneity. But this time we see how it works and why both observers in relative motion can consistently say that they measure one another's rods as contracted—they observe the ends of the moving rods at different times in the rod's rest frame.

PROBLEMS

2-1. A spaceship (frame S') passes Earth (frame S) at velocity $0.6c$. The two frames synchronize their clocks at $x = x' = 0$ to read $t = t' = 0$ when they pass one another (call this event 1). Ten minutes later, as measured by Earth clocks, a light pulse is emitted toward the spaceship (call this event 2). Later, the light pulse is detected on the spaceship (call this event 3).

a. Is the time interval between events 1 and 2 a proper time interval in the spaceship frame? In the Earth frame?

b. Is the time interval between events 2 and 3 a proper time interval in the spaceship frame? In the Earth frame?

c. Is the time interval between events 1 and 3 a proper time interval in the spaceship frame? In the Earth frame?

d. What is the time of event 2 as measured by the spaceship?

e. According to the spaceship, how far away is Earth when the light pulse is emitted?

f. From your answers in parts (d) and (e), what does the spaceship clock read when the light arrives?

g. Find the time of event 3 according to Earth's clock by analyzing everything from Earth's perspective.

h. Are your answers to parts (f) and (g) consistent with your conclusions from parts (a), (b), and (c)?

2-2. Rockets A and B, each having a proper length 100 m, pass each other moving in opposite directions. According to clocks on rocket A, the front end of rocket B takes $1.5 \cdot 10^{-6}$ s to pass the entire length of A.

a. What is the relative velocity of the rockets?

b. According to clocks on rocket B, how long does the front end of A take to pass the entire length of rocket B?

c. According to clocks on rocket B, how much time passes between the time when the front end of A passes the front end of B and the time when the rear end of A passes the front end of B? Does this time interval agree with your answer to part (b)? Should it?

2-3. Short-lived particles are produced in high-energy collisions at accelerator centers such as Fermilab in Batavia, Illinois. The study of such particles teaches us about the fundamental building blocks of matter, how nuclei in atoms such as uranium fission and produce energy, how the universe evolved from its origin as a big bang, and so on. One particularly important particle that is produced copiously in high-energy collisions between protons is the pion. It is the carrier of the nuclear force and has been the subject of much research since the 1950s. Pions decay in their own rest frame according to

$$N(t') = N_o 2^{-t'/T},$$

where T is the half-life, $T \approx 1.8 \cdot 10^{-8}$ s. Imagine that experimenters create pulses of pions at Fermilab and observe that two-thirds of the pions in a particular pulse reach a detector at a distance of 35 m from the point where they were created. All the pions have the same velocity.

a. What is the velocity of the pions?

b. What is the distance from the target to the detector in the rest frame of the pions?

2-4. The distance from Planet X to a nearby star is 12 light-years (a light-year is the distance light travels in 1 year as measured in the rest frame of Planet X). The relative velocity between Planet X and the nearby star is negligible.

a. How fast must a spaceship travel from Planet X to the star to reach the star in 7 years according to a clock fixed on the spaceship?

b. How long would the trip take according to a clock fixed on Planet X?

c. What is the distance from Planet X to the nearby star, according to an astronaut on the spaceship?

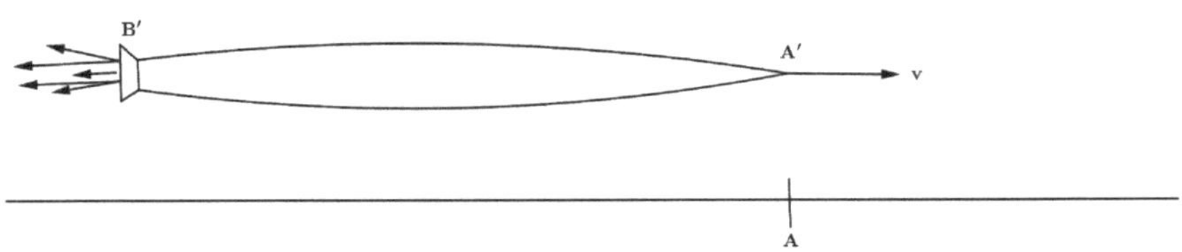

Figure 2.11 A rocket ship travels at velocity v with respect to frame S.

2-5. A spaceship of proper length ℓ_o travels at a constant velocity v relative to a frame S, as shown in Fig. 2.11. The nose of the ship (A') passes point A in frame S at time $t = t' = 0$, when a light signal is sent from A' to B', the tail of the spaceship.

a. When does the signal reach the tail, B', according to spaceship time t'?

b. At what time t_1 in the frame S does the signal reach the tail B'?

c. At what time t_2 in the frame S does the tail of the spaceship, B', pass point A?

2-6. A flash of light is emitted at position x_1 and is absorbed at position $x_2 = x_1 + \ell$. In a reference frame moving with velocity v along the x axis:

a. What is the spatial separation ℓ' between the point of emission and the point of absorption?

b. How much time elapses between the emission and absorption of the light?

2-7. At 1:00, a spaceship passes Earth with a velocity $0.8c$. Observers on the ship and on Earth synchronize their clocks at that moment. Just to make this exercise interesting, answer each question from both the viewpoint of Earth and the spaceship.

a. At 1:30, as recorded in the spaceship's frame, the ship passes another space probe that is fixed relative to Earth and whose clocks are synchronized with respect to Earth. What time is it according to the space probe's clock?

b. How far from Earth is the probe, as measured by Earth's coordinates?

c. At 1:30, spaceship time, the ship sends a light signal back to Earth. When does Earth receive the signal by Earth time?

d. Earth sends another light signal immediately back to the spaceship. When does the spaceship receive that signal according to spaceship time?

2-8.

a. Two particles move along the x axis at velocities $0.8c$ and $0.9c$, respectively. The faster one is 1 m behind the slower 1 at $t = 0$. How many seconds elapse before they collide?

b. A rod of proper length 0.10 m moves longitudinally along the x axis of frame S at speed $0.50c$. How long in frame S does it take a particle moving oppositely at the same speed to pass the rod?

REFERENCE

[1] N.D. Mermin, Space Time in Special Relativity, Waveland Press, Prospect Heights, IL, 1968.

CHAPTER 3

Visualizing Relativity—Minkowski Diagrams and the Twins

Contents

3.1 SPACE AND TIME AXES FOR INERTIAL FRAMES AND THE CONSTANCY OF THE SPEED OF LIGHT

The effects discussed in Chapter 2 follow from the two postulates of relativity. The constancy of the speed limit played a central role in them, and, thinking back, we could argue that the qualitative nature of time dilation, Lorentz contraction, and the relativity of simultaneity followed. All this can be done visually using Minkowski diagrams. These pictures of time and position measurements date to the earliest days of relativity. The relativity of simultaneity follows from these diagrams particularly simply, and some of our quantitative results can be obtained anew with less (!) effort.

To begin, set up a space axis x and a time axis t. The transverse directions y and z will be omitted. We choose to display the time t on a clock at rest in frame S as an axis perpendicular to the x axis, as shown in Fig. 3.1. Note that we plot ct on the vertical axis, so both axes have units of length. The depiction of light rays is particularly convenient if we do this. It will prove convenient, especially when we discuss the Doppler shift and the twin paradox, to use light-years as the units along the ct axis. A light-year is the distance light travels in 1 year. For example, if $ct = 5$ light-years, then t is simply 5 years.

Special Relativity, Electrodynamics, and General Relativity
ISBN 978-0-12-813720-8
https://doi.org/10.1016/B978-0-12-813720-8.00003-9

Figure 3.1 Space and time axes.

Now consider the synchronization of two clocks at rest in this frame, one at point A and the second at point C. We emit light from point B midway between A and C. The light is recorded on the clocks at events A_1 and C_1, as shown in Fig. 3.2.

What do the lines in Fig. 3.2 mean? The vertical line $\overline{AA_1}$ indicates that the clock A is at rest in this frame. The fact that the dashed line $\overline{A_1C_1}$ is horizontal indicates that events A_1 and C_1 are simultaneous in this frame. Note that the two light rays, $x = x_B \pm ct$, are drawn at 45 degrees inclined to the x axis—including the factor c in the time axis produces this handy result. In fact, if we shot off a light ray from the origin of the frame to the right, $x = ct$, then its path, called a world line, appears as shown in Fig. 3.3.

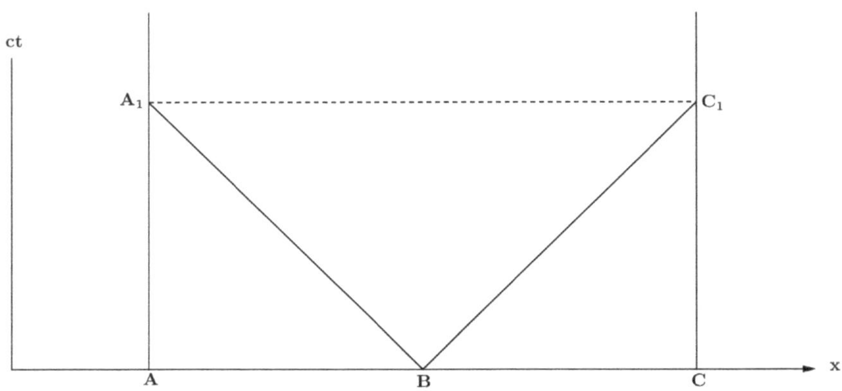

Figure 3.2 Synchronizing two clocks in the frame *S*.

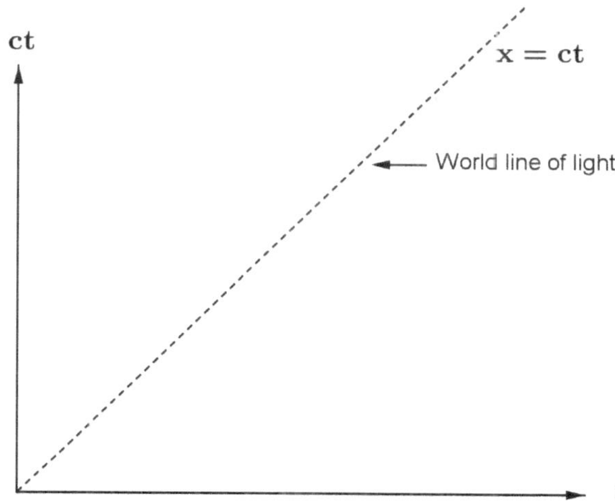

Figure 3.3 The path of a light ray plotted on the space and time axes in frame *S*.

Minkowski diagrams really become useful when we visualize space–time measurements made in two inertial frames, S and S', in relative motion. The ct–x axes are laid down. How shall the ct'–x' axes be put on the same figure in a way that is consistent with the postulates of relativity? The ct' axis is the path (world line) in the ct–x graph of the origin $x' = 0$ of the S' frame. But, any point at rest in S' moves to the right in frame S with velocity v, so $x' = 0$ corresponds to $x = vt$. So, the ct' axis is inclined at an angle θ, $\tan\theta = v/c$, to the ct axis as shown in Fig. 3.4 (no surprises, so far).

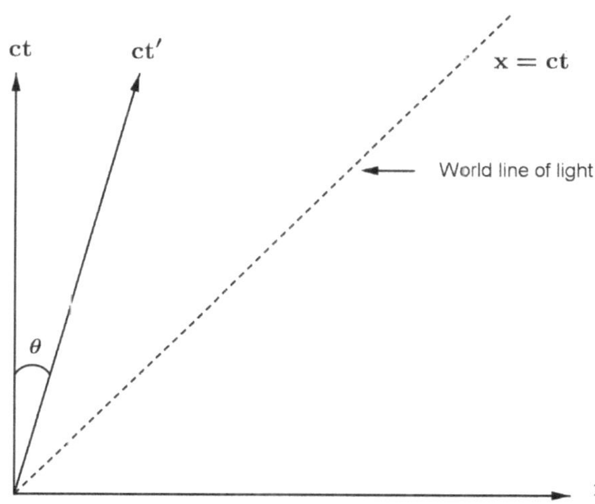

Figure 3.4 Adding the time axis of frame *S'* into Fig. 3.3.

However, now we need to put the x' axis into Fig. 3.4. We have put the world line of a light ray in the figure because the second postulate of relativity, that light travels at the speed limit in any inertial frame, guides us. The light ray bisects the angle between the ct and x axes to guarantee that its velocity is measured as c, $x = ct$. But, it must also bisect the angle between ct' and the x' axis so that its velocity is also c in this frame! Thus, the x' axis must be tilted at an angle θ, $\tan\theta = v/c$, *above* the x axis as shown in Fig. 3.5.

Because the central results of relativity, time dilation, Lorentz contraction, and the relativity of simultaneity, follow simply from this result, Fig. 3.5, let us discuss it in more detail. Note that the ct'—x' coordinate system is not orthogonal. The ct—x and the ct'—x' coordinate systems are not related to one another by a rotation. How do we read off the coordinates of a space—time event in a coordinate system with nonorthogonal axes? Lines of constant ct' are parallel to the x' axis and lines of constant x' are parallel to the ct' axis. So, the event P has the coordinate ct_p' and x_p', as shown in Fig. 3.6.

Recall that we drew the ct and x axes orthogonally. This was convenient and familiar, but the ct'—x' coordinate system shows that we must learn to use nonorthogonal systems. If the relative velocity is negative, so that S' is moving to the left in frame S, then Fig. 3.6 becomes Fig. 3.7. Lines of constant ct', indicating a line of events that are simultaneous in frame S', are

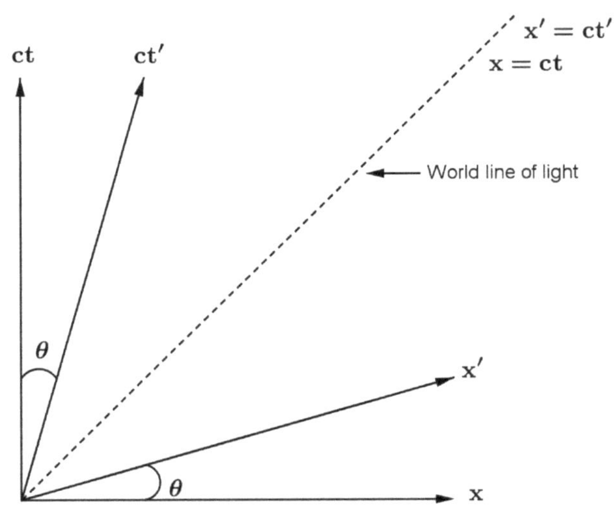

Figure 3.5 Adding the space axis of frame S' into Fig. 3.3.

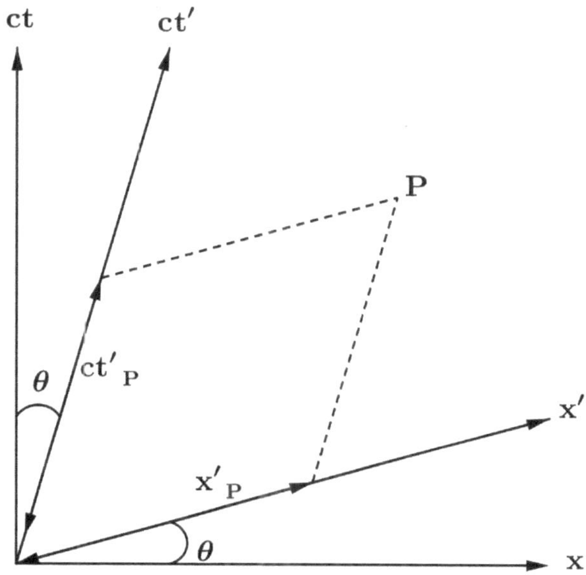

Figure 3.6 Labeling a point P in nonorthogonal coordinates, $ct' - x'$.

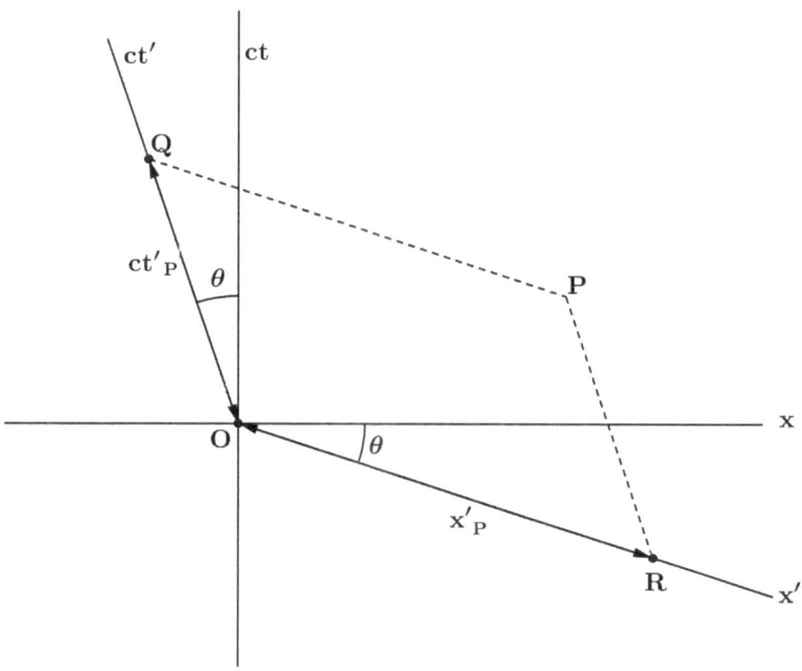

Figure 3.7 Space–time coordinates for frames S and S' f S' moves to the left relative to frame S.

parallel to the x' axis, which has $t' = 0$. The line $\overline{PQ}$ is a line of constant t'. Finally, the line $\overline{PR}$ is a line of constant x' and so is parallel to the t' axis.

Fig. 3.5 is so central to relativity that we should derive it in yet another way. The x' axis was drawn so that the frame S' measures the same speed limit c as does frame S. The x' axis consists of events that are simultaneous in frame S', events with $t' = 0$. Because this axis is tilted with respect to the x axis, events that are simultaneous in one frame are not simultaneous in the other. We know this from considerations in Chapter 2, but Minkowski diagrams bring that fact to the forefront.

We could also have found the x' axis by boosting the clocks A and C and the signal generator B into the S' frame, giving them a common velocity v with respect to the S frame. Then the world lines of the clocks would tilt to the right by an angle θ, $\tan\theta = v/c$ (Fig. 3.8). We have arranged the ct–x world lines so that the signal generator passes $x = 0$ at $t = 0$. The world lines of A, B, and C have a common tilt of θ because they all have a common velocity v with respect to frame S. The dashed line in Fig. 3.8 denotes the light rays that are tilted by 45 degrees, one to the right and one to the left, off the t axis. The intersection of the light rays with the world lines of the clocks synchronizes them in the moving frame—the signal generator lies halfway between the clocks, and all light rays travel at velocity c with respect to the moving objects, so light reaches the clocks simultaneously in the moving frame. So we have drawn the line $t' =$ constant accordingly in Fig. 3.8. It is clearly tilted with respect to the

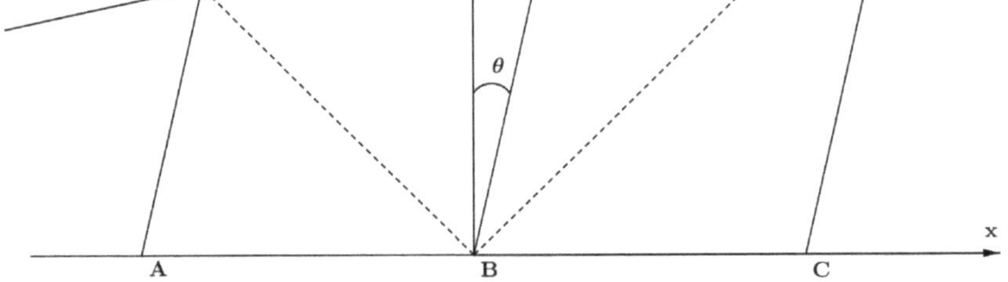

Figure 3.8 Synchronization of two clocks at rest in frame S' from the perspective of frame S.

x axis. In fact, the time it takes light to reach clock C is $ct_c = \ell' + vt_c$, and the time it takes light to reach clock A is $ct_A = \ell' - vt_A$. Here ℓ' is the distance, measured in frame S, between the signal generator and one of the clocks. The slope of the t' axis is then

$$\text{Slope} = (ct_c - ct_A)/(ct_c + ct_A). \tag{3.1}$$

Substituting $ct_c = \ell' + vt_c$ and $ct_A = \ell' - vt_A$ into the numerator of Eq. (3.1) shows that the slope is v/c. This simply confirms that the x' axis, a line of constant t' along which the moving clocks are synchronized, is tilted at the same angle above the x axis as the ct' axis is tilted below the ct axis. The constancy of the speed limit forces this result on us any way we choose to look at it. Note that ℓ' cancels out of these considerations—all we need to know is that the signal generator lies halfway between the clocks.

3.2 VISUALIZING THE RELATIVITY OF SIMULTANEITY, TIME DILATION, AND LORENTZ CONTRACTION

Let us use Minkowski diagrams to understand the three basic relativistic phenomena we have been discussing. The Minkowski diagram, because it shows both the ct–x and ct'–x' axes, will allow us to view all the phenomena from both points of view and see how paradoxes are avoided.

First, consider the relativity of simultaneity. In Fig. 3.9 we show two clocks at rest in frame S and separated by a proper distance ℓ_o. We show the

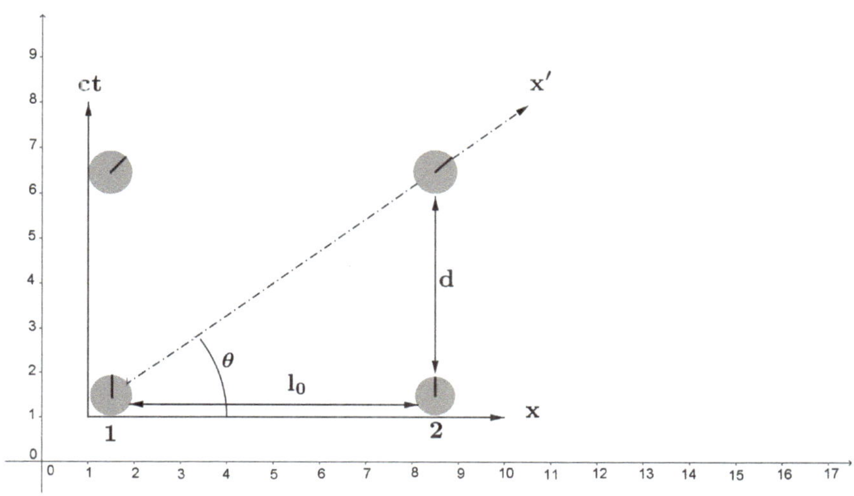

Figure 3.9 Illustration of the relativity of simultaneity: clocks synchronized in frame S are not synchronized in frame S'.

two clocks at $t = 0$ and at some later time t. We also show the x' axis, points of constant t', for the frame S' moving with velocity v, $\tan\theta = v/c$. We see instantly that the clocks at rest in frame S are not synchronized in frame S' if they are separated in the direction x. We also see that clock 1 is behind clock 2 when viewed in frame S'. This confirms the result of Chapter 2, but in a totally transparent fashion. In fact, it is trivial to calculate the basic formula of the relativity of simultaneity from the geometry of Fig. 3.9. We see in the figure that the distance $d = \ell_o \tan\theta = \ell_o v/c$. Because $d = ct$, we have $t = \ell_o v/c^2$, the time difference between clocks 1 and 2 in Fig. 3.9 at a specified t'.

Even more interesting, we can now turn the tables and view clocks at rest in frame S' from the perspective of an observer at rest in frame S. Consider the world line of two clocks at rest in S' in Fig. 3.10. The clocks on the x' axis ($t' = 0$) are synchronized as shown. Therefore, when compared at the same value of t, along the x axis as shown in the figure, clock 2 is behind clock 1. The moving clock that is spatially ahead lags in time. The quantitative effect can also be read off Fig. 3.10. By the law of sines, we read off the triangle,

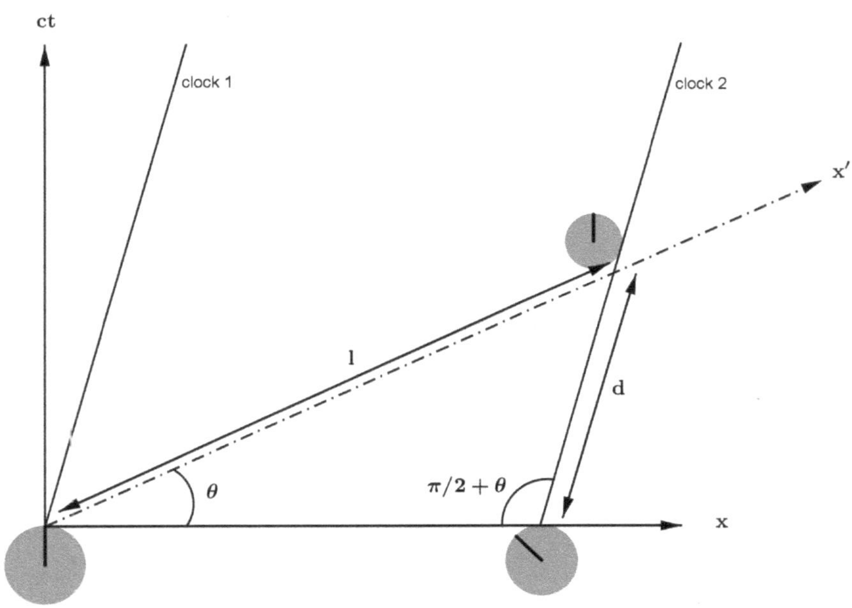

Figure 3.10 Illustration of the relativity of simultaneity: clocks synchronized in frame S' are not synchronized in frame S.

$$\frac{d}{\ell} = \frac{\sin\theta}{\sin(\pi/2+\theta)} = \frac{\sin\theta}{\cos\theta} = \tan\theta = \frac{v}{c}. \tag{3.2}$$

Because $d = ct'$, the time difference between the clocks in frame S', we again have $t' = \ell v/c^2$, where ℓ is the distance between the clocks in their rest frame S'.

Now turn to the visualization of Lorentz contraction using Minkowski diagrams. First, however, we must settle on the units of length and times used on the unprimed and primed axes in the diagram. The relation of space–time measurements in frame S and frame S' does not preserve angles, as is evident from our Minkowski diagram pictures. It also does not preserve lengths as we have seen in our discussions of time dilation and Lorentz contraction. This means that the scale of lengths of the x' and ct' axes relative to the x and ct axes is not unity. To see this, consider two frames having relative velocity v and $\gamma = 1\big/\sqrt{1 - v^2/c^2}$. Consider a rod of unit length at rest in the moving frame S'. By Lorentz contraction, we measure this length if we observe a rod of length γ in the frame S. We show this situation in Fig. 3.11—the rod at rest in frame S lies along $\overline{\mathrm{OA}}$, which has a length γ. The world lines of the ends of the rod are shown as the vertical lines in the Minkowski diagram.

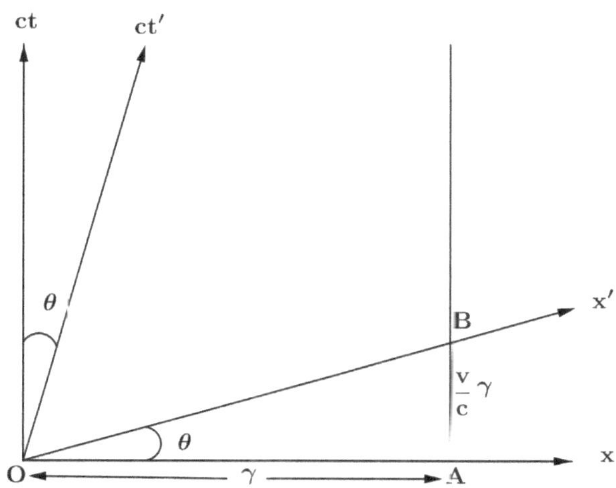

Figure 3.11 Setting the scales of relative lengths and times in a Minkowski diagram.

Because $\overline{\mathrm{OA}} = \gamma$ units of length in the S frame and because $\tan\theta = v/c$, the segment $\overline{\mathrm{AB}} = v\gamma/c$. Therefore, the length $\overline{\mathrm{OB}}$, which is the length of the rod as measured in the frame S', is

$$\overline{\mathrm{OB}} = \sqrt{\frac{v^2}{c^2}\gamma^2 + \gamma^2} = \sqrt{\left(1+\frac{v^2}{c^2}\right)\gamma^2} = \sqrt{\left(1+\frac{v^2}{c^2}\right)\Big/\left(1-\frac{v^2}{c^2}\right)}. \quad (3.3)$$

But $\overline{\mathrm{OB}}$ represents the length of a rod of unit length in the S' frame. So, in the Minkowski diagram there is a change of scale between the frames:

$$S'/S = \sqrt{\left(1+\frac{v^2}{c^2}\right)\Big/\left(1-\frac{v^2}{c^2}\right)}. \quad (3.4)$$

This effect only occurs when we visualize lengths of two frames on a single Minkowski diagram. It will show up in no other parts of our analysis, and we need it just to draw our figures accurately.

To visualize Lorentz contraction, choose $v/c = 3/5$, so the arithmetic works out neatly and $\gamma = 1\big/\sqrt{1 - v^2/c^2}$ is exactly $\gamma = 5/4$. The scale factor is $S'/S = \sqrt{34}/4 \approx 1.46$. So, if we consider a rod of unit length at rest in the frame S', we generate the Minkowski diagram in Fig. 3.12, where the length of the rod of unit proper length is measured as having a length $\Delta x = 1/\gamma = 1/(5/4) = 0.8$ in the frame S, by the Lorentz contraction formula. Turning the tables, a rod of unit length at rest in the frame S, whose ends sweep out the world lines $x = 0$ and $x = 1.0$, the dashed line, is measured to have length 0.8 along the x' axis. Again, we see from Fig. 3.12 that each observer measures the length of a moving rod as contracted by a factor of 1.25. The Minkowski diagram also shows that the reason for this contraction is the fact that simultaneous measurements in one frame are not simultaneous in a moving frame.

The visualization of time dilation is similar. Consider a clock at rest in frame S and let frame S' be moving to the right at relative velocity $v/c = 3/5$, so $\gamma = 5/4$ again. The t and t' axes are shown in Fig. 3.13. We have drawn the tick marks on the ct' axis to scale, $S'/S = \sqrt{34}/4 \approx 1.46$. At $ct = 5$ the observer at rest in frame S notes that the clock at rest in frame S' reads $ct' = 4$ because the dashed horizontal line of $ct = 5$ intercepts the ct' axis at $ct' = 4$. This is time dilation, $ct = \gamma ct'$, which reads $5 = (5/4)\cdot 4$. The observer at rest in frame S states that the moving clock at rest in frame S' is running slowly by a factor of $1/\gamma$. Similarly, the observer at rest in frame S'

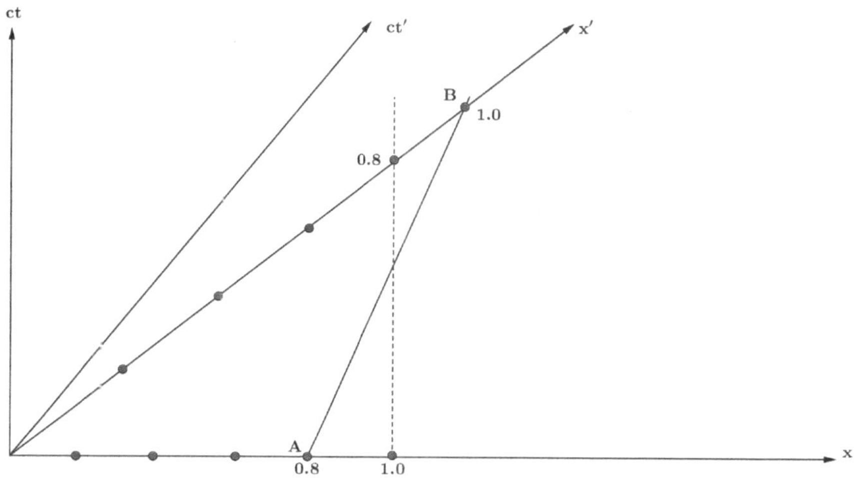

Figure 3.12 Illustration of Lorentz contraction on a Minkowski diagram.

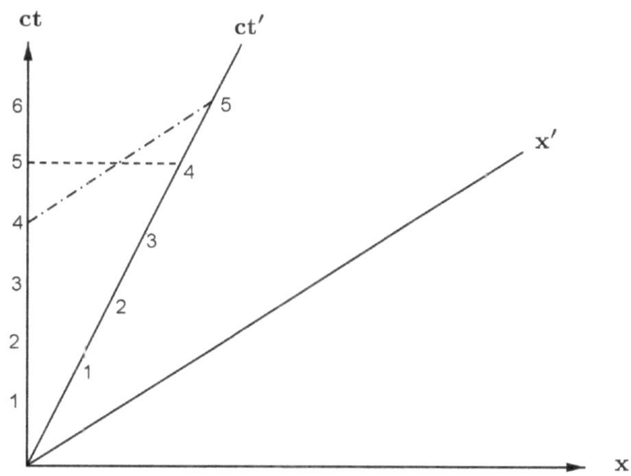

Figure 3.13 Illustration of time dilation on a Minkowski diagram.

notes that where his clock reads $ct' = 5$ the clock in frame S reads $ct = 4$ (follow the dash-dot line in Fig. 3.13).

In summary, both observers agree that moving clocks run slowly. There is no contradiction in this statement because the two observers do not share the same time axis.

3.3 THE DOPPLER EFFECT

Everyone knows that a train whistle sounds higher when it is approaching and lower when it is receding. A similar effect occurs in relativity when we observe a traveling wave. We need the quantitative details of this effect for light before we analyze the twin paradox. The Doppler effect is very important in astronomy, so it is worth examining in some detail.

Suppose that there is a signal generator at rest in frame S', which is moving at velocity v with respect to frame S, and an observer at rest in frame S sees the signal generator approaching her (Fig. 3.14). Let the wave train consist of n cycles that the observer detects in the course of a time interval Δt. Because light moves at velocity c with respect to frame S and the signal generator approaches at velocity v, the length of the wave train, $n\lambda$, observed in frame S is

$$n\lambda = c\Delta t - v\Delta t, \tag{3.5}$$

where λ is the wavelength of the light in frame S. As wavelength and frequency are related by $\nu\lambda = c$, we can write Eq. (3.5) as

$$n = \nu(1 - v/c)\Delta t. \tag{3.6}$$

But we can also write n in terms of quantities observed in frame S'. The source is at rest there and produces a frequency ν_o, say, over a time interval $\Delta t'$, so another expression for n is

$$n = \nu_o\Delta t'. \tag{3.7}$$

But $\Delta t'$ is the proper time interval for the signal generator, so Δt is given by time dilation,

$$\Delta t = \gamma\Delta t', \tag{3.8}$$

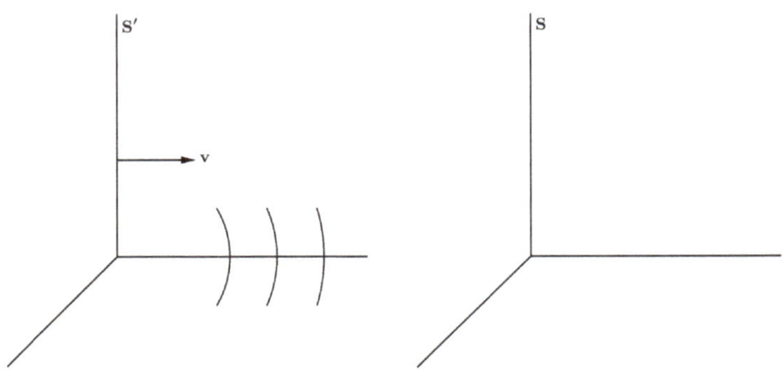

Figure 3.14 An observer at rest in frame S detects a light ray produced by a signal generator moving to the left at velocity v.

and Eqs. (3.6)–(3.8) give

$$\nu_o \Delta t/\gamma = \nu(1 - v/c)\Delta t. \tag{3.9}$$

Solving for the frequency detected in frame S,

$$\nu = \frac{\nu_0}{\gamma\left(1 - \frac{v}{c}\right)} = \nu_0 \sqrt{\frac{1 + v/c}{1 - v/c}} \quad \text{(approaching)} \tag{3.10}$$

So, the observer at rest in frame S measures the frequency of the light to be higher due to the approaching motion of the source.

Clearly, if the source had been receding we would have $v \rightleftarrows -v$, in Eq. (3.10),

$$\nu = \nu_e \sqrt{\frac{1 - v/c}{1 + v/c}} \quad \text{(receding)} \tag{3.11}$$

This effect is particularly important in astronomy where it produces the red shift of distant receding stars—the frequency of light coming from rapidly receding stars is reduced, as described by Eq. (3.11).

Consider a numerical example we use later in a discussion of the twin paradox. Let a source generate a light pulse once a year and suppose it moves away from us at $v/c = 0.8$. How long is the interval between the pulses that we receive? Substituting into Eq. (3.11), we find

$$\nu = \sqrt{\frac{1 - 0.8}{1 + 0.8}} \nu_o = \nu_o / \sqrt{9} = \nu_o/3. \tag{3.12}$$

So, we receive one signal every 3 years. If the source were moving toward us at $v/c = 0.8$, then we would observe $\nu = 3\nu_o$; that is, we would receive a signal every 4 months.

It will prove instructive to visualize the Doppler shift effect on a Minkowski diagram. Let the source be at rest in frame S, and let the frame S' recede at $v/c = \tan\,\theta = 0.8$. The Minkowski diagram is shown in Fig. 3.15.

We have set the scale using $S'/S = \sqrt{(1 + v^2/c^2)/(1 - v^2/c^2)} = \sqrt{41}/3 \approx 2.13$. We see in Fig. 3.15 that the clock at rest in frame S emits a light pulse at $ct = 1$, which is detected by the clock at rest in frame S' at $ct' = 3$. We were careful in the figure to set the relative scales of the time axes when we drew the tick marks, and we were careful to draw the light

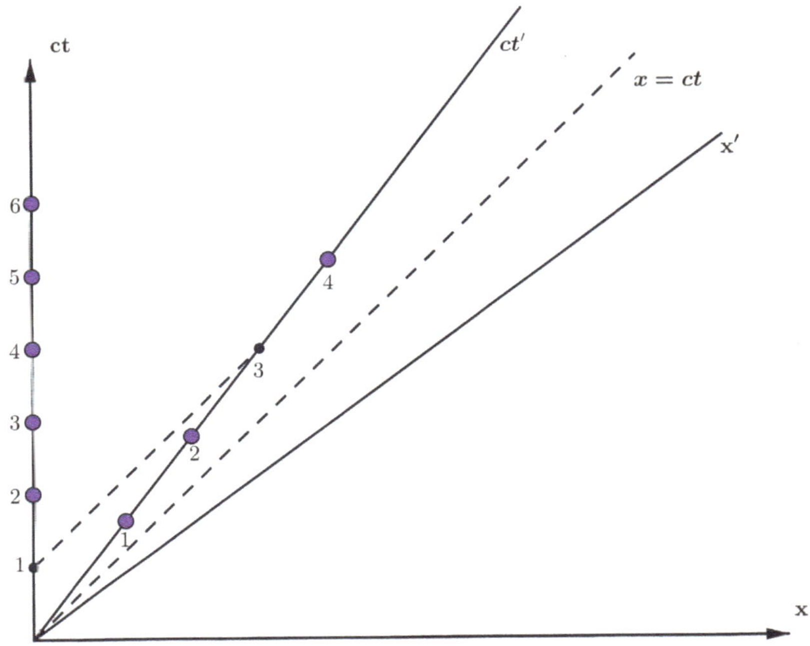

Figure 3.15 An observer at rest in frame S' detects a light beam produced annually in frame S. The observer receives one signal every 3 years.

ray with slope 1. We will see that the Doppler shift gives us a convenient way of monitoring the readings on clocks in relative motion. This will be handy in our discussion of the twin paradox.

3.4 THE TWIN PARADOX

Probably the most famous paradox in all of science is the twin paradox. We will resolve this paradox in several ways because it overthrows the notion of Newtonian time so dramatically. We do so in the context of Minkowski diagrams, although the algebra that could accompany the diagrams would do as well.

The paradox is the following. Suppose there are two identical twins, Mary and Maria. They have a terrible fight and Maria rockets away at a velocity of 80% of the speed of light, $\tan\,\theta = v/c = 4/5$ for 5 years, as measured by Mary, at rest in frame S. Then Maria has a change of heart. She returns as quickly as she departed and throws herself into Mary's arms as

Mary's clock strikes a decade. They embrace and make up, but something is not quite right—Maria looks considerably younger than Mary. In fact, Maria's clock indicates that only 6 years has passed! Yes, indeed, Maria is now 4 years younger than Mary. They are no longer identical twins!

Is this possible? Mary, who knows relativity, tries to understand the result by noting that Maria had a relative velocity of $v/c = 0.8$ throughout a decade, so 10 years must be the time-dilated interval of the time passed on a wristwatch attached to Maria. Because $\gamma = 1\big/\sqrt{1-0.8^2} = 5\big/3$, time dilation implies $\Delta t = \gamma \Delta t'$ or $10 = (5/3)\cdot 6$, Mary feels that she understands the time interval involved. But Maria, distraught at now being Mary's "younger" sister, contends that Mary was, from her perspective, racing away at a relative velocity $0.8c$ and she should have aged more slowly! How can both see one another's clocks as running slowly and a paradoxical situation not be found when their trips are over?

In light of our earlier discussions of time dilation, Lorentz contraction, and the relativity of simultaneity, the reader, hopefully, is not as puzzled by this apparent paradox as the uninitiated. Let us go through the Minkowski diagram analysis of the trips and see that all works out. Yes, Maria ends up 4 years younger than Mary and has to live with that—at least until the next trip.

The trip is shown in Fig. 3.16. Mary and her clock reside at $x = 0$ and 10 years click off. Maria dashes off at $v/c = 0.8$ for 5 years according to

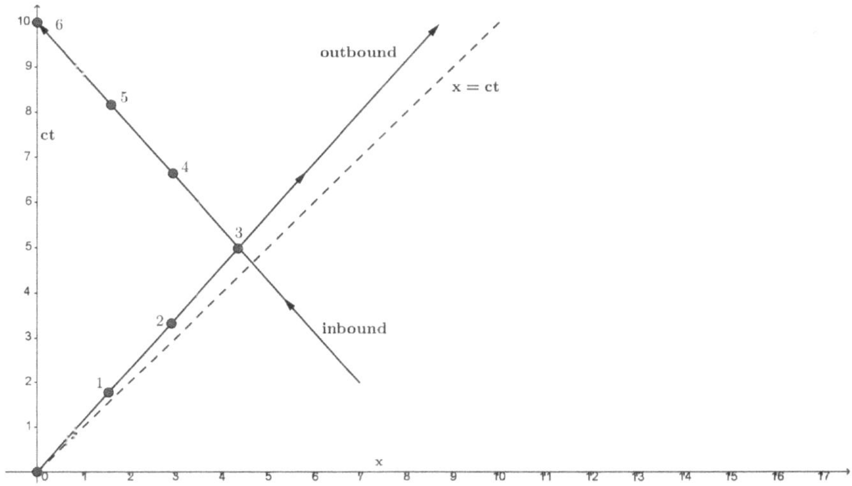

Figure 3.16 The space–time trips of the twins: Mary, who stays home, and Maria, who travels out and back.

Mary and then returns at $v/c = -0.8$ for another 5 years. From Mary's perspective, Maria's clock, her wristwatch, say, is running slowly, and $\Delta t = \gamma \Delta t'$ applies with $\gamma = 1\big/\sqrt{1 - v^2/c^2} = 1\big/\sqrt{1 - 0.8^2} = 5\big/3$, so Maria ages 3 years during her outbound trip and 3 more years during her inbound trip.

All of this seems clear, except the portion of Maria's trip where she turns around. At this point she experiences acceleration and her frame is no longer inertial. There is a simple way to deal with this. We could consider the outbound leg of the trip as made by one rocket with $v/c = 0.8$ and the inbound trip as made by *another* rocket with $v/c = -0.8$. Let the inbound and outbound rockets pass closely by one another at $ct = 5$ and synchronize their clocks. In this way we never have to consider acceleration, and our analysis holds without compromise.

To get the full understanding we seek, we must consider the trip from the perspective of Maria. She travels outbound for 3 years as measured on her wristwatch. From Maria's perspective, Mary is racing away at $v/c = -0.8$. Maria observes Mary's clock running slowly and records that Mary ages by an amount $\Delta t'/\gamma = 3/(5/3) = 9/5 = 1.8$ years. We see this in the Minkowski diagram, as shown more explicitly in Fig. 3.17, where the line $\overline{AB}$ of constant t' is parallel to the x' axis. So, Maria measures that Mary has aged only 1.8 years throughout her outbound trip.

Now, something special happens at the point B where Maria turns back toward her sister—the line of constant t' tilts because we are jumping onto a new inertial frame, one with $v/c = -0.8$. This frame is shown in Fig. 3.18. Thus, during the inbound portion of the trip, the lines of constant t' are parallel to the x' axis in Fig. 3.18, which tilt by angle $-\theta$ *below* the x axis.

The lines of constant t' are shown in Fig. 3.19 for both legs of the trip. We see that the turnaround, where $ct' = 3$ light-years, corresponds to both $ct = 1.8$ light-years *and* $ct = 8.2$ light-years. Then, when $ct' = 4$, $ct = 8.8$ light-years, and when $ct' = 5$, $ct = 9.4$ light-years. It is quite interesting that Maria concludes that $8.2 - 1.8 = 6.4$ years pass on Mary's clock during the turnaround! (When we analyze the twin paradox in the context of general relativity in Chapter 11, our focus will be the turnaround and its acceleration.) Maria can now understand how 10 years pass on Mary's clock, even though Maria measures that Mary's clock runs slower than her own!

Another perspective on this curious sequence of events is afforded by the Doppler effect. This was first illustrated by the mathematician C. G. Darwin [1]. Mary and Maria can monitor one another's clocks by agreeing

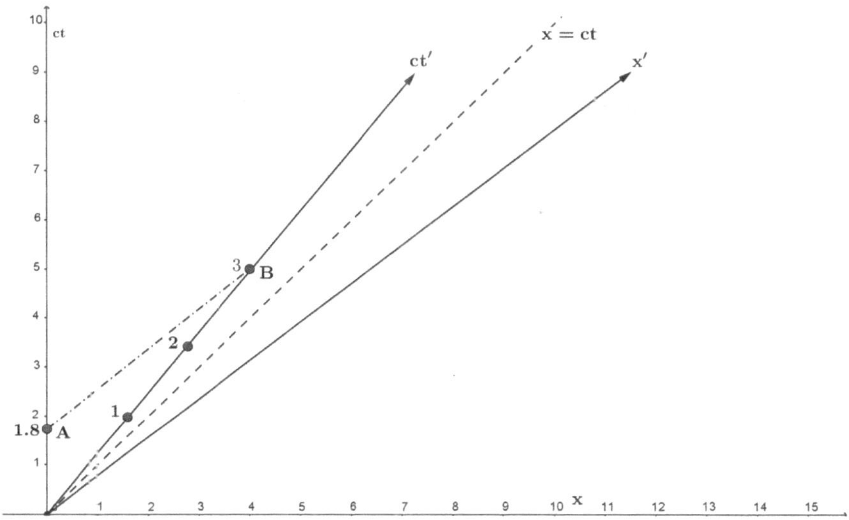

Figure 3.17 Mary's aging from Maria's perspective during her outbound trip.

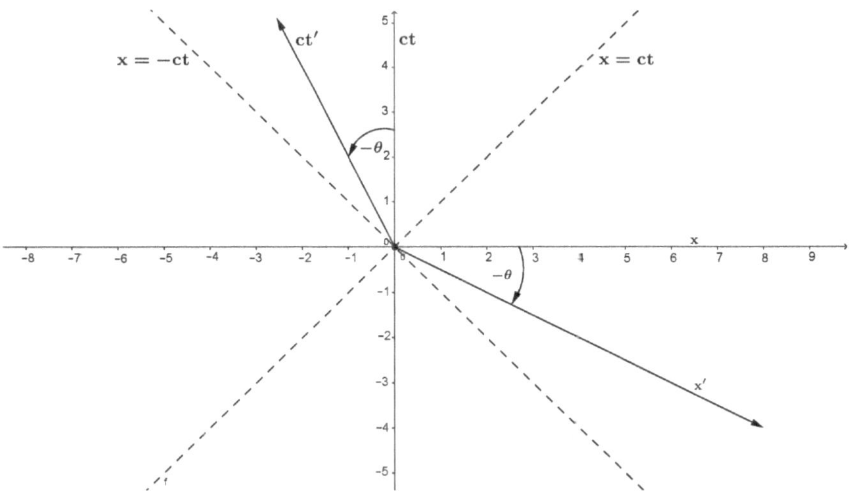

Figure 3.18 The coordinates $ct' - x'$ of Maria's rest frame during her inbound trip back to Mary.

to send one another signals on their birthdays. So, Mary sends a greeting to Maria on each of her 10 birthdays. This is shown in Fig. 3.20.

Mary sends a birthday greeting at $ct = 1$ light-year, and Maria receives it at the turnaround point $ct' = 3$ light-years, as shown. This agrees with the Doppler formula $\nu = \sqrt{(1 - v/c)/(1 + v/c)} = \sqrt{(1 - 4/5)/(1 + 4/5)} = 1/3$. So, 3 years pass in Maria's frame before she receives the birthday greeting from Mary.

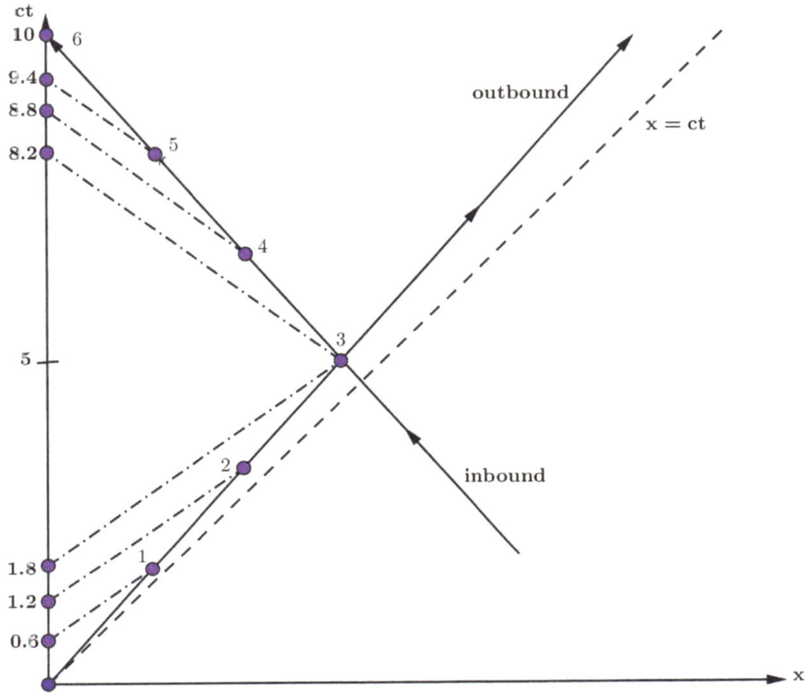

Figure 3.19 The dash-dotted lines are lines of constant t', Maria's time, during the outbound and inbound portions of her trip.

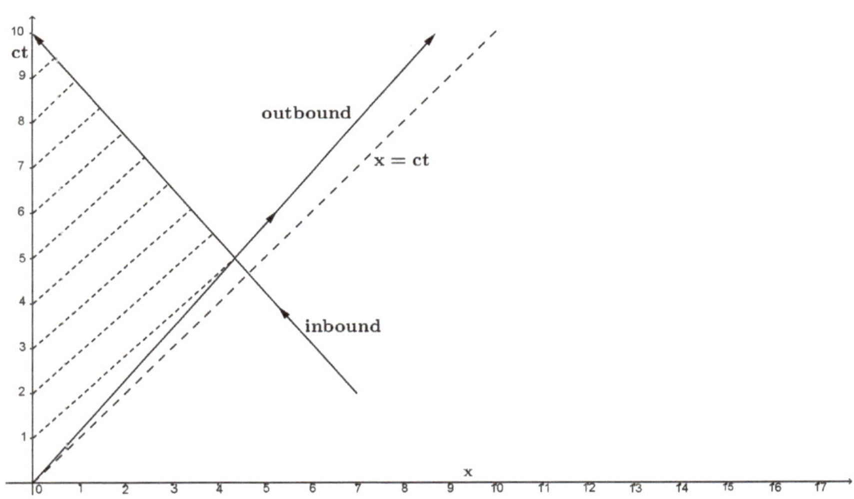

Figure 3.20 Mary sends 10 birthday greetings by light pulses (*dashed lines*) to Maria during the round trip.

Maria's position is $x_m = (4/5)ct$ during the outbound trip, and the light ray sent by Mary on her first birthday travels on $x = c(t - 1)$. So, Maria receives the signal at $(4/5)\ t = t - 1$ or $t = 5$. Maria is now 4 light-years from Earth and is about to turn around. The next eight light rays sent by Mary and received by Maria are shown in the figure. Maria receives nine signals during the second half of her journey (the inbound part) and just one during the first half. She receives 10 signals and knows that Mary ages 10 years while she ages only 6 years. The Doppler formula for the inbound part of Maria's journey reads $\nu = \sqrt{(1 + v/c)/(1 - v/c)} = \sqrt{(1 + 4/5)/(1 - 4/5)} = 3$, so Maria receives three signals each year from Mary for the duration of the inbound journey, 3 years on Maria's wristwatch, accounting for nine signals.

Finally, we turn the tables again and suppose that Maria sends greetings to Mary on each of her birthdays. This situation is shown in Fig. 3.21. During the outbound trip, Maria sends three greetings and Mary receives one every 3 years, at years 3, 6, and 9 on her clock. Then after the turnaround, Maria sends three more greetings and Mary receives them every 4 months between years 9 and 10 on her wristwatch. Again, both Mary and Maria agree that Mary has aged 10 years while Maria has aged 6.

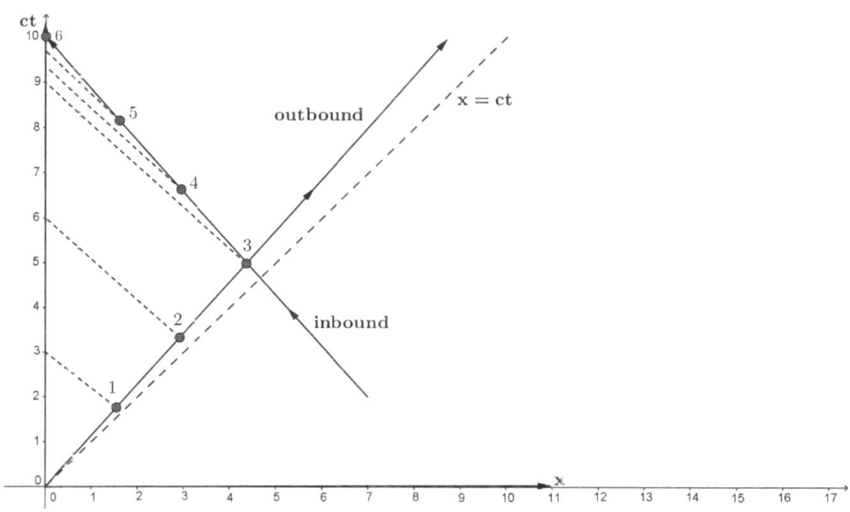

Figure 3.21 Maria sends six birthday greetings by light pulses (*dashed lines*) to Mary during the round trip.

3.5 EINSTEIN MEETS SHAKESPEARE—RELATIVISTIC HISTORY

The twin paradox illustrates how much of present culture and history is tied to the Newtonian concept of absolute time. In a world where space travel at relative speeds approaching the speed limit would be possible, everything would be so different. For example, the ascendancy of the throne in the days of the kings of England would have to be revised. The very idea that the son of the king of England will later become the king who will have another son who will then become king, and so on, is rooted in the notion of absolute time. For example, say that Prince William was jealous of his older brother Prince Harry because Prince William has little chance to rule as king. Well, he could always take a relativistic journey akin to that taken by Maria and time things so that on his return King Harry, who, we suppose, is childless, is on his deathbed while he, Prince William, has aged only a few years. Shortly, then, Prince William would become King William and rule a long and glorious time—barring other surprises!

3.6 REALITY, HORSE RACING, AND THE SPEED LIMIT

The fact that the speed of light is independent of the speed of its source is quite hard to grasp intuitively. It emphasizes that we, humans, have our intuitions rooted in the particulars of our existence on earth where massive bodies typically have relative speeds, which are small compared with the speed of light, and Newtonian kinematics seem adequate.

Because of this, it is worthwhile to demonstrate this fact that the speed of light is independent of the speed of its source, using the principles of relativity directly. Let us do this by considering a "horse race" where the fastest horse, "Lightening," can run arbitrarily close to c. Lightening runs a race in the frame S where the track is at rest and beats his two competitors, Georgie and Dickie, in that order. Track officials have found that Georgie and Dickie have a history of cheating, so they set up a camera at the finish line, which takes pictures of each horse's front paw as it touches the finish line. There is also a clock at the finish line to record the time of the event in frame S. The horses run the race and try as they may; Georgie and Dickie finish in second and third place. Each horse is awarded a picture of their finish showing the time and place so there can be no doubt.

Now let us observe the same race in a frame S', which moves at a velocity $\boldsymbol{v}$ with resect to S. Of course the results of the race are the same in every frame, so any observer at rest in S' finds that Lightening beats Georgie who beats Dickie. The order of the three finishes does not depend on the frame in which the race is viewed: the reality of the race does not depend on how you view it. The pictures taken by the camera cannot lie and can be passed from frame to frame.

All this is patently self-evident. But now realize that the same result applies to *any* race: Lightening always wins because he runs arbitrarily close to the speed limit c in the original frame S. Try as they may, Georgie and Dickie cannot beat him. But this means that he, Lightening, must be running arbitrarily close to c in *any* frame as he wins all possible races.

Returning to experiments with light, we learn that the principles of relativity alone dictate that a beam of "light" travels at the speed limit c in all frames if it does so in any particular frame. It does not matter how you created the beam.

It will be interesting to see how these matters of principle work out when we derive the generalization of the addition of velocities for Einstein's relativity. Somehow that formula must have the property that when the speed of light is added to any ordinary velocity, the result will be just the speed of light!

When stated this way, our little exercise at the track looks quite perplexing indeed!

PROBLEMS

3-1. Two inertial coordinate systems S and S' are moving with respect to one another at a relative velocity $c/2$. Draw a Minkowski diagram relating these two frames.

- **a.** Calibrate the axes and mark a unit length along each following Eq. (3.4).
- **b.** Plot the events on the diagram: (1) $ct = 1$, $x = 1$; (2) $ct = 2$, $x = 0$; (3) $ct' = 1$, $x' = 1$; and (4) $ct' = 2$, $x' = 0$.
- **c.** Take each event in part (b) and determine from the Minkowski diagram its coordinates in the other frame.

3-2. Radar trap! A policeman aims his stationary radar transmitter backward along the highway toward oncoming traffic. His radar detector picks up the reflected waves and analyzes their frequencies. Suppose that his transmitter generates waves at a frequency ν_o and detects the waves reflected by an approaching speeding car at frequency ν_r.

a. Draw a Minkowski diagram showing the world lines of the stationary policeman, the approaching speeding car, and several transmitted and reflected radar waves.

b. From the geometry of your Minkowski diagram, relate the time between transmitted radar waves to the time between reflected waves, as detected by the stationary policeman, and derive $\nu_r = [(1 + v/c)/(1 - v/c)]\nu_o$.

c. Use this result to derive the Doppler shift formula obtained in the chapter above. Call the frequency that an observer in the speeding car measures ν'. Argue on the basis of the first principle of relativity that $\nu'/\nu_o = \nu_r/\nu'$. Combine this result with part (b) to derive the standard Doppler shift formula, $\nu' = \sqrt{(1 + v/c)/(1 - v/c)}\nu_o$.

d. Suppose that the car's speed is 100 mph and $\nu_o = 10^{10}$ cps. Predict ν_r approximately by linearizing the formula in part (b). (Because 100 mph is tiny compared with the speed of light, this approximate treatment is very good.) Can handheld radar equipment carried by your local law enforcement department detect such a small fractional frequency shift? [2].

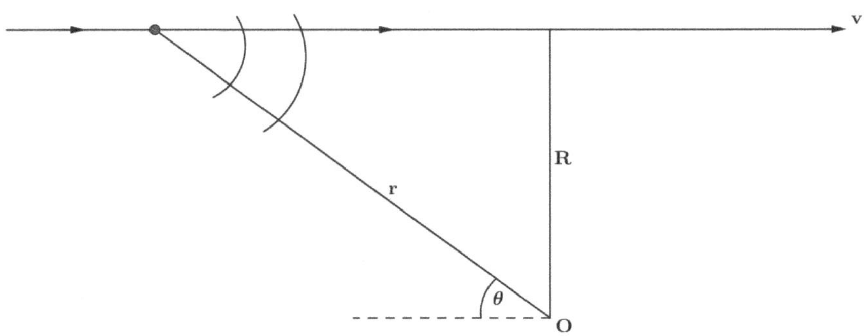

Figure 3.22 Set up of transmitter and observer *O* for the transverse Doppler effect.

3-3. Transverse (quadratic) Doppler shift. The Doppler shift we have discussed so far occurs when the transmitter and the detector of the waves are heading directly toward or away from one another. It might happen, however, that the transmitter and the detector approach one another at an angle θ as shown in Fig. 3.22. The figure shows the transmitter with a velocity v traveling past an observer with the distance of closest approach, labeled R. When $\theta = 90$ degrees, we have the transverse Doppler shift.

a. Show that there is a Doppler shift in the case $\theta = 90$ degrees, which is given just by time dilation. Calling ν_o the frequency of light in the rest frame of the transmitter and ν' the frequency of light in the rest frame of the observer, at the origin in the figure, derive $\nu' = \nu_o/\gamma$, where $\gamma = 1/\sqrt{(1 - v^2/c^2)}$, as usual. Note that for small v^2/c^2, $\nu' \approx (1 - v^2/2c^2)\nu_o$ and hence the name quadratic Doppler shift.

b. Armed with the result of part (a) and the standard Doppler shift formula for head-on motion, obtain the Doppler shift formula for the general case shown in the figure, $\nu' = \nu_o/[\gamma(1 - (v/c)\cos\theta)]$. (*Hint*: Resolve the relative motion between the transmitter and the observer into radial motion, along the radius r shown in the figure, and transverse motion.)

3-4. Two photons travel along the x axis of frame S with a constant distance d between them. Prove that in frame S' the distance between the photons is $d\sqrt{(1 + v/c)/(1 - v/c)}$. (Hint: One approach is to write the Doppler shift formula for the wavelength of light instead of its frequency.)

3-5. When we resolved the twin paradox, we noted that from Maria's perspective, Mary ages 6.4 years as Maria jumps from the outgoing to the incoming rocket. Obtain this numerical result from our relativity of simultaneity formula, vx/c^2, by noting that v changes by $2v$ as Maria jumps between the rockets. Choose $v = 0.8c$.

3-6. Here is a relatively easy paradox based on an incomplete understanding of time dilation—your task is to resolve it.

Suppose that frame S', a huge transport rocket, moves in the x direction relative to frame S, a space station, at velocity v, as usual. Suppose that there is an astronaut in the huge transport rocket, who has a rocket-powered backpack and is traveling in the $-x$ direction at speed $-v$ relative to the transport rocket. Therefore, our astronaut is actually at rest with respect to the space station. A critic of relativity might argue that time dilation states that clocks at rest in the transport rocket run slowly compared with those at rest in the space station, and the astronaut's clock runs slowly compared with those at rest in the transport rocket. Therefore, the astronaut's clock is predicted to run slow-squared compared with those at rest in the space station! But the astronaut is actually at rest in the space station, so we have an apparent contradiction!

What's wrong with this argument? Clarify it with a Minkowski diagram showing the three relevant time axes and spatial axes of the three frames involved.

3-7. Consider Problem 2.1 once more. Diagram the events and world lines of that problem in a Minkowski diagram. Establish the scales on the time and spatial axes so you can make quantitative predictions from the Minkowski diagram itself in either frame. Answer Problem 2.1(a)−(h) directly from your Minkowski diagram.

3-8. The wavelength of a spectral line measured to be λ on earth is found to increase by 50% on a far distant galaxy. What is the speed of the galaxy relative to the earth?

REFERENCES

[1] C.G. Darwin, The clock paradox in relativity, Nature 180 (1957) 976.
[2] T.M. Kalotas, A.R. Lee, Am. J. Phys. 58 (1990) 187.

CHAPTER 4

Lorentz Transformations (Boosts), Addition of Velocities, and Invariant Intervals

Contents

4.1 LORENTZ TRANSFORMATION (BOOSTS)

Now that we understand how relativity works, let us derive the generalization of the Galilean transformation relating the time and position of an event measured in one inertial frame to those measurements in another inertial frame. Recall the Galilean result. We have a frame S' moving along the x axis of frame S at velocity v. Then if a point x' is observed at time t' in frame S', an observer in frame S would assign the values

$$x = x' + vt \tag{4.1a}$$

$$t = t' \tag{4.1b}$$

to the point, assuming that the origins of the two frames, $x = x' = 0$, coincided at $t = t' = 0$.

In relativity, Eqs. (4.1a) and (4.1b) are replaced by the Minkowski diagram shown in Fig. 4.1. Instead of the picture, we want explicit formulas for x and t, given x' and t' for an event in Fig. 4.1. Armed with our knowledge of Lorentz contraction, time dilation, and the relativity of simultaneity, this is a snap.

Special Relativity, Electrodynamics, and General Relativity
ISBN 978-0-12-813720-8
https://doi.org/10.1016/B978-0-12-813720-8.00004-0

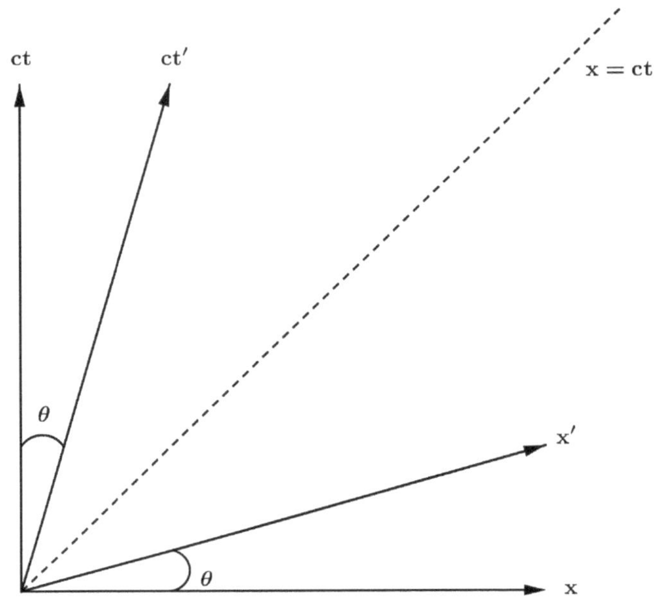

Figure 4.1 Minkowski diagram showing the space–time coordinates of inertial frames S and S' in relative motion.

Consider the relativistic generalization of Eq. (4.1a) in Fig. 4.2. Galileo states that the distance vt and x' add up to x. However, because x' is measured in frame S', Lorentz contraction shrinks it to x'/γ when observed in frame S. So, Eq. (4.1a) is replaced by

$$x = x'/\gamma + vt,$$

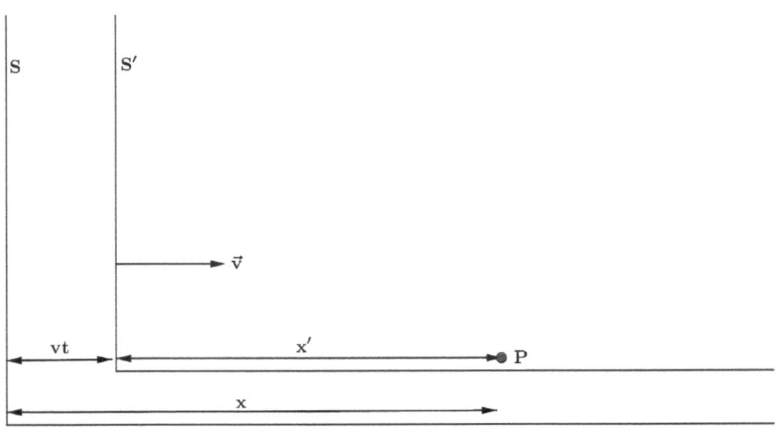

Figure 4.2 Space–time visualization of the $x - x'$ transformation law.

which can be solved for x',

$$x' = \gamma(x - vt). \tag{4.2a}$$

This tells us how the x' axis in the Minkowski diagram Fig. 4.1 is related to the x and ct axes there.

Next we need the generalization of Eq. (4.1b). In other words, we need to know how the ct' axis in the Minkowski diagram Fig. 4.1 is related to the ct and x axes there. But this relation must have the same form as Eq. (4.2a), in order that both frames S and S' observe the same speed of light. In other words, take Eq. (4.2a) and replace x' with ct', x with ct and t with x/c. So,

$$ct' = \gamma(ct - vx/c)$$

or

$$t' = \gamma(t - vx/c^2). \tag{4.2b}$$

Once more with emphasis—the central idea in this derivation is the constancy of the speed of light—once we know how x' is related to x and t, t' must be related to x and t so that both frames measure the identical speed of light, as expressed pictorially in the Minkowski diagram.

Eqs. (4.2a) and (4.2b) constitute the Lorentz transformations, often called "boosts." It is a central result of special relativity. It incorporates all we know about time dilation, Lorentz contraction, and the relativity of simultaneity. To see how the formulas work, let us obtain our three basic results by taking special cases of Eqs. (4.2a) and (4.2b).

4.1.1 Time Dilation

Consider two times t_2 and t_1 that occur on a clock at rest in frame S. From Eq. (4.2b) an observer in S' would measure the time interval

$$t_2' - t_1' = \gamma\left[t_2 - t_1 - \frac{v}{c^2}(x_2 - x_1)\right] = \gamma(t_2 - t_1),$$

where we used the fact that the clock is at rest in frame S, so $x_2 = x_1$. This is our usual time dilation formula because $t_2 - t_1 = \Delta\tau$, the proper time interval.

4.1.2 Lorentz Contraction

Consider a rod at rest in frame S' extending from x_1' to x_2'. Its length is measured in frame S by noting the corresponding x_1 and x_2 at a given time $t_2 = t_1$. Eq. (4.2a) predicts

$$x_2' - x_1' = \gamma(x_2 - x_1).$$

Because $x_2' - x_1' = \ell_o$ is the proper length of the rod, we have Lorentz contraction, $\Delta x = \ell_o/\gamma$.

4.1.3 Relativity of Simultaneity

Consider five clocks at rest and synchronized in frame S [1]. Let each clock be separated from its neighbor by ℓ_o (Fig. 4.3). Now measure the time of each clock in frame S', as shown in the Minkowski diagram (Fig. 4.4). We observe the times on each clock at a given single time t' in frame S', as shown in the diagram. Let us apply Eq. (4.2b) and note that clock C at $x = 0$ gives the value of t' directly, $t' = \gamma t_c$. So,

$$\gamma t_c = \gamma\left(t - \frac{v}{c^2}x\right),$$

which we solve for t,

$$t(x) = t_c + \frac{v}{c^2}x \quad t' = \text{constant}. \tag{4.3}$$

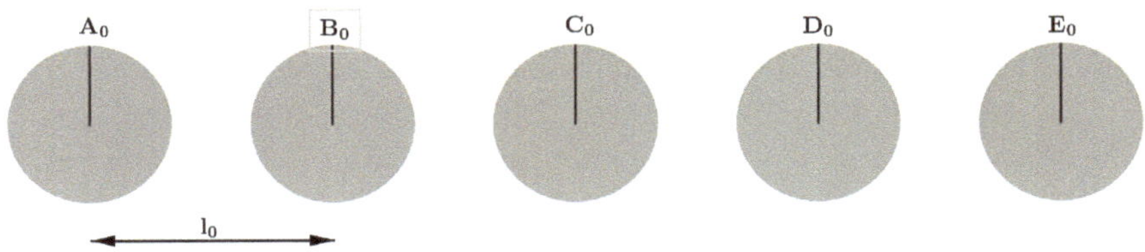

Figure 4.3 Five clocks at rest in frame S each separated in the x direction by a proper distance l_0.

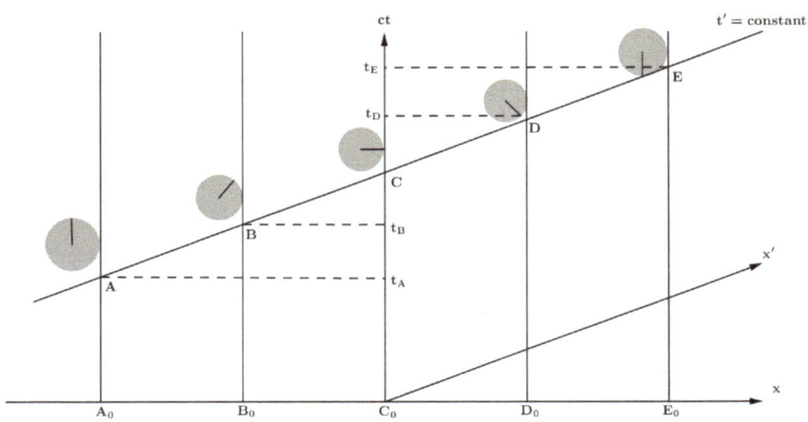

Figure 4.4 The five clocks of Fig. 4.3 as perceived in frame S' at a given time t'.

This result expresses the relativity of simultaneity effect—the clocks that are moving in frame S are not synchronized, with the leading clock A behind the trailing clock E by an amount $v\Delta x/c^2$.

The Lorentz transformation formulas are easy to apply, but they frequently do not elucidate as much as more elementary but logically challenging arguments as presented in Chapter 2.

4.2 RELATIVISTIC VELOCITY ADDITION

One of the hallmarks of Newtonian physics is the "obvious" fact that velocities add. Consider the Galilean transformation Eq. (4.1a) and let the particle move at velocity $v_p{}'$ relative to frame S'. What then is its velocity, $v_p = x/t$, with respect to frame S? Substituting into Eq. (4.1a) we have

$$v_p t = v_p{}' t' + vt.$$

Using the absolute nature of Newtonian time, $t = t'$, we see that

$$v_p = v_p{}' + v. \tag{4.4}$$

This result is used in day-to-day physics problems and serves us well as long as all the velocities are small compared to the speed limit c. It is clearly inconsistent with special relativity because it does not respect the existence of a speed limit—by adding velocities linearly we can generate a velocity for a physical particle as high as we like.

We need to know what replaces Eq. (4.4) in special relativity. To answer this, we go back to the Lorentz transformations, Eqs. (4.2a) and (4.2b) and substitute in $x' = v_p{}'\, t'$ and $x = v_p t$ to describe a particle in uniform motion. Now Eq. (4.2a) becomes

$$v_p{}' t' = \gamma(v_p t - vt) = \gamma(v_p - v)t. \tag{4.5a}$$

Eq. (4.2b) becomes, upon substituting, $x = v_p t$,

$$t' = \gamma\left(t - \frac{v}{c^2} v_p t\right) = \gamma\left(1 - \frac{v v_p}{c^2}\right)t. \tag{4.5b}$$

where the second term expresses the relativity of simultaneity effect. Substituting back into Eq. (4.5a), we have

$$v_p{}' = \frac{v_p - v}{1 - v v_p / c^2}. \tag{4.5c}$$

We can also solve Eq. (4.5c) for v_p,

$$v_p = \frac{v_p{}' + v}{1 + v v_p{}'/c^2}. \tag{4.5d}$$

Note that this result is nonlinear in v and $v_p{}'$ in just the right way to enforce the speed limit! For example, let the particle move at the speed limit in frame S', $v_p{}' = c$. Then,

$$v_p = \frac{c + v}{1 + vc/c^2} = c \cdot \frac{c + v}{c + v} = c.$$

In other words, light has velocity c with respect to both frames S and S' even though they are in relative motion! This result was built into our formalism by the second postulate of relativity, but it is still interesting to see how it is enforced. The reader should check the even more peculiar case of letting $v_p{}' = -c$ in Eq. (4.5d). Again we find $v_p = -c$ for *all* v, no matter how large!

This derivation also suggests that although transverse distances y and z are uneffected by the boost between frames S and S',

$$y' = y$$
$$z' = z,$$

it is not true that transverse velocities are the same in both frames. Let the particle again have a velocity of v_p in the x direction in frame S but let it also have a y component of velocity, $y = u_p t$. In the frame S' we have $y' = u_p{}'t'$, and $u_p{}'$ can be computed using Eq. (4.5b),

$$u_p{}' = \frac{y'}{t'} = \frac{y}{t'} = \frac{u_p t}{t'} = \frac{u_p}{\gamma(1 - v v_p/c^2)}. \tag{4.6}$$

This complicated formula involves both the velocities in the x direction, v and v_p, and those in the y direction, u_p and u'_{p}. We will need Eq. (4.6) later when we discuss dynamics and relativistic momentum.

In summary, it is interesting to revisit Eq. (4.5d) and reiterate the source of the crucial nonlinearity, the denominator, $1 + v v_p/c^2$. Which of the three effects, Lorentz contraction, time dilation, or the relativity of simultaneity, is responsible here? The γ factors in Eqs. (4.5a) and (4.5b) cancel out of (4.5c), so Lorentz contraction and time dilation are not the culprits here. Inspecting Eq. (4.5b), we see the relativity of simultaneity at work—the fact that a moving clock that is attached to the particle is increasing its x' coordinate as $v_p' t'$, produces the nonlinearity that makes the speed of light truly universal.

4.3 CAUSALITY, LIGHT CONES, AND PROPER TIME

When observers measure physical phenomena, such as the position and time of events, they record space–time coordinates whose values depend on their coordinate systems. Some quantities are the same in all reference frames and they are particularly useful and significant. The speed of light is such an invariant—it is measured to be the speed limit, $3.0 \cdot 10^{10}$ cm/s, in all inertial frames. The proper times of clocks and proper lengths of rods are also invariants. This last remark might sound trivial—to measure the proper time of a clock, we must boost ourselves to the rest frame of the clock of interest and note a proper time interval. Similarly, for a proper length of a rod. The term "proper" indicates exactly how the measurement is to be done.

All this would be more interesting if we could infer the proper time interval passing on a clock by making measurements in another inertial frame. Suppose the clock is at rest in frame S and a proper time $\Delta\tau$ passes. In another inertial frame S', a time interval $\Delta t'$ will pass while the clock moves a distance $\Delta x' = -v\Delta t'$. $\Delta t'$ is related to $\Delta\tau$ by time dilation: $\Delta t' = \Delta\tau / \sqrt{1 - v^2/c^2}$. However, we would like to have a relation between $\Delta\tau$, $\Delta t'$, and $\Delta x'$ so we can calculate $\Delta\tau$ from $\Delta t'$ and $\Delta x'$ without knowing v. This would be an expression of the invariance of $\Delta\tau$.

This problem is analogous to a familiar one in Euclidean geometry. We consider a rotation by angle θ between two frames (Fig. 4.5). In the frame O, the point P has coordinates (x, y) and in O' it has coordinates (x', y'). They are related by a rotation,

$$x' = x \cos\theta + y \sin\theta$$
$$y' = y \cos\theta - x \sin\theta.$$

So, (x, y) is related to (x', y') by a linear transformation. We want to know what quantities are the same in both frames O and O'. In other words, what quantities are preserved by rotation? Intrinsic geometrical relations are preserved, such as the angles between vectors and the lengths of vectors. For example, the distance P from the origin is independent of the choice of coordinates,

$$x^2 + y^2 = x'^2 + y'^2.$$

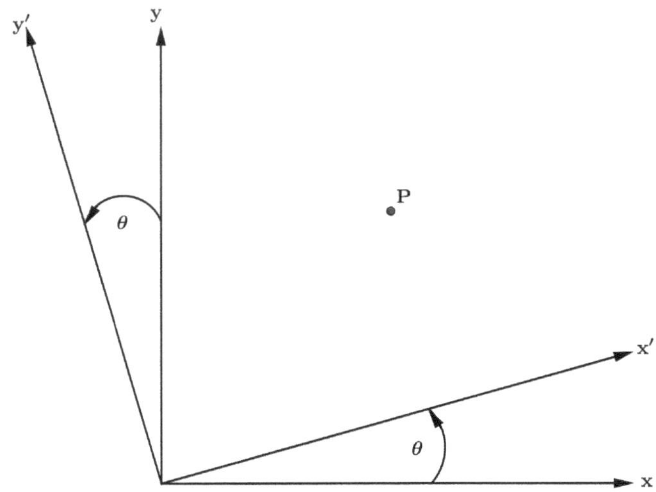

Figure 4.5 A rotation in two-dimensional Euclidean geometry.

This relation is independent of θ—it is true for any two frames related by rotation.

What is the analogous result for Lorentz transformations? These transformations, as is clear from Minkowski's diagrams, do not preserve angles or Euclidean distances. So, $c^2(\Delta t')^2 + (\Delta x')^2$ will not work. Using $\Delta t' = \Delta\tau/\sqrt{1 - v^2/c^2}$ and $\Delta x' = -v\Delta t' = -v\Delta\tau/\sqrt{1 - v^2/c^2}$, we can calculate $c^2(\Delta t')^2 + (\Delta x')^2$ and check that it is not an invariant. But in so doing, we note that a change of sign does the trick:

$$c^2\left(\Delta t'\right)^2 - \left(\Delta x'\right)^2 = (\Delta\tau)^2\frac{c^2}{1 - v^2/c^2} - (\Delta\tau)^2\frac{v^2}{1 - v^2/c^2} = c^2(\Delta\tau)^2.$$

We learn from this that we can tell time in frame S by using measurements in another inertial frame S',

$$c\Delta\tau = \sqrt{\left(c\Delta t'\right)^2 - \left(\Delta x'\right)^2}. \tag{4.7}$$

We also learn that $(c\Delta t)^2 - (\Delta x)^2$ is an invariant (i.e., it is the same in all inertial frames), and so is called a "Lorentz invariant."

This analysis also applies to the proper length of a rod, l_0. Let the rod be at rest in the frame S and orient the rod so that one end lies at $x = 0$ and the other is at $x = l_0$, so $\Delta x = l_0$ and $\Delta t = 0$. In the frame S', $\Delta x' = \gamma l_0$ and $\Delta t' = -\gamma v \Delta x/c^2 = -\gamma v l_0/c^2$. Therefore the invariant interval is,

$$c^2(\Delta t')^2 - (\Delta x')^2 = \gamma^2 v^2 l_0^2/c^2 - \gamma^2 l_0^2 = \gamma^2 l_0^2(v^2/c^2 - 1) = -l_0^2$$

which shows the invariant interval also allows us to measure proper distances without knowledge of v, the speed between the inertial frames.

These results are more general than our discussion so far. Take two events, one at $P_1 = (ct_1, x_1, y_1, z_1)$ and the second at $P_2 = (ct_2, x_2, y_2, z_2)$. Then defining the differences $c\Delta t = ct_2 - ct_1$, and so on, we can check using the Lorentz transformation formulas Eqs. (4.2a) and (4.2b) that

$$(\Delta s)^2 \equiv (c\Delta t)^2 - (\Delta x)^2 - (\Delta y)^2 - (\Delta z)^2$$

is the same in all other frames. Unlike the invariant distance of Euclidean space, $x^2 + y^2 + z^2$, this invariant interval can be either positive, negative, or zero.

Let us illustrate these three possibilities and choose $P_1 = (0, 0, 0, 0)$ and $P_2 = (ct, x, 0, 0)$ for easy visualization. Clearly if P_1 and P_2 are connected by a light ray,

$$(\Delta s)^2 = 0 \quad (\text{light} - \text{like}).$$

The fact that $(\Delta s)^2 = 0$ in all inertial frames means that if light travels at the speed limit c in one frame, then it does so in all. This follows from writing out $(\Delta s)^2 = 0$ in frames S and S',

$$(\Delta s)^2 = (c\Delta t)^2 - (\Delta x)^2 = (c\Delta t')^2 - (\Delta x')^2 = 0 \tag{4.8}$$

where we took the x direction as the direction of light propagation. Eq. (4.8) states that if $\Delta x = \pm c\Delta t$, then $\Delta x' = \pm ct'$.

The light ray $(\Delta s)^2 = 0$ separates our space−time plot into a region I where $(\Delta s)^2 > 0$ and a region II where $(\Delta s)^2 < 0$ (Fig. 4.6). For P_2 in region I, we can boost to a frame where P_1 and P_2 occur at the same x'. In other words, the two events could be the ticks on a clock at rest in a frame S'. In that frame, $\Delta x' = 0$ and $\Delta t'$ is just the proper time $\Delta\tau$. Similarly, if P_2 lies in region II, we can boost to a frame S' where P_1 and P_2 occur simultaneously, $\Delta t' = 0$, so $-(\Delta s)^2$ then represents the proper length squared of a rod

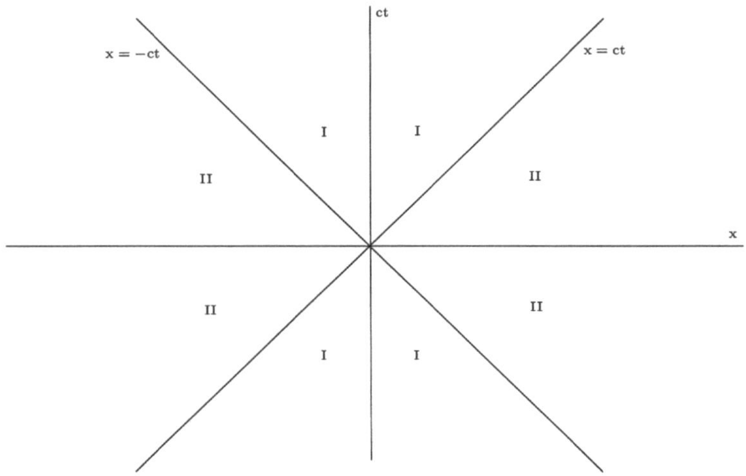

Figure 4.6 A space−time diagram showing regions causally connected to the origin (I) and those not (II).

extending from P_1 to P_2. It is conventional to label region I time-like because $(\Delta s)^2 > 0$ and to label region II space-like because $(\Delta s)^2 < 0$.

Another important invariant distinction between regions I and II involves causality. If the event P_2 lies in region I for $t > 0$, one can find a frame S' where it occurs at the same x' as P_1 but at a later time t'. Clearly then, an action at P_1 could cause a result at P_2. We say that the space−time points P_1 and P_2 are causally connected and that P_2 lies in P_1's "forward light cone." Similarly, if P_2 lies in the region I for $t < 0$, then an action at P_2 could cause a result at P_1 and the points are again causally connected. However, if P_2 lies in region II, one can find a physical frame S' where P_1 and P_2 are simultaneous. There is no way for them to communicate without inventing a signal that violates the speed limit c, and so they are not causally connected. In fact, we can also find a frame S'' where P_2 occurs *before* P_1, as shown in Fig. 4.7. Space−time points separated by a negative interval $(\Delta s)^2$ do not have a unique time ordering and cannot influence one another. The region II is sometimes called "elsewhere" because of this.

We return to some of these notions when we discuss dynamics, a topic in which causality plays a particularly important role.

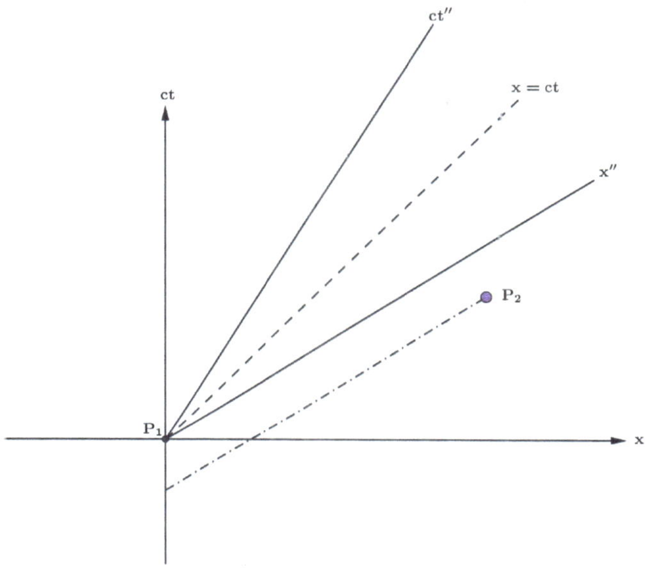

Figure 4.7 Space–time events P_1 and P_2 lie in region II and have different time orderings in different frames.

PROBLEMS

4-1. Two events occur at the same place in the frame S and are separated by a time interval of 5 s. What is the spatial separation between these two events in frame S' in which the events are separated by a time interval of 7 s? Frame S' moves at a constant velocity along the x direction of frame S.

4-2. Two events occur at the same time in the frame S and are separated by a distance of 2 km along the x axis. What is the time difference between these two events in frame S' in which their spatial separation is 4 km? Frame S' moves at a constant velocity along the x direction of frame S.

4-3. An event occurs at $x' = 100$ m, $t' = 9 \cdot 10^{-8}$ in frame S'. If this frame moves with velocity $4c/5$ along the x axis of frame S, what are the space–time coordinates of the event in frame S?

4-4. The space and time coordinates of two events are measured in frame S to be

Event 1: $x_1 = L,\ t_1 = L/c$

Event 2: $x_2 = 2L,\ t_2 = L/2c$

a. Find the velocity of a frame S' in which both events occur at the same time.

b. What is the time t' that both events occur in S'?

4-5. Frame S' has a speed of $0.8c$ relative to S.

a. An event occurs at $t = 5 \cdot 10^{-7}$ s and $x = 100$ m in frame S. Where and when does it occur in S'?

b. If another event occurs at $t = 7 \cdot 10^{-7}$ s and $x = 50$ m, what is the time interval between the events in frame S'?

4-6. In parts (b) and (c) of this problem, we measure the times and locations of events by looking at them. In other words, instead of recording locations and times by being at the event, the event will send us a light signal that takes some additional time to propagate to us. This extra time must be taken into account accordingly. Astronomical data are of this type. For example, we see the Sun approximately 8 min in its past.

Suppose a meter stick, aligned along the x direction, moves with velocity $0.8c$ and its midpoint passes through the origin at $t = 0$. Let there be an observer at $x = 0$, $y = 1$ m, $z = 0$.

a. Where in the observer's frame are the end points of the meter stick at $t = 0$?

b. When does the observer see the midpoint pass through the origin?

c. Where do the end points appear to be at this time, as seen by the observer?

4-7. Frame S' has velocity $0.6c$ relative to frame S. At $t' = 10^{-7}$ s, a particle leaves the point $x' = 12$ m, traveling in the negative x' direction with a velocity $u' = -c/3$. It is brought to rest suddenly at time $t' = 3 \cdot 10^{-7}$ s.

a. What was the velocity of the particle as measured in the frame S?

b. How far did it travel as measured in the frame S?

4-8. Frame S' has velocity v relative to frame S. At time $t = 0$, a light ray leaves the origin of S, traveling at a 45 degrees angle with the x axis.

a. What angle does the light ray make with respect to the x' axis in the frame S'?

b. Repeat part (a) replacing the light ray with a particle of mass m and speed u.

c. Repeat part (a) replacing the light ray with a rod that is stationary in frame S.

4-9. The Fizeau Experiment. Light propagates more slowly through a material medium than through a vacuum. If v_m is the speed of light in a medium and the medium is moving with respect to the frame S at the velocity v, find an expression for the velocity of light in the frame S, assuming that the light ray propagates in the same direction as v. Show that when v is much smaller than c, this general expression reduces to $v_m + v\left(1 - v_m^2/c^2\right)$, to leading order in v. (This expression was tested by H. Fizeau in the mid-19th century using interferometry techniques.)

4-10. The Čerenkov Radiation. Although light travels at the speed limit in empty space, its speed is diminished as it propagates through materials, as discussed in the Fizeau experiment. The process by which light is slowed is complicated—the waves scatter from the atoms making up the material and a light wave can lose speed and coherence in the process. We learn about these scattering processes in quantum mechanics. Nonetheless, light may not even be the fastest racer in a particular material. When a charged particle moves through the material faster than the speed of light in that material, it radiates coherent light in a cone that trails behind it, as shown in Fig. 4.8. Note in the figure that the charged particle is shown to radiate light at several points, and the resulting wave front of Čerenkov radiation, the conical surface BC, is perpendicular to the direction of propagation of the light rays.

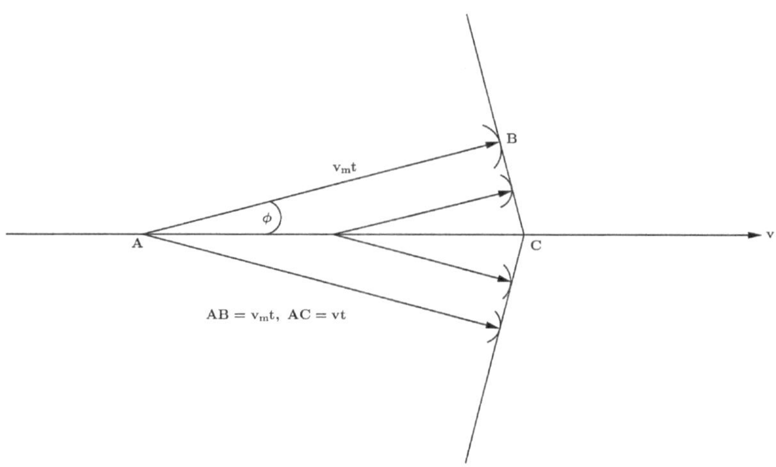

Figure 4.8 Illustration of Čerenkov radiation.

a. Show that the half-angle of the cone of Čerenkov radiation, ϕ, is $\cos\phi = v_m/v$, where v_m is the speed of light in the material and v is the speed of the charged particle. (In high energy physics and cosmic ray experiments, Čerenkov radiation and this formula are used to measure the speeds of exotic charged particles).

b. Compare this phenomenon to these well-known ones: When a speed boat travels at high velocity over the water, it leaves a sharp wake behind it; and when a jet plane travels faster than the speed of sound, it makes a sonic boom.

4-11. Stellar Aberration. Imagine that Earth were at rest with respect to a distant star and that an earthbound telescope had to be pointed at an angle θ above the horizon to view the star. Now suppose that there is a relative velocity v between Earth and the star. The angle θ would change to θ'.

a. Show that θ' is given by $\cos\theta' = (\cos\theta + v/c)/[1 + (v/c)\cos\theta]$.

b. If v/c is very small, show that the formula in part (a) reduces to $\cos\theta' \approx \cos\theta + (v/c)\sin^2\theta$.

c. Because the difference between θ and θ' is very small under the conditions of part (b), it is convenient to introduce the angle $\alpha \equiv \theta' - \theta$. Show that part (b) reduces to the prediction $\alpha \approx -(v/c)\sin\theta$.

This result is called stellar aberration and is significant in astronomical observations. It is particularly important when we compare the position of a star in the sky in winter to its position in summer, because the velocity of Earth reverses relative to the star every 6 months.

4-12. Use the velocity transformation formulas to derive the "particle aberration formula,"

$$\tan\alpha' = \frac{\sin\alpha}{\gamma(v)(\cos\alpha - v/u)}$$

where the particle moves in frame S with a velocity $\boldsymbol{u}$ making an angle α with the x axis and the frame S' moves in the x direction with velocity v, as usual, and the angle α' is the corresponding angle in the frame S'.

4-13. Headlight Effect. Suppose that a car's headlight sends out light rays into a forward hemisphere. Use the result of Problem 4.11a to deduce that the maximum angle of the light with respect to the line of motion of the car is $\cos\theta' = v/c$ in the rest frame of the road. Explain. (This result is called the headlight effect.)

4-14. A frame S' moves along the x axis of frame S at a velocity v. A particle moves with a velocity $v_x' = 0$ and $v_y' \neq 0$ in the S' frame. (a) What are v_x, and v_y in the frame S? (b) Why are v_y' and v_y different?

4-15. Two particles are shot out simultaneously from a given point with equal speed u in orthogonal directions. Show that the speed of each particle relative to the other is $u\sqrt{2 - u^2/c^2}$.

4-16. Show that relativistically "adding" a velocity $\boldsymbol{u}$ to a velocity $\boldsymbol{v}$ is not generally the same as "adding" velocity $\boldsymbol{v}$ to velocity $\boldsymbol{u}$. [Hint. Take an example such as $\boldsymbol{u} = (0,u,0)$ and $\boldsymbol{v} = (v,0,0)$]

4-17. A frame S' moves along the x axis of frame S at a velocity v. In frame S there is a meter stick parallel to the x axis and moving in the y direction with a velocity v_y, as shown in Fig. 4.9. The center of the meter stick passes the point $x = y = x' = y' = 0$ at $t = t' = 0$.

a. Argue, on the basis of the relativity of simultaneity, and without calculation, that the meter stick is measured as tilted upward in the positive x' direction in the frame S'!

b. Calculate the angle of the tilt, ϕ', in frame S', as shown in the figure, by answering several easy questions. Where and when does the right end of the meter stick cross the x axis as observed in the frame S'? (You might answer this question first in the frame S and then transform this information to the frame S'.).

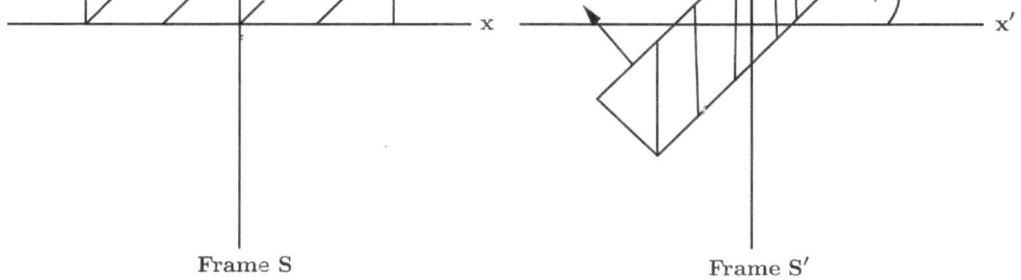

Figure 4.9 A meter stick parallel to the x axis in frame S but moving vertically is measured to be tilted in frame S'.

Referring to Problem 4.14 for the velocity of the meter stick in the frame S', determine where the right end of the meter stick is at time $t' = 0$ when the center is at the origin. Your formula for ϕ' follows from this Ref. [2].

4-18. A stick of length l_0 in its rest frame is situated at angle θ to the x axis.

a. What is the length of the rod as measured by an observer at rest in the frame S' which moves along the x axis at speed v?

b. What is the orientation of the rod as measured in frame S'?

4-19. A race car speeds past two markers spaced a distance 100 m apart in a time 0.04 μs as measured by an observer on the ground.

a. How far apart are the markers as measured by the racer?

b. What time does the racer measure when he/she passes the second marker?

c. What speeds do the racer and observer on the ground measure?

REFERENCES

[1] A.P. French, Special Relativity, W. W. Norton, New York, 1968.
[2] Shaw, Am. J. Phys. 30 (1972) 72.

CHAPTER 5

Illustrations and Problems in Space–Time Measurements

Contents

Let us apply the principles we have learned so far to some interesting problems. Some of these problems are solved in several different ways to illustrate pitfalls and strategies.

5.1 A SPACESHIP RENDEZVOUS

A spaceship of length 100 m has a radio receiver in its nose [1]. It travels at a velocity $v = 0.6c$ relative to a space station. A radio pulse is emitted from the space station just as the tail of the spaceship passes by the transmitter.

1. How far from the space station is the nose of the spaceship when it receives the radio signal?
2. How long did it take the radio pulse to reach the nose of the spaceship from the perspective of the space station?
3. How long did it take the radio pulse to reach the nose of the spaceship from the perspective of the spaceship?

Consider Fig. 5.1, showing the emission and reception of the radio pulse. From the perspective of the space station, the radio pulse travels a distance ct and goes a distance vt plus the length of the spaceship in its (the space station's) rest frame. This distance is $100\ \text{m}/\gamma$. So,

$$ct = \frac{100}{\gamma} + vt.$$

The relevant distance is $100\ \text{m}/\gamma$ because the spaceship's proper length, 100 m, is being measured in a frame at relative velocity v. Since $v = 0.6c$,

Special Relativity, Electrodynamics, and General Relativity
ISBN 978-0-12-813720-8
https://doi.org/10.1016/B978-0-12-813720-8.00005-2

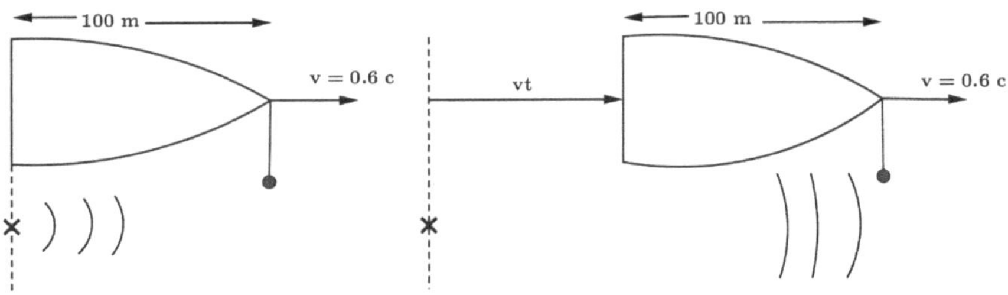

Figure 5.1 Emission of a radio pulse from a space station and its reception on a moving rocket ship.

we compute $\gamma = 1/\sqrt{1 - v^2/c^2} = 5/4$ and solve for ct, finding $ct = 200$ m. This answers part a. For part b, $t = 200\ \mathrm{m}/c = 200/(3\cdot 10^8) \approx 6.67\cdot 10^{-7}$ s. Now we need the time elapsed in the spaceship's frame t'. Over this time interval, light travels from the rear of the spaceship to its nose, a distance of 100 m, at velocity c. So, $t' = 100/c = 3.37\cdot 10^{-7}$ s, which answers part c.

Where did we use the principles of relativity? First, we had Lorentz contraction in part a, and second, in part c we used the fact that light travels at the speed limit c in any inertial frame. Easy problem—but wait, why are the times t and t' not related by time dilation, a factor of $\gamma = 5/4$? Instead, we found in a most elementary way that $t' = t/2$. This is puzzling, until you notice that the events, the emission and absorption of the radio pulse, occur at *separate* points in the spaceship frame. From the perspective of the space station, a clock in the tail of the spaceship and one in the nose are not synchronized, and this time difference contributes to t'.

To see the effect, let us redo the problem by plugging into our Lorentz transformation formulas. First, we need the space–time coordinates of the two events in the spaceship frame. A radio pulse is emitted at $t'_1 = x'_1 = 0$ and is received in the nose of the spaceship at $x'_2 = 100$ m, $t'_2 = x'_2/c = 100\ \mathrm{m}/(3\cdot 10^8)\ \mathrm{m/s} = 3.33\cdot 10^{-7}$ s. So, in the space station's frame the second event occurs at

$$x_2 = \gamma\left(x'_2 + vt'_2\right) = \frac{5}{4}(100 + 0.6\cdot 100) = \frac{5}{4}\cdot 1.6\cdot 100 = 200\ \mathrm{m}$$

as we calculated before from the space station's perspective. The reception of the radio pulse occurs at

$$t_2 = \gamma\left(t'_2 + \frac{v}{c^2}x'_2\right) = \frac{5}{4}\left(\frac{100}{c} + 0.6\frac{100}{c}\right) = \frac{5}{4}\cdot 1.6\cdot\frac{100}{c} = \frac{200}{c}$$
$$= 6.67\cdot 10^{-7}\ \mathrm{s}$$

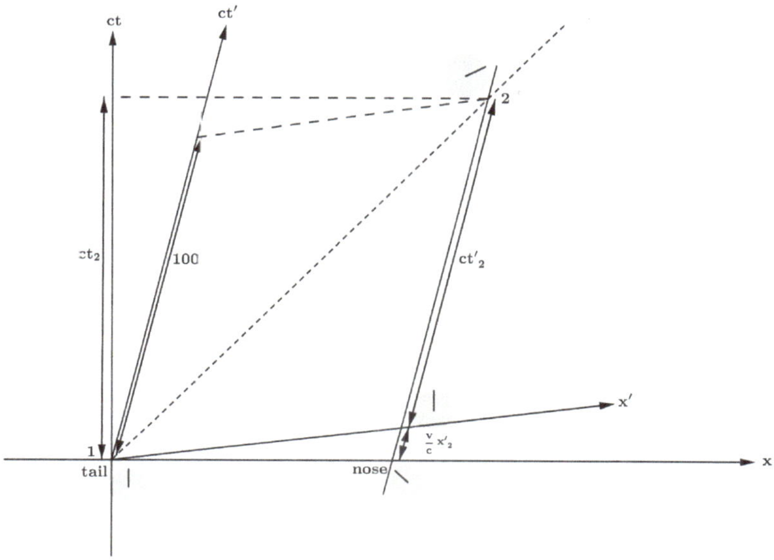

Figure 5.2 Emission and reception of a radio pulse in a Minkowski diagram, showing measurements in the rocket and the space station frames.

as before. Here we see the relativity of simultaneity at work. The formula states that t_2 is *not* time dilation applied to t_2'; it is time dilation applied to $t_2' + (v/c^2)x_2'$. We recognize this as the time that the space station says passes on a clock at the nose of the spaceship between the two events. Multiplying by γ gives the time in the space station's frame. The term vx_2'/c^2 is the extra time that the space station says must pass on the clock at the nose of the spaceship because that clock is vx_2'/c^2 seconds *behind* the clock at the tail.

All this becomes evident if we make a Minkowski diagram of the situation (see Fig. 5.2). Note that the clocks in the tail and the nose are synchronized in the spaceship frame, S'. We see that, from the point of view of the space station, the clock in the nose is vx_2'/c^2 behind the clock in the tail and that ct_2 is related by time dilation to the *sum* of $ct_2' = 100$ and $vx_2'/c = 0.6 \cdot 100$ m.

5.2 A HOLE IN THE ICE

A relativistic skater with 15-inch-long blades on his skates travels at $v = 0.8c$ over an icy surface. There is a hole in the ice of diameter 10 in. before the finish line. The skater decides to skate over the hole to win the race. He

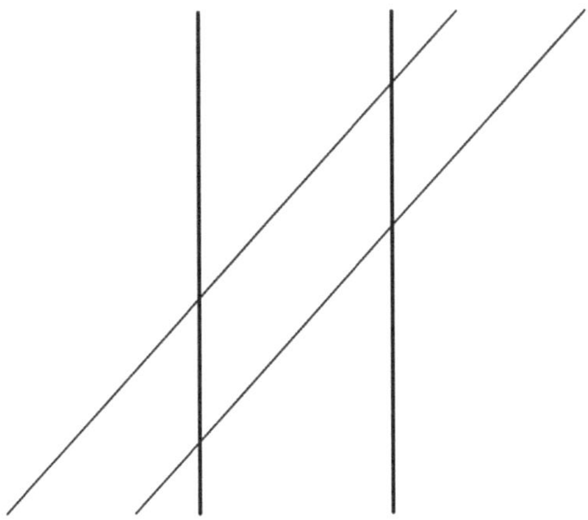

Figure 5.3 The world lines (vertical) of the edges of the hole and the ends (tilted) of the skater's blade.

figures that this is safe because the diameter of the hole is Lorentz contracted in his frame to $10/\gamma = 10/(5/3) = 6$ in., which is less than the length of his blade. A judge at the finish line sees the skater approaching and tries to wave him off because in *her* frame the length of the skater's blade is $15/\gamma = 15 \cdot (3/5) = 9$ in., so the entire blade will fit into the hole and the skater will fall through the ice and be injured. Which person is right? We better get the right answer because a serious accident lies in the balance!

This is a paradox of relativity almost as famous as the twin paradox. Let us describe it in a Minkowski diagram in which frame S is the rest frame of the ice (or the judge) and S' is the rest frame of the skater's blade (or the speed skater). The world lines of the edges of the hole and the ends of the blade are shown in Fig. 5.3 [2].

Now we want to describe these events in various frames where we lay down lines of constant time at various angles. From the perspective of the ice (or the judge), the lines of constant time t are horizontals as shown in Fig. 5.4. At t_1, the front of the blade reaches the edge of the hole. At t_2, the back of the blade does the same. Between t_2 and t_3, the entire blade is inside the hole. At time t_3, the front end of the blade hits the other edge of the hole and the skater falls! Fig. 5.4 ignores this and shows the blade emerging from the hole, and finally, at t_4, the back end of the blade comes out of the hole.

From the perspective of the skater, things are quite different because his lines of constant time t' are inclined to the horizontals by an angle tan $\theta = v/c = 0.8$. These lines are shown in Fig. 5.5. Inspecting this diagram,

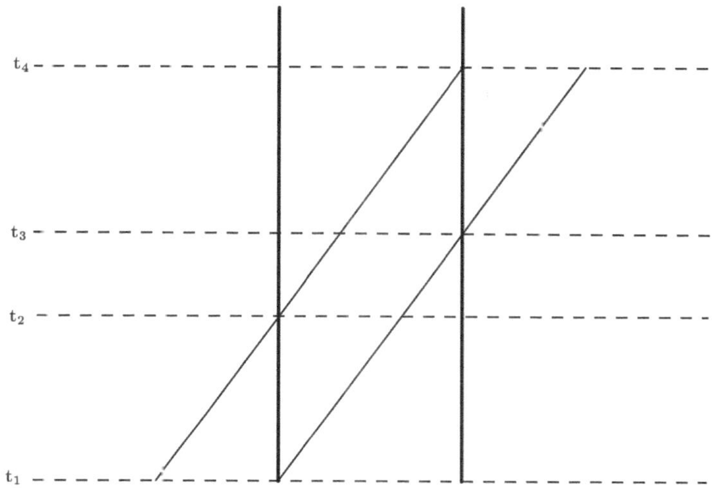

Figure 5.4 Fig. 5.3 with lines of constant time t in the frame of the ice (stationary judge).

we see that the skater's impressions were also correct—at no time t' are both ends of his blade inside the hole—and even the time order of the events is different than that given in frame S. In S', the front of the blade emerges from the hole (at t_2') *before* the back end comes in (at t_3'). We can easily check

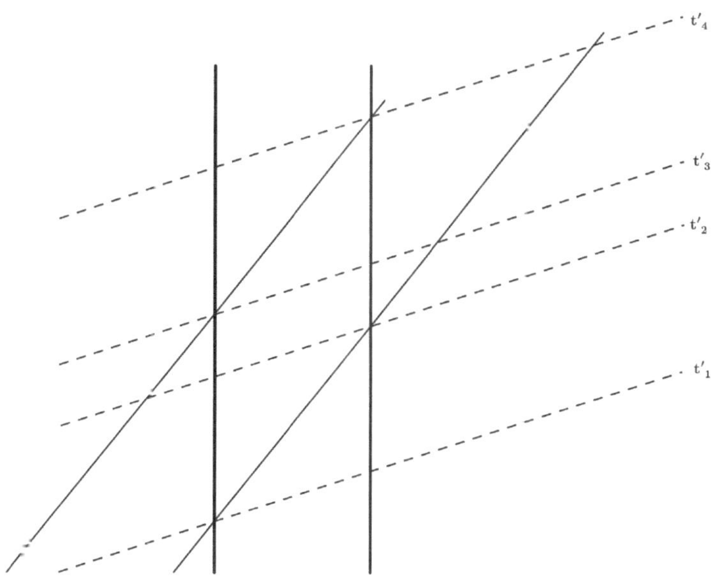

Figure 5.5 Fig. 5.3 with lines of constant time t' in the frame of the skater's blade.

that the invariant interval $(\Delta s)^2$ for the two events, the back end of the skater's blade reaches the left edge of the hole and the front end of the blade reaches the right edge of the hole, is negative. So, the events are not casually connected and their time ordering is frame dependent. This is obvious from the Minkowski diagram. We could also imagine that the back end of the skate emits a light pulse when it reaches the left edge of the hole. The pulse reaches the right end of the hole after a time interval of $10/c$ in the rest frame of the hole. The front end of the skate reaches the right edge of the hole after a much shorter time interval, $(10 - 9)/v = 5/4c$.

So, does the skater fall or not? Yes, he falls, and our discussion of causality should remind us of a flaw in the skater's description of his blade. When he claimed that his skate could not fit into the hole in the ice, he was thinking nonrelativistically, where rigid bodies exist. But in a relativistic world where information cannot travel faster than light, the blade is not rigid and the front end is pulled down by the force of gravity when $t_2' > t' > t_1'$. It is easy to check that if the front end of the skater's blade sent a radio pulse to the back end at time t_1', when the front end reaches the left edge of the hole, the pulse does not reach the back end of the blade until well after the front end has hit the right edge of the hole. In other words, the back end of the blade does not even know that there is a hole in the ice until after the accident!

This example suggests that Lorentz invariant methods would be the best tools to analyze dynamic, relativistic problems involving extended objects. Our frame-dependent discussion comes complete with biases and arbitrariness that can be misleading. In fact, high-energy theorists who study relativistic strings as a framework for grand unification almost exclusively use invariant methods so that they concentrate on the physical content of processes involving strings—those aspects that all observers in any inertial frame can agree on.

5.3 A VELOCITY GREATER THAN THE SPEED LIMIT?

Consider three frames of reference. An observer O sees an observer S moving to her right at speed $v = 0.8c$ and another observer S' moving to her left at the same rate.

1. The observer O says, "The velocity difference between S and S' is $1.6c$, so the distance between them, $d = 1.6ct$, is growing at a rate in excess of

Figure 5.6 An observer *O* with two frames receding to the left and right at velocities $+v$ and $-v$.

the speed of light c." Is this statement correct? Does it contradict Postulate 2 of special relativity?

2. If S' measures the velocity of S, what does he find?
3. If a transmitter at rest with S' broadcasts light with a frequency $\nu' = 10$ cps, what frequency is measured by O and what frequency is measured by S?

Begin with a picture of the three observers shown in Fig. 5.6. There is no doubt that O measures the distance between S and S' to be $d = 1.6ct$. This is just arithmetic. But $1.6c$ does not represent the speed of one object relative to another, so there is no contradiction with relativity. No physical object is moving in an inertial frame with a speed in excess of c. We see this explicitly by considering the next part of the problem. Using our relativistic addition of velocities formula, the speed of S in the frame of S' consists of two pieces: the speed of S in the rest frame of O, v, and the speed of O relative to S', v again. So, the speed of S measured by S' is

$$u = \frac{(v+v)}{\left(1+\frac{v^2}{c^2}\right)} = \frac{1.6c}{(1+0.8)^2} = \frac{1.6c}{1.64} = 0.9756c,$$

which is, indeed, slightly less than c. All is well.

It is amusing to note in passing that the situation depicted here approximates several natural phenomena studied in astrophysical research journals. Some researchers have presented analyses of "superluminal" jets of gases streaming from spinning black holes, much as S and S' are racing away from O. When observed through an Earth-bound telescope, the jets appear to be diverging at a rate exceeding c. These and other optical effects are sometimes raised in the popular science literature as threats to relativity. They do not last long.

Back to our problem. The relation between the frequency ν' and the frequency ν_o, observed by O, is given by the Doppler formula,

$$\nu_o = \sqrt{\frac{1-v/c}{1+v/c}}\nu' = \sqrt{\frac{1-0.8}{1+0.8}}\nu' = \frac{10}{3}\text{ cps}.$$

Finally, we can calculate the frequency ν measured by S by noting that he observes the wave train in O's reference frame as Doppler shifted by another factor of $\sqrt{(1-v/c)/(1+v/c)}$ because the relative speed between O and S is v,

$$\nu = \sqrt{\frac{1-v/c}{1+v/c}}\nu_o = \left(\frac{1-v/c}{1+v/c}\right)\nu' = \frac{1-0.8}{1+0.8}\cdot 10 = \frac{10}{9} \text{ cps.}$$

Alternatively we could calculate ν directly in terms of ν' by again applying the Doppler formula, but using the relative velocity u between S and S'. Call $\beta = v/c$, so

$$\nu = \sqrt{\frac{1-u/c}{1+u/c}}\nu' = \sqrt{\frac{1-2\beta/(1+\beta^2)}{1-2\beta/(1+\beta^2)}}\nu' = \left(\frac{1-v/c}{1+v/c}\right)\nu',$$

where we did some algebra in the previous step and found agreement with the earlier result.

PROBLEMS

5-1. The scissors paradox. A long straight rod that is inclined at an angle θ to the x axis has a velocity v downward as shown in Fig. 5.7.

a. Derive a formula for the speed v_r of the intersection of the rod with the x axis in terms of v and θ.

b. Can you choose values for v and θ so that v_r is greater than the speed limit c?

c. Can you transmit information along the x axis at speeds greater than c using this trick?

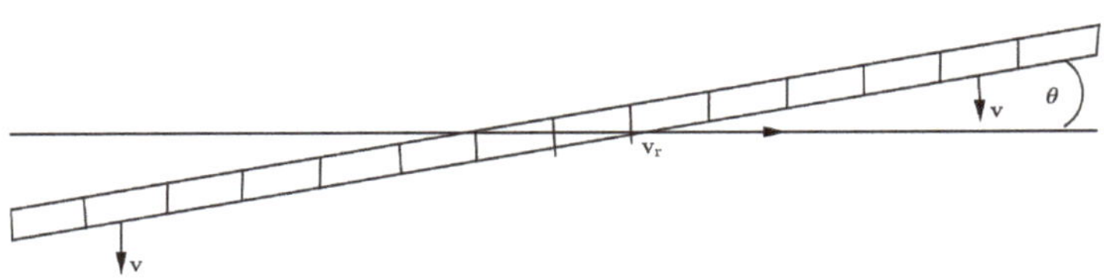

Figure 5.7 The scissors paradox. The blade of the scissors is tilted with respect to the horizontal and moves vertically down with velocity v. Its point of intersection with the horizontal moves to the right at velocity v_r.

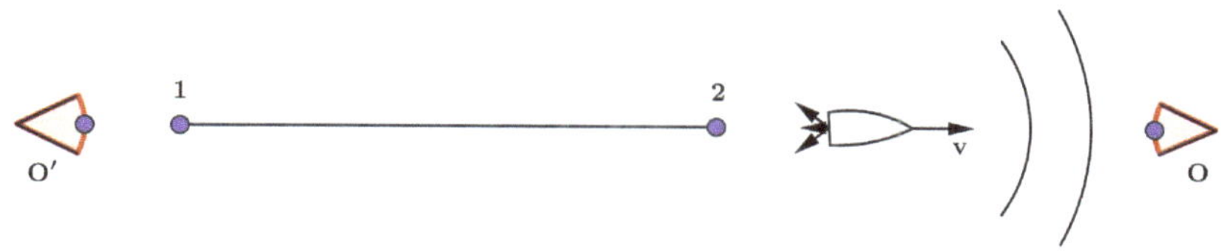

Figure 5.8 A rocket goes from point 1 to point 2, and at both points it radiates a pulse of light, which are detected by an observer *O* at rest to the right of point 2. Another observer *O*′ is at rest to the left of point 1.

5-2. Tricks your eyes can play. Consider a rocket that goes from point 1 to point 2, a total distance d, in time t at velocity v. At both points it radiates a pulse of light, and these pulses are detected by an observer O at rest to the right of point 2, as shown in Fig. 5.8.

a. Show that the time interval between the reception of the pulses at O is $T = [1 - (v/c)]t$, where t is the time it took the rocket to travel from point 1 to point 2, $d = vt$.

b. Because the light pulses provide observer O with two images of the rocket, one at point 1 and the other at point 2, a distance d to the right of point 1, argue that the observer sees the rocket moving with an apparent speed $d/T = v/[1 - (v/c)]$, which is greater than c, whenever $c/2 < v < c$!

c. Suppose that there is a stationary observer O' to the left of point 1. Find a formula for the apparent speed that he attributes to the rocket. What is its maximum value?

This problem illustrates the visual effects that occur in astrophysical observations of moving and exploding stars. Interpreting astronomical data can get pretty tricky! If you pursue astrophysics, you will see more such effects in more realistic settings. This problem also illustrates why we are so particular about how measurements are made in relativity—why we phrase measurements in terms of events, specifying the time an event occurs on a clock at that exact spot.

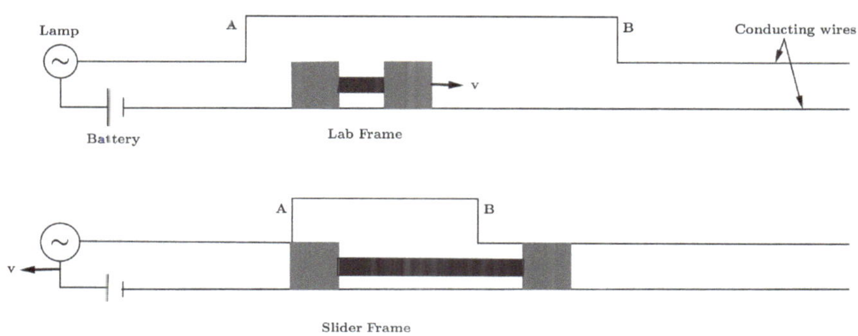

Figure 5.9 Two long parallel conducting rails are open at one end, but connected electrically at the other end through a lamp and battery. An H-shaped slider, whose vertical pieces are made of copper but whose horizontal piece is an insulator, moves along the rails. The upper portion of the figure shows the lab frame view of the experiment, and the lower portion shows the slider view.

5-3. A project to elucidate the physics of Section 5.2. Two long parallel conducting rails are open at one end, but connected electrically at the other end through a lamp and battery, as shown in Fig. 5.9. An H-shaped slider, whose vertical pieces are made of copper but whose horizontal piece is an insulator, moves along the rails. Let the length of the slider be 2 m, matching the rest length of the region AB, and let the height of the slider match the distance between the rails, except in the special region AB where the rails are further apart. Suppose that the H-shaped slider moves at a large velocity, so that it contracts by a factor of 2 relative to the region AB. In the rest frame of the rails, an observer concludes that there is a period of time when the slider does not complete the circuit between the battery and the lamp, so the lamp must go dark for a moment. However, from the perspective of the slider, the region AB is contracted by a factor of 2, so an observer on the slider predicts that the circuit is always completed and the lamp will always shine. Analyze this system as we analyzed the ice skate and the hole problem in Section 5.2 and resolve the paradox. Does the lamp go dark for an instant or not?

This project assumes that the reader knows how signals, current, and voltages propagate along a wire. The time delay between the arrival of the right end of the slider at point B and the transmission of that information from point B back to point A on the wire is an important part of the resolution of this puzzle. (It is the analogue of the time it takes the back end of the skate to know that the front end crashed into the far side of the hole.) This challenging problem can be used as a fascinating research project involving the reading and mastery of the material in Ref. [3].

REFERENCES

[1] A.P. French, Special Relativity, W. W. Norton, New York, 1968.
[2] N.D. Mermin, Space Time in Special Relativity, Waveland Press, Prospect Heights, IL, 1968.
[3] G.P. Sastry, Am. J. Phys. 55 (1987) 943.

CHAPTER 6

Relativistic Dynamics: First Steps

Contents

6.1 ENERGY, LIGHT, AND $E = MC^2$

We have spent all our time so far discussing how measurements of space–time coordinates transform between inertial frames in such a way that there is a universal speed limit. Has this been an empty exercise? We need to see that we can write down a scheme of relativistic *dynamics* that satisfies Postulate 1 of relativity—that the relative speed between inertial frames has no material influence on the dynamics in either frame. We know that the transformation laws between inertial frames in a Newtonian world, the Galilean transformations we have discussed in previous sections, are consistent with Newton's laws of motion. Now we need the relativistic laws of motion to underpin the Lorentz transformations.

The hallmarks of Newtonian mechanics are the second law (that force equals inertia times acceleration) and the third law (that action equals reaction; i.e., if I push on you, you push on me in an equal and opposite fashion). These results lead to momentum conservation. The Galilean transformations then imply that if momentum conservation is true in one inertial frame, it is true in them of all. These transformations implement the idea that all inertial frames are equivalent.

Let us briefly review these Newtonian developments before moving on to relativistic dynamics. The logic here is important. Consider two point

Special Relativity, Electrodynamics, and General Relativity
ISBN 978-0-12-813720-8
https://doi.org/10.1016/B978-0-12-813720-8.00006-4

particles of mass m_1 and m_2 and let them collide in an inertial frame where, by definition, there is no external force. So, if $\mathbf{f}_{12}$ is the force that particle 1 exerts on particle 2 and $\mathbf{f}_{21}$ is the force that particle 2 exerts on particle 1, Newton's second and third laws imply

$$\frac{d}{dt}(\mathbf{p}_1 + \mathbf{p}_2) = \mathbf{f}_{21} + \mathbf{f}_{12} = 0,$$

where $\mathbf{p}_1$ is the momentum of particle 1, $\mathbf{p}_1 = m_1\mathbf{v}_1$, and $\mathbf{p}_2 = m_2\mathbf{v}_2$. So, the total momentum is conserved, $m_1\mathbf{v}_1 + m_2\mathbf{v}_2 = \text{constant}$. This means that the center of mass $\mathbf{R} = (m_1\mathbf{x}_1 + m_2\mathbf{x}_2)/(m_1 + m_2)$ travels with a constant velocity, $(d/dt)\,\mathbf{R} = \text{constant}$, and we can boost to a frame where $\mathbf{R}$ is the origin and remains there forever. But we can also view the collision in another frame S', which moves with velocity $\mathbf{v}$ with respect to frame S. By Galilean invariance, we have the nonrelativistic rule of addition of velocities, so the velocities of the particles in frame S' read

$$\mathbf{v}_1' = \mathbf{v}_1 - \mathbf{v}, \quad \mathbf{v}_2' = \mathbf{v}_2 - \mathbf{v}.$$

So, in the frame S',

$$\begin{aligned} \mathbf{p}_1' &= m_1\mathbf{v}_1' = m_1\mathbf{v}_1 - m_1\mathbf{v} = \mathbf{p}_1 - m_1\mathbf{v} \\ \mathbf{p}_2' &= m_2\mathbf{v}_2' = m_2\mathbf{v}_2 - m_2\mathbf{v} = \mathbf{p}_2 - m_2\mathbf{v} \end{aligned}.$$

So,

$$\frac{d}{dt'}(\mathbf{p}_1' + \mathbf{p}_2') = \frac{d}{dt}(\mathbf{p}_1 + \mathbf{p}_2) = 0$$

because $\mathbf{v}$ is a constant. So, the same dynamics holds in either frame and, if we have momentum conservation in one inertial frame, we have it in all of them.

In short, Newtonian dynamics works together with Galilean invariance to assure that physics satisfies Newton's version of relativity. The extension of this consistency to the relativistic domain will take some work because, although relativistic dynamics must reduce to Newtonian dynamics when all the relative velocities involved are small compared with the speed limit c, Lorentz transformations are more intricate than Galilean transformations. We can anticipate that the Newtonian notion of momentum $\mathbf{p} = m\mathbf{v}$ will not work in Einstein's world because the frame independence of momentum conservation used the addition of velocities rule, which is not true relativistically.

So, we will follow Einstein and rethink the notions of momentum, mass, and energy. To begin, let us present one of Einstein's famous thought experiments, which shows that relativistic energy and inertia must be unified into one concept in the new dynamics.

Consider a box of mass M and length L, and suppose that radiant energy (light) E is emitted from one end and absorbed by the other (Fig. 6.1). We need to know one fact about light, which we will understand in more detail later: if a wave carries energy E, it also carries momentum p and they are related by $E = pc$. So, when light is emitted from the left end of the box, as shown in the top half of Fig. 6.1, the box recoils with a momentum $-E/c$. The box then moves to the left with velocity $v = -E/Mc$, supposing that M is so large that v is very small, $v \ll c$, and the box's motion is well described by Newtonian nonrelativistic considerations. Then, the light reaches the right end of the box in time $\Delta t = L/c$ and is absorbed. Now, the box should be at rest again, but apparently it has moved by a distance $\Delta x = v\Delta t = -EL/Mc^2$! But this is crazy—the center of mass of the system cannot move because there are no external forces to push it. If our expression for Δx were correct, we could repeat this process as often as we wished and move the box as far as we liked to the left. Something is wrong! M is large and Newtonian mechanics should work to describe it.

It must be that when energy E moves from one end of the box to the other it must deposit some mass m on the right-hand end of the box, so the center of mass of the heavy box does not move. How much should m

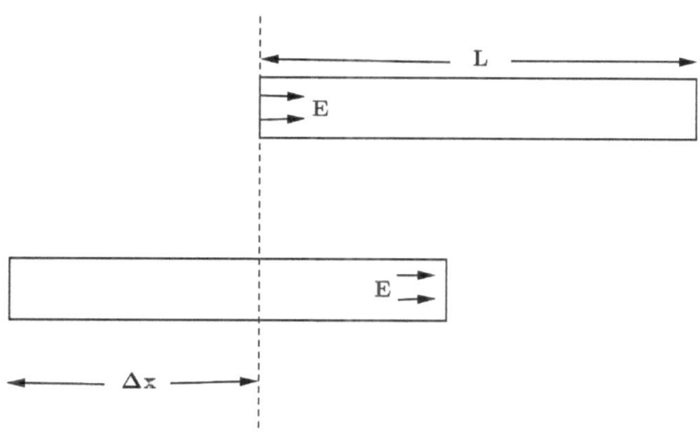

Figure 6.1 Upper portion of the figure shows radiation being emitted from the left-hand side of a box, and the lower portion shows that radiation being absorbed on the right-hand side as the box recoils.

be to do the trick? Certainly $m \ll M$. The condition that $\Delta\overline{x}$ (the change in the position of the center of mass) vanishes reads

$$\Delta\overline{x} = 0 = mL + M\Delta x.$$

Solving for m,

$$m = -\frac{M}{L}\Delta x = \frac{M}{L}\frac{EL}{Mc^2} = E/c^2.$$

In other words,

$$E = mc^2. \tag{6.1}$$

This means that the energy carried by the light wave results in a mass increase of $m = E/c^2$ when it is absorbed as heat on the end of the box. Because c is so large, m is extraordinarily tiny, and this thought experiment is not practical. However, in the realm of nuclear and high-energy physics, dramatic illustrations of Eq. (6.1) can be found. Some will be discussed later.

Eq. (6.1) is one of the most famous equations in physics. When we study elementary particle collisions later, we shall see that it predicts that we can convert the rest mass of heavy particles into the kinetic energy of lighter ones. The equation underlies nuclear power.

Because mass can be converted into energy and vice versa, according to Eq. (6.1), one of the sacred conservation laws of Newton must be modified, that conservation law states that mass, as well as total momentum, is conserved in a collision of particles in an inertial frame,

$$\begin{aligned} \mathbf{p}_1 + \mathbf{p}_2 &= \mathbf{p}_1' + \mathbf{p}_2' \\ m_1 = m_1', &\quad m_2 = m_2' \end{aligned}.$$

According to Newton, there is separate overall momentum conservation and mass conservation for each elementary particle. Our next task is to find the relativistic generalization of these statements.

6.2 PATCHING UP NEWTONIAN DYNAMICS—RELATIVISTIC MOMENTUM AND ENERGY

Can we invent a formula for momentum, a relativistic analogue to $\mathbf{p} = m\mathbf{v}$, so that relativistic momentum is conserved in the collision of particles in an inertial frame, and the conservation law is truly relativistic (i.e., it holds in all inertial frames if it holds in one)? Let us take a simple collision.

Figure 6.2 A two-body inelastic collision.

First, consider a Newtonian inelastic collision, shown in Fig. 6.2. Initially, there is a particle of mass m at rest in the lab, and a particle of the same mass but with velocity u collides and sticks to it. The final composite particle of mass $2m$ recoils with velocity $\overline{u}$, determined by momentum conservation,

$$mu = 2m\overline{u}.$$

So,

$$\overline{u} = \frac{1}{2}u$$

Because the initial kinetic energy is $T_i^{\mathrm{NR}} = (1/2)mu^2$ (NR denotes "nonrelativistic," or Newtonian) and the final kinetic energy is $T_f^{\mathrm{NR}} = (1/2)(2m)\overline{u}^2 = (1/4)mu^2$, heat Q is generated,

$$Q = T_i^{NR} - T_f^{NR} = \frac{1}{4}mu^2.$$

In this discussion, we assumed that the mass of the composite particle was twice the mass of its parts. Actually, the additivity of mass is forced on us by Postulate 1 and the form of the Newtonian momentum. To see this, call the mass of the composite particle M_o and view the collision in a frame S', which has a *transverse* velocity $\mathbf{v}$ (Fig. 6.3).

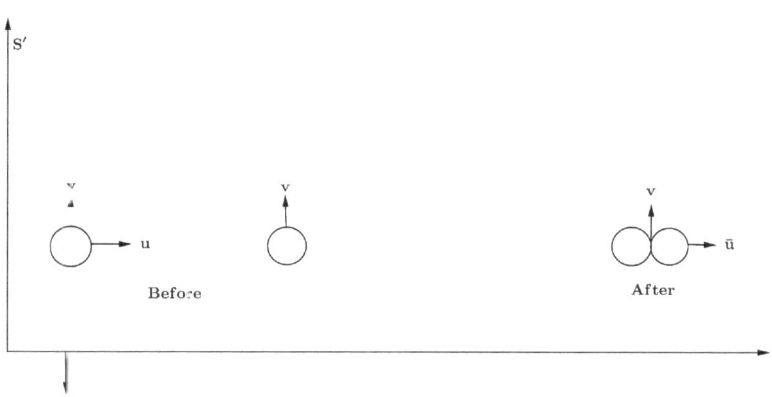

Figure 6.3 The collision of Fig. 6.2 viewed from a frame with a transverse velocity v down.

Conservation of transverse momentum reads

$$m\mathbf{v} + m\mathbf{v} = M_c\mathbf{v}.$$

So, we read off $M_c = 2m$, as expected. What is the point of this discussion? Conservation of momentum must hold in any inertial frame, and it will do so only if masses add in a Newtonian world.

Now we want a *relativistic* description of this process that will satisfy Postulate 1—the law of conservation of momentum must be the same in all frames [1]. Our task is to find an expression for the relativistic momentum that accomplishes this. The expression must reduce to its nonrelativistic cousin, $\mathbf{p} = m\mathbf{v}$, when the velocity is small compared with the speed limit c. The beam particle in the initial state in Fig. 6.2 has relativistic momentum, which must have the mathematical form

$$\mathbf{p} = f(u/c)m\mathbf{u}.$$

This expression should accommodate what we know about the momentum. A study of the collision should determine the function f. Why have we written this form for $\mathbf{p}$? First, we have chosen $\mathbf{p}$ to point in the direction $\mathbf{u}$ because that is the only vector in the problem. We have taken $\mathbf{p}$ to be proportional to the inertia m so its dimensions are correct. Finally, there is the possibility that f depends on the dimensionless ratio u/c, the ratio of the magnitude of $\mathbf{u}$ to the speed limit. The dimensionless function f incorporates features of $\mathbf{p}$ that we need to determine. It is conventional to write $m(u) \equiv f(u/c)m$ and call $m(u)$ the relativistic mass.

Because there are no external forces, we require that relativistic momentum be conserved,

$$m(u)\boldsymbol{u} = M(\overline{u})\overline{\boldsymbol{u}} \tag{6.2}$$

where $M(\overline{u})$ will be determined by momentum conservation and Postulate 1. We do *not* assume $M(\overline{u}) = 2m(\overline{u})$, as nonrelativistic reasoning might suggest. In fact, this is *not* true!

To see how Postulate 1 constrains all the apparent arbitrariness here, view the collision in a frame S', where the beam particle of Fig. 6.2 is at rest as in Fig. 6.4. In this frame the collision is just turned around. But frame S' is obtained from frame S by a boost (Fig. 6.5). So, the velocity of the composite particle, as measured in frame S, must also be given by an application of the addition of velocities formula,

$$\overline{u} = \frac{-\overline{u} + u}{1 - u\overline{u}/c^2},$$

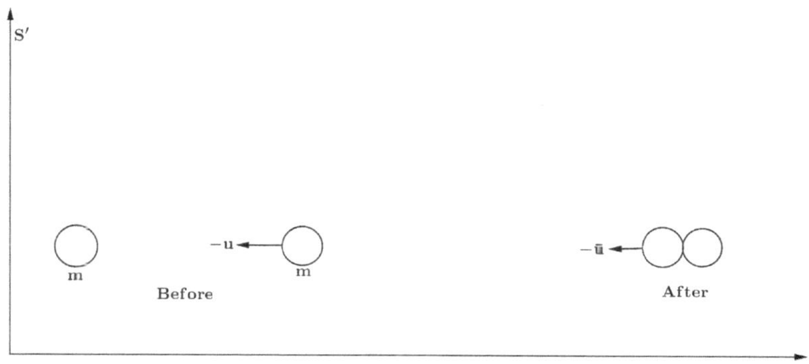

Figure 6.4 The collision of Fig. 6.2 in the rest frame of the beam particle.

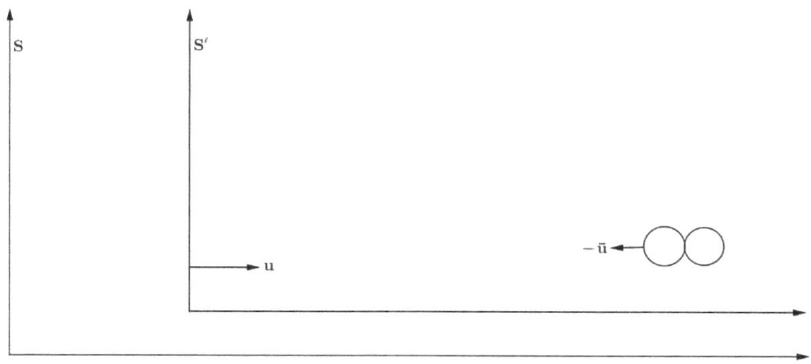

Figure 6.5 The final state of the collision in the frame *S′*. *S′* is related to frame *S* by a boost through velocity *u*.

which we can solve for u,

$$u = \frac{2\overline{u}}{1 + \overline{u}^2/c^2}. \tag{6.3}$$

Is this result reasonable? Well, if $\overline{u} \ll c$, then $u \approx 2\overline{u}$, and we are back to the Newtonian result we had before. All is well, so far.

To determine $M(\overline{u})$ by Postulate 1 of relativity, view the collision in a frame with a small transverse velocity in the $-y$ direction (Fig. 6.6). In the frame S', the beam particle (see "Before" in Fig. 6.7) then has $v'_y = v$, the transverse velocity between the frames. The beam particle had velocity u in the x direction in the lab frame, so in frame S'

$$u'_x = \frac{x'}{t'} = \frac{x}{\gamma(v)t} = \frac{1}{\gamma(v)}u,$$

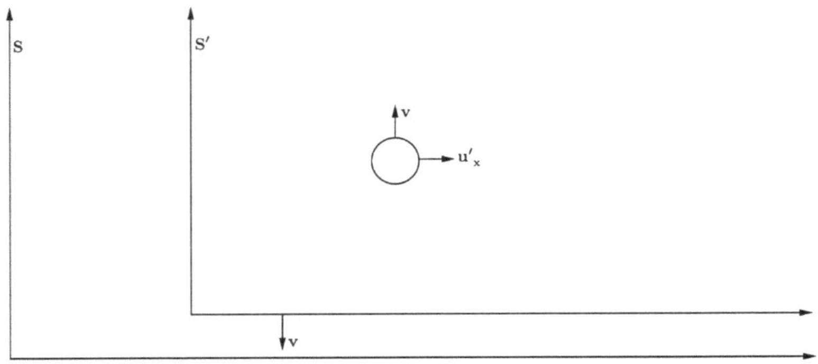

Figure 6.6 View of the collision in a frame S' with a small transverse velocity v in the $-y$ direction.

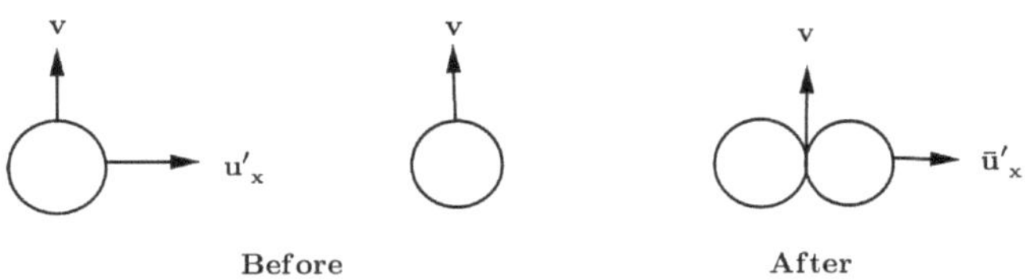

Figure 6.7 The collision in the frame S' of Fig. 6.6.

where $\gamma(v) = 1\big/\sqrt{1 - v^2/c^2}$, using our time dilation formula. Now we can write momentum conservation for the collision in frame S' (Fig. 6.7).

But in the frame S' the mass of the beam particle is a function of the length of the velocity $\mathbf{u}'$, which is $\mathbf{u}'^2 = v^2 + u_x'^2 = v^2 + u^2(1 - v^2/c^2)$. The same remark applies to $\overline{\mathbf{u}}'$, the velocity of the composite particle, $\overline{\mathbf{u}}'^2 = v^2 + \overline{u}_x'^2 = v^2 + \overline{u}^2(1 - v^2/c^2)$. So, conservation of the y component of momentum in frame S' reads

$$m(u')\mathbf{v} + m(v)\mathbf{v} = M(\overline{u}')\mathbf{v},$$

which gives the composite mass $M(\overline{u}')$. We only need this general formula for the special situation $v \to 0$,

$$m(u) + m(0) = M(\overline{u}). \tag{6.4}$$

This formula tells us how to add the relativistic masses of the colliding particles to make up the mass of the composite. It is a statement of mass conservation for relativistic systems where mass must depend on velocity. Without Eq. (6.4) we could not have conservation of relativistic momentum. Substituting into Eq. (6.2), momentum conservation in the lab frame S,

$$m(u)\mathbf{u} = M(\overline{u})\overline{\mathbf{u}} = (m(u) + m(0))\overline{\mathbf{u}},$$

we can now determine the velocity dependence of the relativistic mass,

$$m(u) = m(0) \cdot \frac{\overline{u}}{u - \overline{u}}.$$

This result is useful when we express $\overline{u}$ in terms of u, the velocity of a beam particle in the lab frame. This can be done using Eq. (6.3), giving

$$\frac{\overline{u}}{u - \overline{u}} = \frac{\overline{u}}{\dfrac{2\overline{u}}{1 + \overline{u}^2/c^2} - \overline{u}} = \frac{1 + \overline{u}^2/c^2}{1 - \overline{u}^2/c^2}, \tag{6.5}$$

where we also did some algebra to get the last expression. The right-hand side of Eq. (6.5) must be written in terms of u. Using the identity

$$\left(1 - \overline{u}^2/c^2\right)^2 = \left(1 + \overline{u}^2/c^2\right)^2 - 4\overline{u}^2/c^2$$

we have

$$\left(\frac{1 - \overline{u}^2/c^2}{1 + \overline{u}^2/c^2}\right)^2 = 1 - \frac{4\overline{u}^2/c^2}{\left(1 + \overline{u}^2/c^2\right)^2} = 1 - u^2/c^2,$$

where we identified Eq. (6.3), $u = 2\overline{u}/\left(1 + \overline{u}^2/c^2\right)$, in the last step. So, we end with the elegant result,

$$\frac{\overline{u}}{u - \overline{u}} = \gamma(u),$$

which gives the relativistic mass

$$m(u) = \gamma(u)m(0) = m(0)\Big/\sqrt{1 - u^2/c^2}.$$

In summary, we have the very important result for the relativistic momentum of a particle of rest mass $m(0) = m$ and velocity $\mathbf{u}$,

$$\mathbf{p} = \gamma m \mathbf{u}, \quad \gamma = 1\Big/\sqrt{1 - u^2/c^2}. \tag{6.6}$$

What do we learn from all this? What are the special features of $\mathbf{p}$? $\mathbf{p}$ has a factor of $\gamma = 1\Big/\sqrt{1 - u^2/c^2}$, which crept in from the exercise in addition of relativistic velocities. The factor of γ, therefore, is forced on us by the way positions and times transform from one inertial frame to another.

Because Eq. (6.6) is much simpler than its derivation, we must find a more fundamental derivation for it. We will return to this later.

One of the satisfying features of Eq. (6.6) is that it is consistent with the speed limit c. Because γ grows without bound as u approaches c, the relativistic mass $m(u)$ and the relativistic momentum $\mathbf{p} = m(u)\mathbf{u}$ do so as well.

There is still more to be learned from our inelastic collision. Let us look at the mass of the composite particle in more detail. To do this, boost to yet another frame, the center-of-momentum frame where the total momentum vanishes. This frame is clearly obtained from the lab frame by a boost to the left through velocity $\overline{u}$, just sufficient to bring the composite particle to rest. The collision is shown in Fig. 6.8.

Just as we found $m(u) + m(0) = M(\overline{u})$ by considering momentum conservation in the lab frame, we now find $m(\overline{u}) + m(-\overline{u}) = M(0)$. Because m depends only on the magnitude of its argument, $m(-\overline{u}) = m(\overline{u})$, we have, more simply,

$$M = 2m(\overline{u}) = \frac{2m}{\sqrt{1 - \overline{u}^2/c^2}}. \tag{6.7}$$

This result replaces the Newtonian conservation law in which M would be just twice m. If $\overline{u}^2/c^2 \ll 1$, Eq. (6.7) will not deviate much from its Newtonian limit. Using the expansion from Appendix C,

$$\frac{1}{\sqrt{1 - \overline{u}^2/c^2}} \approx 1 + \frac{1}{2}\frac{\overline{u}^2}{c^2} + O\left(\frac{\overline{u}^4}{c^4}\right),$$

we have

$$M \approx 2m\left(1 + \frac{1}{2}\frac{\overline{u}^2}{c^2}\right) = 2m + 2\cdot\frac{m\overline{u}^2}{2c^2}. \tag{6.8}$$

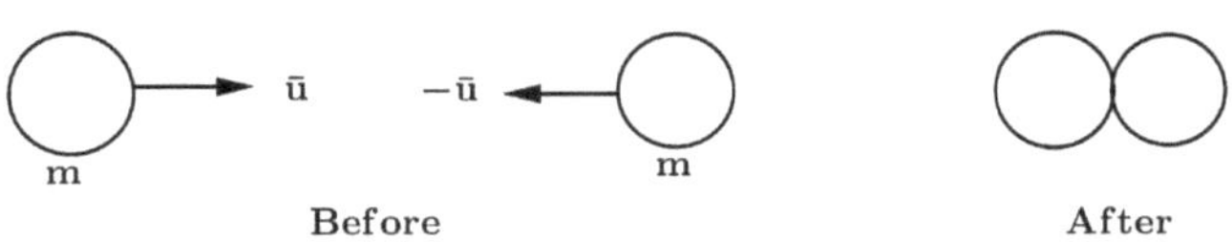

Figure 6.8 The inelastic collision in the center-of-momentum frame.

Here we can identify the nonrelativistic kinetic energy $T^{NR} = m\bar{u}^2/2$. The total kinetic energy, $2\cdot\left(m\bar{u}^2/2\right)$, is converted to heat Q in this inelastic collision, so Eq. (6.8) can be written

$$M \approx 2m + Q/c^2.$$

Identifying the inertial mass in this inelastic collision,

$$\Delta M \equiv M - 2m = Q/c^2,$$

we obtain Einstein's famous formula,

$$Q = \Delta M c^2,$$

which states that heat Q is equivalent to rest mass ΔM through a conversion factor of c^2. This is the same result as in Section 6.1, obtained from a different perspective.

To assimilate all these statements about energy and mass, we define the relativistic energy to be

$$E = \gamma mc^2 \tag{6.9}$$

for a rest mass m having velocity u, $\gamma = 1\Big/\sqrt{1 - u^2/c^2}$, in a certain inertial frame. This definition then uses the concept of relativistic mass $m(u)$ that has proved so convenient and fundamental in our study of the inelastic collision. The fact that the relativistic mass is conserved, and this conservation law satisfies Postulate 1 (it retains its form and is true in all inertial frames) means that the relativistic energy is a conserved quantity. This is the relativistic generalization of the conservation of rest mass that is an essential aspect of Newtonian mechanics.

Eq. (6.9) also suggests a definition of relativistic kinetic energy that generalizes the Newtonian quantity, $m\,v^2/2$. Because E reduces to mc^2 for a body at rest, the difference, $E - mc^2$, is a relativistic measure of the energy due to velocity. Call this difference the relativistic kinetic energy T,

$$T \equiv E - mc^2 = (\gamma - 1)mc^2.$$

When $v^2/c^2 \ll 1$, T reduces to the familiar quantity $mu^2/2$, because

$$T = (\gamma - 1)mc^2 \approx \left[\left(1 + \frac{1}{2}\frac{u^2}{c^2}\right) - 1\right]mc^2 = \frac{1}{2}mu^2.$$

T proves to be a handy quantity in relativistic kinematics problems.

Most textbooks on relativity work with the relativistic energy $E = \gamma mc^2$ and rest mass m rather than with the relativistic mass, a velocity-dependent quantity, $m(u) = \gamma m$. We shall do the same from here on to avoid possible confusion. Mass means rest mass m, just as in Newton's world. Relativistic energy is $E = \gamma mc^2$ and relativistic momentum is $\mathbf{p} = \gamma m\mathbf{u}$. The factor of γ will be written out explicitly.

6.3 RELATIVISTIC FORCE AND ENERGY CONSERVATION

To do problems in relativity with a given applied force that generates acceleration, we need the relativistic generalization of Newton's second law, $\mathbf{F} = (d/dt)\mathbf{p}$, $\mathbf{p} = m\mathbf{v}$. In many situations, $\mathbf{F}$ varies from point to point in a simple way. For example, to describe gravitational attraction between masses m and M a distance r apart, we know from Newton's law of gravity that

$$\mathbf{F} = -\frac{GmM}{r^2}\widehat{\mathbf{r}},$$

where $\widehat{\mathbf{r}}$ is a unit vector in the direction pointing from particle M to m, G is Newton's constant, and $\mathbf{F}$ is the force that M exerts on m. In the case of a harmonic oscillator,

$$\mathbf{F} = -kr\widehat{\mathbf{r}},$$

where a particle is attracted to the origin by a force proportional to its distance away from the origin. If the particle is constrained to move along the positive x axis, then the harmonic force reads $-kx$. In all these cases, we can introduce the concept of potential energy and write the force as the spatial rate of change of the potential energy. Because the problems we deal with here are essentially one dimensional, we illustrate this just for motion along the x axis.

In the case of the harmonic oscillator, then, it proves useful to introduce the potential energy $U(x) = (1/2)kx^2$, so for positive x,

$$F(x) = -kx = -\frac{d}{dx}U(x).$$

The harmonic oscillator illustrates a general strategy: Given a general function U(x), we can calculate the force by computing the spatial rate of change $F(x) = -dU/dx$. We introduce the potential energy because it simplifies problem solving and leads to energy conservation. For example, the work that the force does on a particle of mass m when it is moved from point 1 to point 2 along the positive x axis is

$$W_{12} = \int_1^2 F(x)dx = -\int_1^2 \frac{dU(x)}{dx}dx = U(1) - U(2). \tag{6.10}$$

But, using the equation of motion, we can calculate W_{12} in another way,

$$W_{12} = \int_1^2 Fdx = \int_1^2 \left(\frac{d}{dt}\right)pdx = \int_1^2 m\frac{d}{dt}v \cdot vdt,$$

where we noted that distance is velocity times time, $dx = vdt$. But the integral can be done exactly, noting that $(d/dt)\ (mv^2/2) = mvdv/dt$,

$$W_{12} = \int_1^2 \frac{d}{dt}\left(\frac{1}{2}mv^2\right)dt = T_2^{\mathrm{NR}} - T_1^{\mathrm{NR}}, \tag{6.11}$$

where T^{NR} (the nonrelativistic kinetic energy) $= (1/2)mv^2$. Combining Eqs. (6.10) and (6.11), we have

$$U(1) - U(2) = T_2^{\mathrm{NR}} - T_1^{\mathrm{NR}},$$

or

$$T_1^{\mathrm{NR}} + U_1 = T_2^{\mathrm{NR}} + U_2.$$

In other words, the total energy $E^{NR} = T^{\mathrm{NR}} + U$ is conserved in Newton's world whenever an x-dependent static potential energy $U(x)$ exists.

In summary, using Newton's second law, $\mathbf{F} = d\mathbf{p}/dt$, we derive energy conservation under appropriate conditions. These two concepts are central in the solution of nonrelativistic mechanics problems.

The issue here is the generalization of these ideas to the relativistic world of Einstein. Certainly the momentum in Newton's second law must become its relativistic cousin,

$$\mathbf{F} = \frac{d}{dt}(\gamma m\mathbf{v}). \tag{6.12}$$

In a force-free environment, this equation predicts conservation of the relativistic momentum, as desired. If several particles are interacting among themselves, but the total of the forces sum to zero, then the total relativistic momentum is conserved. In this way, Eq. (6.12) becomes the underpinning of our previous discussion of the inelastic collision.

Let us do a simple practice problem with Eq. (6.12) before continuing [2]. Place a charged particle of mass m in a constant electric field. The electric force will accelerate the particle and, in a Newtonian description, the particle's velocity will increase without bound. But in relativity there is a speed limit, so v will increase arbitrarily closely to c, never attaining it because of the factor of $\gamma = 1\Big/\sqrt{1 - v^2/c^2}$ in Eq. (6.12), which grows without bound as v approaches c. In more detail, let the particle have an initial velocity of 0, $v(t = 0) = 0$, and let it be subjected to a constant force F in the x direction. Then,

$$F = \frac{d}{dt}p.$$

So, p will increase linearly with t,

$$\gamma m v = Ft. \tag{6.13}$$

Because $\gamma = 1/\sqrt{1 - v^2(t)/c^2}$, we can solve Eq. (6.13) for $v(t)$ and find, after some algebra,

$$v(t) = \frac{c}{\sqrt{1 + (mc/Ft)^2}}. \tag{6.14}$$

We see, as expected, that $v(t)$ is always less than c, but approaches it when $Ft \gg mc$. At the other extreme, if $Ft \ll mc$, then $v(t) \approx (c/(mc/Ft)) = (F/m) t \ll c$ and we retrieve Newton's result that the particle experiences a constant acceleration F/m that produces a velocity that grows linearly with time.

In summary, our relativistic dynamics has passed an important test—we cannot accelerate a massive body beyond the speed limit. Furthermore, Eq. (6.14) predicts $v(t)$ for a particle in a linear accelerator, so its functional form can be compared with experimental data. Of course, in the real world there are complications, such as the fact that accelerated charged particles radiate energy, which must be accounted for in quantitative tests. After all such effects are dealt with, relativistic dynamics with all the trimmings proves to be an unparalleled success.

Our next big task is to find the relativistic generalization of energy conservation for forces that are obtained from a static potential. As in the nonrelativistic world,

$$W_{12} = \int_1^2 F dx = -\int_1^2 \frac{dU}{dx} dx = U(1) - U(2).$$

Now we need to use the equation of motion to see how the force changes the particle's relativistic energy:

$$W_{12} = \int_1^2 F dx = \int_1^2 \frac{d}{dt}(\gamma m\, v) v dt. \tag{6.15}$$

Can we write the integrand as a total time derivative? Inspired by the nonrelativistic derivation, we hope that the answer will involve the relativistic energy,

$$E = \gamma mc^2.$$

Let us write Eq. (6.15) in terms of just the relativistic momentum and energy by writing the velocity as,

$$v = c^2\frac{\gamma mv}{\gamma mc^2} = c^2\frac{p}{E}$$

so Eq. (6.15) reads,

$$W_{12} = c^2\int_1^2 \frac{p}{E}\frac{dp}{dt}dt = c^2\int_1^2 \frac{1}{2E}\frac{dp^2}{dt}dt$$

Now we must relate p^2 to E^2. Using $p = \gamma mv$ and $E = \gamma mc^2$, we calculate,

$$\frac{1}{c^2}E^2 - p^2 = \gamma^2 m^2 c^2 - \gamma^2 m^2 v^2 = \gamma^2 m^2 c^2\left(1 - \frac{v^2}{c^2}\right) = m^2c^2$$

where we used $\gamma^2 = \left(1 - \frac{v^2}{c^2}\right)^{-1}$ in the last step. The important point in this "Energy—momentum" relation is the fact that the right-hand side is a constant, Lorentz invariant quantity, the rest mass of the particle squared. Now p^2 can be replaced by E^2/c^2 in the expression for W_{12},

$$W_{12} = \int_1^2 \frac{1}{2E}\frac{dE^2}{dt}dt = \int_1^2 \frac{dE}{dt}dt = E_2 - E_1 = T_2 - T_1 \tag{6.16}$$

which is the relativistic analogue of Eq. (6.11)! Combining Eqs. (6.15) and (6.16),

$$T_1 + U_1 = T_2 + U_2, \tag{6.17}$$

and the total energy, relativistic kinetic and potential, is conserved under these conditions.

Note that since $T = E - mc^2$, we could put either T or E into Eq. (6.17). We used the relativistic kinetic energy just to mirror the nonrelativistic discussion.

6.4 ENERGY AND MOMENTUM CONSERVATION, AND FOUR-VECTORS

Our formulas for the relativistic momentum and energy, $\mathbf{p} = \gamma m\mathbf{v}$ and $E = \gamma mc^2$, are much simpler than their derivations. We aim to remedy this problem here. Both of these quantities were determined to satisfy Postulate 1. The factors of γ are forced on us by the transformation laws of space and time measurements between inertial frames. In Newton's world, the momentum inherits its transformation properties from those of space and time,

$$\mathbf{p} = m\frac{d\mathbf{x}}{dt} \quad \text{(Newton)}. \tag{6.18}$$

$\mathbf{x}$ transforms according to Galileo, $x = x' + vt$, $y' = y$, $z' = z$, and t is universal, $t = t'$. Eq. (6.18) then implies that $\mathbf{p}$ transforms as a velocity, and because velocities add in Newton's world, we can show that if there is momentum conservation in one inertial frame, there is momentum conservation in all.

Can we write a relativistic generalization of Eq. (6.18)? The key in Eq. (6.18) is that the numerator is a distance that transforms simply and the denominator is an invariant. Because the proper time τ is an invariant in a relativistic world, we should try

$$\mathbf{p} = m\frac{d\mathbf{x}}{d\tau} \quad \text{(Einstein)}. \tag{6.19}$$

Eq. (6.19) reproduces $\mathbf{p} = \gamma m\mathbf{v}$ because $d\tau$ is related to the time interval dt by time dilation, $d\tau = \sqrt{1 - v^2/c^2}\,dt$. The truly crucial feature of Eq. (6.19) is that $\mathbf{p}$ transforms between inertial frames in the same way as $\mathbf{x}$ because $d\tau$ is an invariant.

There is still a puzzle here. When $\mathbf{x}$ transforms between frames, the time variable mixes in. In other words, it takes *four* quantities, the three components of $\boldsymbol{x}$ and t, to write an expression for the transformation law of any one of them. So, Eq. (6.19) must be supplemented by an expression involving t. Clearly, $dt/d\tau$ is the first candidate to come to mind and because this derivative is γ, we are led to

$$E = mc^2\frac{dt}{d\tau} = \gamma mc^2,$$

where hindsight led us to include the factor of c^2 and identify the relativistic energy. So, E gives the zeroth component of a four-vector of energy–momentum. We write,

$$p^\mu = \left(\frac{E}{c}, p^1, p^2, p^3\right)$$

where μ is an index that can take the values 0, 1, 2, or 3 and $p^o = E/c$, and so on.

Our original four-vector is space–time,

$$x^\mu = (ct, x^1, x^2, x^3)$$

and the crucial feature about this quantity is its transformation law under boosts, the Lorentz transformation,

$$\begin{aligned} t' &= \gamma\left(t - \frac{v}{c^2}x^1\right) \\ x'^1 &= \gamma(x^1 - vt) \\ x'^2 &= x^2 \\ x'^3 &= x^3 \end{aligned}$$

The transformation laws for $\mathbf{p}$, E are now immediate:

$$\begin{aligned} E' &= mc^2\frac{dt'}{d\tau} = \gamma mc^2\left(\frac{dt}{d\tau} - \frac{v}{c^2}\frac{dx^1}{d\tau}\right) = \gamma(E - vp^1) \\ p'^1 &= m\frac{dx'^1}{d\tau} = \gamma m\left(\frac{dx^1}{d\tau} - v\frac{dt}{d\tau}\right) = \gamma\left(p^1 - \frac{v}{c^2}E\right) \\ p'^2 &= p^2 \\ p'^3 &= p^3 \end{aligned} \tag{6.20}$$

Knowing the transformation law Eq. (6.20) will help us solve problems in relativistic collisions and help us formulate how charged particles respond to electric and magnetic fields.

The reader should be careful not to confuse the γ factors in the expression for $\mathbf{p}$ and E with the γ factors in Eq. (6.20). In the first case, γ contains the velocity of the particle in a given frame, and in the transformation formulas Eq. (6.20), γ contains the relative velocity between two inertial frames.

We learn several points from Eq. (6.20). First, momentum and energy conservation are unified. Because momentum and energy mix under a boost, we must have conservation of all *four* components of the energy–momentum four-vector together. It would be inconsistent with Postulate 1 to have fewer. Second, because **p** and E form a four-vector, we should be able to construct an invariant frame-independent quadratic form in analogy to the invariant interval

$$c^2t^2 - \mathbf{x}^2 = c^2t'^2 - \mathbf{x}'^2$$

For energy–momentum, we have

$$\frac{1}{c^2}E^2 - \mathbf{p}^2 = \frac{1}{c^2}\gamma^2 m^2 c^4 - \gamma^2 m^2 \mathbf{u}^2 = \gamma^2 m^2 (c^2 - \mathbf{u}^2)$$
$$= \frac{m^2c^2}{1 - u^2/c^2}(1 - \mathbf{u}^2/c^2) = m^2c^2.$$

by the same algebra we used in Section 6.3. So, we have rederived the energy–momentum relation for a particle of mass m,

$$E^2 = \mathbf{p}^2c^2 + m^2c^4. \tag{6.21}$$

This is the relativistic version of the energy–momentum relation of Newtonian physics,

$$T^{\mathrm{NR}} = \frac{1}{2}m\mathbf{u}^2 = \frac{\mathrm{p}^2}{2m}. \tag{6.22}$$

Note that Eq. (6.21) is quite different from Eq. (6.22). Substituting the relativistic kinetic energy $T = E - mc^2$ into Eq. (6.21) gives

$$T^2 + 2mc^2T = \mathbf{p}^2c^2. \tag{6.23}$$

If $v/c \ll 1$, then we observed earlier that $T \approx (1/2)\,mv^2$ so $T \ll mc^2$ and Eq. (6.23) reduces to

$$T \approx \mathbf{p}^2/2m$$

as expected.

But for $v \approx c$, $pc \gg mc$, so Eq. (6.21) becomes

$$E \approx pc.$$

Therefore, E and p, the magnitude of $\mathbf{p}\left(p \equiv \sqrt{p_1^2 + p_2^2 + p_2^2}\right)$, become linearly related.

Finally, note that if the rest mass of the particle vanishes, it can still carry momentum and energy, and we have exactly,

$$E = pc. \tag{6.24}$$

This is the exact energy–momentum relation for light, as we mentioned early in this chapter. Light travels at the speed limit because it has no rest mass. Maxwell's wave theory of light predicts Eq. (6.24) from first principles as we shall derive in Chapter 9. This relation is particularly important in the quantum theory of light, as we illustrate in Section 6.6.

6.5 FOCUSING ON FOUR-VECTORS, TENSORS, AND NOTATION

In the previous section, we discovered that the energy–momentum of a free particle in relativity has the property that it transforms under boosts in the same fashion as the space–time coordinates. This makes the four-vector $p^\mu = (E/c, p_x, p_y, p_z)$ a particularly significant and useful quantity. We should think more generally about four-vectors for this reason.

First, we noted that the "length" of a four-vector is invariant under Lorentz transformations ("Lorentz invariant"). You will see how powerful this point is in problem solving in Section 6.6. We will write the length of a four-vector a^μ as $a \cdot a = a_0^2 - a_x^2 - a_y^2 - a_z^2$. If we have two four-vectors, a^μ and b^μ, we can take their "inner product," $a \cdot b = a_0 b_0 - a_x b_x - a_y b_y - a_z b_z$. This is also a Lorentz invariant quantity as straightforward algebra shows. Note that it is also invariant under ordinary rotations in three space.

Take a useful example, $p \cdot x = Et - p_x x - p_y y - p_z z$. Consider a light ray propagating in the x direction. In this case $E = p_x c$, so $p \cdot x = Et - p_x x = E(t - x/c)$. We recognize this as the phase of the light ray. The phase depends on the combination $t - x/c$ indicating that points on the wave of constant phase move at the speed limit c. Also, since the phase is Lorentz invariant, observers measure the phase to be the same in all frames, $p \cdot x = p' \cdot x' = E(t - x/c) = E'(t' - x'/c)$, which states that the wave travels at the speed limit in *all* frames.

It will help us make further progress in special relativity, if we modernize our notation for Lorentz transformations, four-vectors, etc. Better notation will also prepare us for general relativity where we will go beyond the requirement that physics satisfies the principle of relativity within inertial frames to the requirement that physics satisfies the principle of relativity in a general noninertial frame including curved space–times. We start with the Minkowski metric $g_{\mu\nu}$, which relates our coordinate mesh to invariant intervals,

$$ds^2 = c^2dt^2 - dx^2 - dy^2 - dz^2 \equiv \sum_{\mu\nu} g_{\mu\nu} dx^\mu dx^\nu \tag{6.25}$$

Here the metric $g_{\mu\nu}$ acts as a 4×4 matrix. In fact, it is a "second rank" tensor as we will discuss in Chapters 11 and 12. In the case of Minkowski space, its only nonzero elements are on the diagonal, $g_{00} = -g_{11} = -g_{22} = -g_{33} = +1$. Of course we could choose spherical or cylindrical coordinates and then the appearance of the metric will change. In spherical coordinates,

$$ds^2 = c^2dt^2 - dr^2 - r^2\left(d\theta^2 + \sin^2\theta d\phi^2\right)$$

In these coordinates (ct, r, θ, ϕ) the metric reads,

$$g_{00} = 1, \quad g_{11} = -1, \quad g_{22} = -r^2, \quad g_{33} = -r^2\sin^2\theta$$

But we really want to consider more possibilities where $g_{\mu\nu}$ has off diagonal elements because this is unavoidable in curvilinear coordinates that occur in general relativity. To accommodate these possibilities we introduce "upper" and "lower" indices. *Covariant* four-vectors A_μ have lower indices and *contravariant* four-vectors A^μ have upper indices. Given A^μ we convert it to a covariant four-vector by lowering its indices using the metric,

$$A_\mu = \sum_\nu g_{\mu\nu} A^\nu \tag{6.26}$$

For the Minkowski metric in Cartesian coordinates this is simple,

$$A^0 = A_0, \quad A^1 = -A_1, \quad A^2 = -A_2, \quad A^3 = -A_3$$

Why should we make these definitions? Because the inner product of four-vectors can now be written more simply! Note that

$$A \cdot B = A_0B_0 - A_1B_1 - A_2B_2 - A_3B_3 = A^0B_0 + A^1B_1 + A^2B_2 + A^3B_3$$

$$A \cdot B = \sum_\mu A^\mu B_\mu \tag{6.27}$$

So, using the upper and lower indices we have a much briefer way of writing invariants without the explicit appearance of the metric. Take an example, the "length" of the energy–momentum four-vector,

$$p^2 = p \cdot p = \sum_\mu p^\mu p_\mu = E^2/c^2 - \boldsymbol{p}^2 = m^2c^2$$

The real payoff occurs when we meet generalizations of four-vectors, second-rank tensors. These are quantities such as the energy–momentum stress tensor $T_{\mu\nu}$, which depends on two four-vector indices and the metric tensor $g_{\mu\nu}$. The energy—momentum stress tensor $T_{\mu\nu}$ is the source for space—time curvature in general relativity, so we will see more of this tensor later. The term tensor means that it transforms from one inertial frame to another like the "outer" product of two four-vectors, $A_\mu B_\nu$. Let us write the Lorentz transformation of a four-vector in matrix notation,

$$ct' = \gamma(ct - vx/c), \quad x' = \gamma(x - vct/c), \quad y' = y, \quad z' = z$$

Equivalently,

$$\begin{pmatrix} ct' \\ x' \end{pmatrix} = \begin{pmatrix} \gamma & -\frac{v}{c}\gamma \\ -\frac{v}{c}\gamma & \gamma \end{pmatrix} \begin{pmatrix} ct \\ x \end{pmatrix} \tag{6.28a}$$

Or, writing out all four members of $(x^0, x^1, x^2, x^3) = (ct, x, y, z)$ in matrix notation, which respects covariant and contravariant indices,

$$x'^\mu = \sum_\nu L^\mu{}_\nu x^\nu \tag{6.28b}$$

where the sum over ν ranges over 0, 1, 2, 3, and we read off Eq. (6.28a) that $L^0_0 = \gamma$, $L^0{}_1 = -\frac{v}{c}\gamma$, $L^1_0 = -\frac{v}{c}\gamma$, $L^1_1 = \gamma$, $L^2_2 = 1$, $L^3_3 = 1$ and all other elements in the Lorentz transformation L^μ_ν are zero in this Cartesian basis for boosts in the x direction.

Let us illustrate this notation further. First, the outer product of two four-vectors transforms under boosts as,

$$A'^\mu B'^\nu = \left(\sum_\sigma L^\mu{}_\sigma A^\sigma\right)\left(\sum_\lambda L^\nu{}_\lambda B^\lambda\right) = \sum_{\sigma\lambda} L^\mu{}_\sigma L^\nu{}_\lambda A^\sigma B^\lambda \tag{6.29}$$

We will find mathematical objects with two four-vector indices as we delve deeper into special and general relativity. They are called "second-rank tensors," and they transform by this rule even if they can not be expressed as the outer product of two four-vectors,

$$T'^{\mu\nu} = \sum_{\sigma\lambda} L^\mu{}_\sigma L^\nu{}_\lambda T^{\sigma\lambda} \tag{6.30}$$

The metric is a second-rank tensor, but a very simple and special one in special relativity,

$$g'^{\mu\nu} = \sum_{\sigma\lambda} L^\mu{}_\sigma L^\nu{}_\lambda g^{\sigma\lambda} \tag{6.31}$$

In this case the metric is unaffected by boosts. We know this because of the invariance of the invariant interval under boosts,

$$ds^2 = c^2dt^2 - dx^2 - dy^2 - dz^2 = c^2dt'^2 - dx'^2 - dy'^2 - dz'^2$$

So, $g'^{\mu\nu} = g^{\mu\nu}$. We can check this explicitly from Eq. (6.31),

$$g'^{00} = \sum_{\sigma\lambda} L^0{}_\sigma L^0{}_\lambda g^{\sigma\lambda} = \gamma^2 g^{00} + \left(\frac{v}{c}\gamma\right)^2 g^{11} = \gamma^2 - \gamma^2\frac{v^2}{c^2} = 1$$

$$g'^{11} = \sum_{\sigma\lambda} L^1{}_\sigma L^1{}_\lambda g^{\sigma\lambda} = \gamma^2 g^{11} + \left(\frac{v}{c}\gamma\right)^2 g^{00} = -\gamma^2 + \gamma^2\frac{v^2}{c^2} = -1$$

$$g'^{01} = \sum_{\sigma\lambda} L^0{}_\sigma L^1{}_\lambda g^{\sigma\lambda} = -\gamma^2\frac{v}{c}g^{00} - \gamma^2\frac{v}{c}g^{11} = \gamma^2\frac{v}{c} - \gamma^2\frac{v}{c} = 0$$

$$g'^{22} = \sum_{\sigma\lambda} L^2{}_\sigma L^2{}_\lambda g^{\sigma\lambda} = g^{22} = -1$$

$$g'^{33} = \sum_{\sigma\lambda} L^2{}_\sigma L^2{}_\lambda g^{\sigma\lambda} = g^{33} = -1$$

and so on. Of course, all of this arithmetic "worked out" because Lorentz transformations were designed to leave ds^2 invariant, $ds^2 = ds'^2$. A fuller discussion can be found in Problem 6.42.

There are many observations we could make about the algebra of $L^\mu{}_\sigma$, $g^{\mu\nu}$, four-vectors and tensors. For example, $L^\mu{}_\sigma$ is a function of the velocity of the boost v. If we boost in the x direction by v and then boost in the x direction by $-v$, we should arrive where we started. In other words, $L^\mu{}_\sigma(-v)$ should be the inverse of $L^\mu{}_\sigma(v)$,

$$\sum_\mu L^\sigma{}_\mu(-v)L^\mu{}_\nu(v) = \delta^\sigma_\nu \tag{6.32}$$

where the "Kronecker" symbol δ^σ_ν is short hand for the identity matrix, $\delta^\sigma_\nu = 1$ if $\sigma = \nu$ and zero otherwise. The explicit verification of this identity is left to the problem set.

Tensors as well as four-vectors can be written with upper or lower indices. The metric illustrates this formalism. Start with the invariant interval, and write it in terms of dx^μ instead of dx_μ,

$$ds^2 = \sum_{\mu\nu} g^{\mu\nu}dx_\mu dx_\nu = \sum_{\sigma\rho} g_{\sigma\rho}dx^\sigma dx^\rho$$

Clearly $g^{\mu\nu} = g_{\mu\nu}$ in Cartesian coordinates, and if $A^\mu = \sum_\nu g^{\mu\nu}A_\nu$, then $A_\sigma = \sum_\lambda g_{\sigma\rho}A^\rho$, and $\sum_\nu g^{\mu\nu}g_{\nu\rho} = \delta^\mu_\rho$. Of course, since $g_{\mu\nu}$ is just on-diagonal (1, −1, −1, −1) in Cartesian coordinates, this notation is rather excessive! It gains value, however, when we work in curvilinear coordinates.

We have written Lorentz transformations for contravariant four-vectors x^μ and p^μ, etc. We can write these transformation laws for x_μ and p_μ, etc., as well. Recall that we raise or lower indices with the metric tensor,

$$x^\mu = \sum_\nu g^{\mu\nu} x_\nu, \quad x_\mu = \sum_\nu g_{\mu\nu} x^\nu \tag{6.33}$$

We see that $g^{\mu\nu}$ is the inverse of $g_{\mu\nu}$,

$$\sum_\nu g^{\mu\nu} g_{\nu\rho} = \delta^\mu_\rho \tag{6.34}$$

as we noted earlier. Since x^μ transforms from frame S to S' by a Lorentz transformation,

$$x'^\mu = \sum_\sigma L^\mu{}_\sigma x^\sigma \tag{6.35}$$

and since,

$$x_0 = x^0, \quad x_1 = -x^1, \quad x_2 = -x^2, \quad x_3 = -x^3 \tag{6.36}$$

we have sufficient information to see how x_μ transforms under a boost. Write out Eq. (6.28a),

$$x'^0 = \gamma(x^0 - vx^1/c), \quad x'^1 = \gamma(x^1 - vx^0/c), \quad x'^2 = x^2, \quad x'^3 = x^3$$

Substituting Eq. (6.36),

$$x'_0 = \gamma(x_0 + vx_1/c), \quad x'_1 = \gamma(x_1 + vx_0/c), \quad x'_2 = x_2, \quad x'_3 = x_3 \tag{6.37}$$

Using the metric to raise and lower indices in Eq. (6.35), we can write the boost for x_μ,

$$x'_\mu = \sum_\sigma L_\mu{}^\sigma x_\sigma \tag{6.38}$$

and read off from Eq. (6.37), $L_0{}^0 = \gamma$, $L_0{}^1 = \frac{v}{c}\gamma$, $L_1{}^0 = \frac{v}{c}\gamma$, $L_1{}^1 = \gamma$, $L_2{}^2 = 1$, and $L_3{}^3 = 1$.

Next, let us turn to the transformation laws of operators. It will serve us well to find the transformation laws of $\partial/\partial x^\mu$ and $\partial/\partial x_\mu$. To begin, write the inverse relation to Eq. (6.37),

$$x_0 = \gamma(x'_0 - vx'_1/c), \quad x_1 = \gamma(x'_1 - vx'_0/c), \quad x_2 = x'_2, \quad x_3 = x'_3$$

so we can compute the partial derivatives,

$$\frac{\partial x_0}{\partial x'_0} = \gamma, \quad \frac{\partial x_1}{\partial x'_0} = -\frac{v}{c}\gamma, \quad \frac{\partial x_0}{\partial x'_1} = -\frac{v}{c}\gamma, \quad \frac{\partial x_1}{\partial x'_1} = \gamma$$

So, using the chain rule,

$$\frac{\partial}{\partial x_0'} = \sum_{\mu} \frac{\partial x_\mu}{\partial x_0'} \frac{\partial}{\partial x_\mu} = \frac{\partial x_0}{\partial x_0'} \frac{\partial}{\partial x_0} + \frac{\partial x_1}{\partial x_0'} \frac{\partial}{\partial x_1} = \gamma\left(\frac{\partial}{\partial x_0} - \frac{v}{c} \frac{\partial}{\partial x_1}\right)$$

$$\frac{\partial}{\partial x_1'} = \sum_{\mu} \frac{\partial x_\mu}{\partial x_1'} \frac{\partial}{\partial x_\mu} = \frac{\partial x_0}{\partial x_1'} \frac{\partial}{\partial x_0} + \frac{\partial x_1}{\partial x_1'} \frac{\partial}{\partial x_1} = \gamma\left(\frac{\partial}{\partial x_1} - \frac{v}{c} \frac{\partial}{\partial x_0}\right)$$

which we identify as the Lorentz transformation for x^μ. So we learn that $\partial/\partial x_\mu$ transforms as an *upper* index four-vector, so we write,

$$\partial^\mu = \partial/\partial x_\mu$$

Similarly we find,

$$\partial_\mu = \partial/\partial x^\mu$$

A simple illustration of this formalism displays its usefulness. Consider the inner product of ∂_μ and ∂^μ,

$$\sum_{\mu} \partial^\mu \partial_\mu = \frac{\partial^2}{c^2 \partial t^2} - \nabla^2$$

This is an invariant operator because ∂^μ transforms as x^μ and ∂_μ transforms as x_μ and $\sum_{\mu} x^\mu x_\mu$ is an invariant. The operator character of ∂^μ and ∂_μ is not important here. The transformation law is important. So,

$$\sum_{\mu} \partial^\mu \partial_\mu = \sum_{\mu} \partial'^\mu \partial'_\mu = \frac{\partial^2}{c^2 \partial t^2} - \nabla^2 = \frac{\partial^2}{c^2 \partial t'^2} - \nabla'^2$$

This is the wave operator of electrodynamics. We shall see it again in Chapter 9.

The meaning of contravariant and covariant vectors is discussed from the perspective of linear algebra and differential geometry in Appendix C, Section D. The student interested in space–time curvature should have a look.

6.6 COLLISIONS AND CONSERVATION LAWS—CONVERTING MASS TO ENERGY AND ENERGY TO MASS, PRODUCING AND DESTROYING PARTICLES

It is interesting to consider relativistic collisions and decay processes that illustrate the energy–momentum conservation laws. Some of these processes involve light that satisfies the energy–momentum relation,

$E = pc$. In addition, if we deal with individual quanta of light, then the energy comes in a packet $E = h\nu$, where ν is the frequency of the light wave and h is Planck's constant, $h = 6.627 \cdot 10^{-34}$ J-s, which sets the scale of quantum physics. We certainly will not be doing any quantum mechanics here, so we just borrow $E = h\nu$ to illustrate relativity in interesting settings involving elementary particles.

As we pursue dynamics further in the following chapters, we will see that we must carefully distinguish between particles that are outside the range of nontrivial forces of other particles and those that are not. The kinematics of collisions are greatly simplified by the assumption that the particles in the initial and final states are so well separated that they experience no forces and move with constant velocities. It is only when the particles come within range of one another and interact that energy and momentum are transferred between them, and they experience forces that change their velocities. The interaction mechanisms must conserve energy and momentum *locally* in a relativistic, causal theory. That description will require us to replace the nonrelativistic notion of forces with the relativistic notion of dynamical fields. We will learn more about this in later chapters, but a full accounting requires the development of field theory. Nonetheless, we will learn a lot about relativistic collisions, the equivalence of mass and energy, particle creation and destruction and conservation laws in this section.

Let us take an example. Consider a nucleus of rest mass M_o that absorbs a photon of energy Q. The final state consists of an excited nucleus of rest mass M' and recoil velocity v. We want to find M' and v in terms of M_o and Q. The process is shown in Fig. 6.9. Energy conservation reads

$$Q + M_o c^2 = \gamma M' c^2$$

and momentum conservation reads

$$Q/c = \gamma M' v.$$

Combining these equations, we find

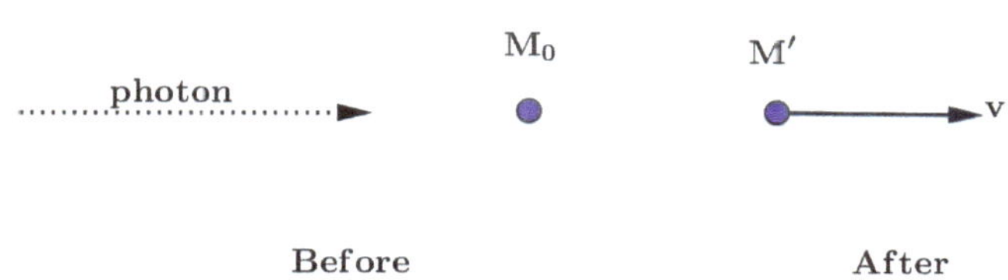

Figure 6.9 Photon absorption by a nucleus of mass M_0 at rest.

$$\frac{v}{c} = \frac{Q}{\gamma M' c^2} = \frac{Q}{Q + M_o c^2},$$

which gives the recoil velocity. Finally, we can solve for M',

$$M' = M_o \sqrt{1 + 2Q/M_o c^2}. \tag{6.39}$$

Note that $M'c^2$ is less than the sum of the initial relativistic energies, $M_o c^2 + Q$, because some energy goes into M's recoil. This is how the kinematics works out. It is interesting that this process may not be possible and may not occur. The point is that the laws of nuclear physics (quantum mechanics) predict a *discrete* list of possible states and possible M' values for each nucleus. If the M' of Eq. (6.39) does not match one of these allowed values, the initial photon will not be absorbed. Scattering of the photon might be the actual physical event in that case.

It is interesting to turn this process around and consider photon emission from an excited atom. The process is shown in Fig. 6.10. Conservation of energy and momentum reads

$$\begin{aligned} M_o c^2 &= E' + Q \\ 0 &= p' - Q/c, \end{aligned} \tag{6.40}$$

where E' and p' label the relativistic energy and momentum of the final (unexcited) atom of mass M'_o. Let us say that we detect the photon and measure its energy Q. So, we need to eliminate E' and p' from Eq. (6.40). A slick way to do this is to use the energy–momentum relation $E'^2 = p'^2 c^2 + M_o'^2 c^4$. Rearranging Eq. (6.40) gives

$$\begin{aligned} E' &= M_o c^2 - Q \\ p' &= Q/c. \end{aligned}$$

So,

$$E'^2 - p'^2 c^2 = (M_o c^2 - Q)^2 - Q^2 = M_o'^2 c^4.$$

Finally,

$$M_o^2 c^4 - 2M_o c^2 Q = M_o'^2 c^4. \tag{6.41}$$

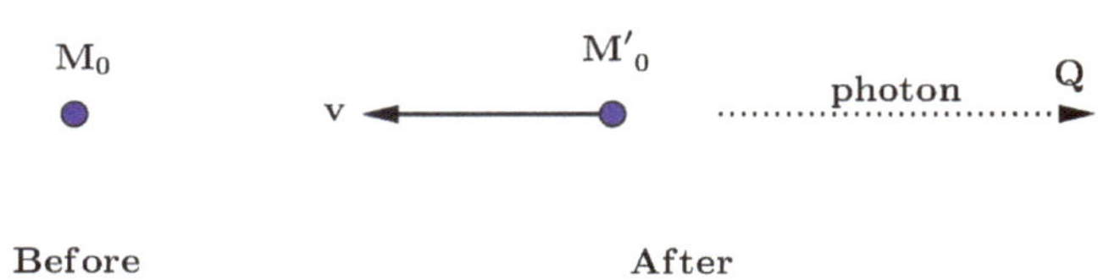

Figure 6.10 Photon emission from an excited atom of mass M_0 at rest.

Let us write this in terms of the energy difference between the initial and final atoms, taken at rest,

$$\Delta E \equiv M_o c^2 - M'_o c^2. \tag{6.42}$$

We focus on ΔE because this would be the energy difference between discrete quantum energy levels of the atom. One could calculate such a difference from first principles in quantum mechanics. Solving Eq. (6.42) for $M'_o c^2$,

$$M_o c^2 - \Delta E = M'_o c^2.$$

Squaring gives

$$M_o^2 c^4 - 2M_o c^2 \Delta E + (\Delta E)^2 = M'^2_o c^4.$$

Combining this with Eq. (6.41) gives

$$Q = \Delta E\left(1 - \frac{\Delta E}{2M_o c^2}\right). \tag{6.43}$$

This is the desired result. We learn that recoil, a consequence of energy–momentum conservation, has reduced the energy of the photon from ΔE to $\Delta E(1 - \Delta E/2M_o c^2)$. If we observe the photon and measure Q, we must use Eq. (6.43) to predict ΔE to compare with predictions of quantum mechanics. For heavy atoms, the recoil is a small effect. For example, consider ^{198}Hg, which emits photons with an energy 412 keV. Because $M_o = 198$ amu $= 3.28 \cdot 10^{-25}$ kg, we compute $\Delta E/2M_o c^2 \approx 10^{-6}$. This is small but not really negligible in our quantum world. For example, if the emitted photon were incident on another ^{198}Hg, it could not be absorbed because its energy does not quite match the needed ΔE—it is not quite energetic enough to cause the transition.

There is a way around this dilemma, first developed by Mössbauer. Let the atom that emits the photon be part of a large regular crystal. Then, when it emits the photon, the recoiling momentum, if it is sufficiently small, is imparted to the entire crystal! The recoil term, $\Delta E/2M_o c^2$, is now reduced by the number of atoms in the crystal and becomes truly negligible, less than the intrinsic energy spread of each spectral line. Using this observation, Mössbauer studied spectral lines in detail. His experimental apparatus employed two crystals, one emitting and the other absorbing, at a relative velocity v, as in Fig. 6.11. If v is too large, then the energy of the photons in the rest frame of the atoms in crystal B is too large to match the energy difference between the quantum states, even accounting for

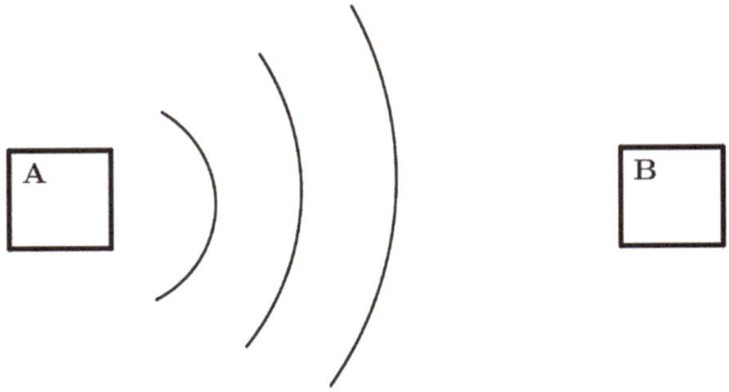

Figure 6.11 The two crystals at relative velocity v in a Mössbauer experiment.

their intrinsic uncertainties (widths). In fact, a tiny v on the order of a few centimeters per second is sufficient to produce a photon energy that misses ΔE. By varying v, the detailed structure of spectral lines can be mapped out.

Now let us turn to several illustrations of particle creation in high-energy physics collisions. At an accelerator center, researchers convert kinetic energy into mass and search for new elementary particles. Colliding beams are a particularly effective way of achieving this goal, and important discoveries of new resonances, evidence for the existence of heavy quarks, were made this way. Consider proton–proton collisions in the center-of-momentum frame and suppose we wish to create a pion, the particle responsible for the nuclear force as you will learn in quantum mechanics:

$$P + P \rightarrow P + N + \pi^{+}.$$

The collision is shown in Fig. 6.12. We want to know the minimal velocity u that will make it possible to create the extra particle, a pion. The rest masses involved are

$$m_{\pi}/m_P \approx 0.149, \quad m_P c^2 \approx 938 \text{ MeV}$$

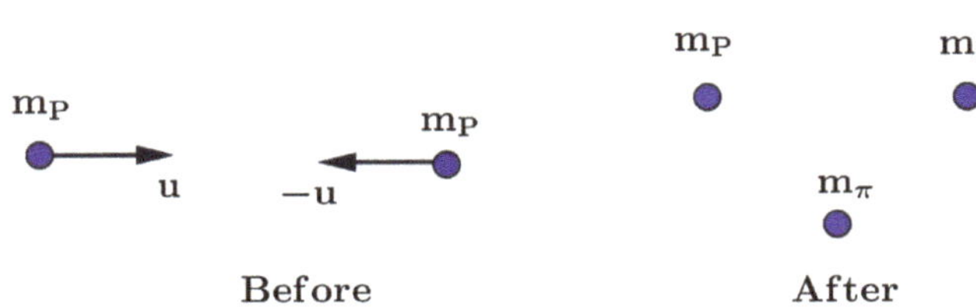

Figure 6.12 A proton–proton collision that produces a pion, a proton, and a neutron in the center-of-momentum frame.

and the proton and neutron have approximately equal masses (actually the neutron is slightly heavier than the proton, approximately 0.1%). We can compute the required u, and thus know how powerful an accelerator is required. Using relativistic energy conservation,

$$2\gamma m_P c^2 = 2m_p c^2 + m_\pi c^2.$$

Therefore,

$$\gamma = \frac{1}{\sqrt{1 - u^2/c^2}} = 1 + \frac{m_\pi}{2m_P} \approx 1.074.$$

Doing the arithmetic gives

$$u/c \approx 0.37.$$

To put this result into perspective, imagine a collision at another accelerator center where one proton is speeding along in a beam and the other proton is in a stationary target. How fast would the proton in the beam have to be to create a pion again? Call this velocity v. We can get this answer by boosting our center-of-momentum analysis. Consider a frame S moving to the left in Fig. 6.12 at velocity u, so that the proton with velocity $-u$ in the center-of-momentum frame is brought to rest (Fig. 6.13). The left proton has a velocity

$$v = \frac{u + u}{1 + u^2/c^2}$$

in the lab frame, using the addition of velocity formula. Since $u/c = 0.37$, we learn that $v = 0.65c$ and the relativistic kinetic energy of the beam proton is $(\gamma - 1)mc^2 = \left(1/\sqrt{1 - v^2/c^2} - 1\right)mc^2 \approx (1.31 - 1)$ 938

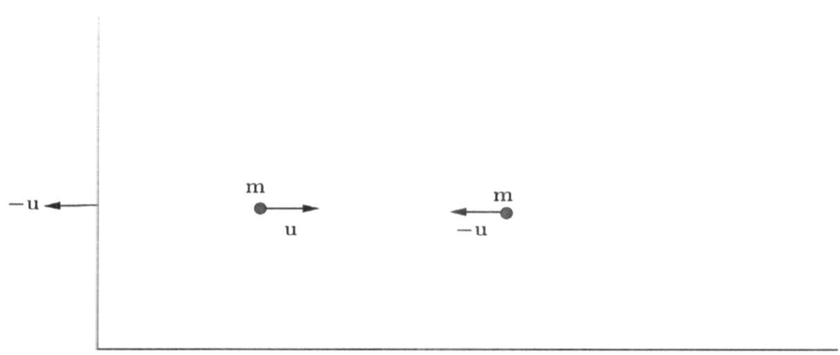

Figure 6.13 Boosting the collision of Fig. 6.12 to the lab frame.

MeV = 290 MeV. We learn from this that the proton in the beam needs to have more than twice the kinetic energy of the rest mass of the pion we are trying to create. Mass creation in this sort of experiment is much less efficient than that in the center-of-momentum frame because of momentum conservation—the three particles in the final state must have net momentum to the right to match the momentum of the initial proton in the beam, so we cannot convert all the kinetic energy to mass. This calculation illustrates why colliding beam experiments are best in exploring the world of new, higher energy states of matter. Unfortunately, it is harder to make colliding beam machines with beams intense enough to make such experiments practical than it is to make fixed target accelerators.

Finally, let us consider a scattering process in three dimensions. Consider Compton scattering, the elastic scattering of a photon off an atom or any charged particle. Studies of this process were important in establishing the quantum theory of light. A photon of energy Q_o scatters off a stationary electron, and the photon emerges at an angle θ with diminished energy Q, while the electron recoils through angle φ with final energy E and momentum $\mathbf{p}$. This is shown in Fig. 6.14, where we have labeled the direction of the initial photon with the unit vector $\widehat{\mathbf{n}}_o$ and the direction of the final photon with the unit vector $\widehat{\mathbf{n}}$. Writing out the conservation laws,

$$\begin{aligned} Q_o + mc^2 &= E + Q \\ \widehat{\mathbf{n}}_o Q_o/c &= \widehat{\mathbf{n}} Q/c + \mathbf{p} \end{aligned}. \tag{6.44}$$

Suppose our experiment just detects the final photon, so we want to eliminate E and φ from the kinematics. Solving Eq. (6.44) for E and $\mathbf{p}c$ gives

$$\begin{aligned} E &= (Q_o - Q) + mc^2 \\ \mathbf{p}c &= (\widehat{\mathbf{n}}_o Q_o - \widehat{\mathbf{n}} Q) \end{aligned}.$$

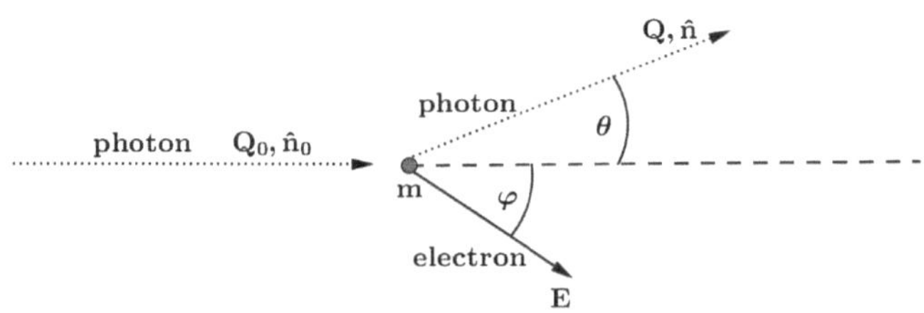

Figure 6.14 Compton scattering: the elastic scattering of a photon off an electron initially at rest in the lab.

Squaring each equation gives

$$(Q_o - Q)^2 + 2(Q_o - Q)mc^2 + m^2c^4 = E^2$$
$$Q_o^2 - 2Q_oQ\cos\theta + Q^2 = c^2\mathbf{p}^2$$

where we identified $\widehat{\mathbf{n}} \cdot \widehat{\mathbf{n}}_o = \cos\theta$. Now subtract these equations using the energy—momentum relative for the recoiling electron, $E^2 = \mathbf{p}^2c^2 + m^2c^4$:

$$2Q_oQ(1 - \cos\theta) - 2(Q_o - Q)mc^2 = 0.$$

Dividing through by $2Q_oQmc^2$ gives

$$\frac{1}{Q} = \frac{1}{Q_0} + \frac{1}{mc^2}(1 - \cos\theta) \tag{6.45}$$

So, if the photon scatters through an angle θ, its energy is diminished from Q_o to Q according to the Compton formula. This result is usually quoted in the context of quantum mechanics where the energy of a single quantum of light is related to its frequency ν by Planck's constant h, $h = 6.627 \cdot 10^{-34}$ J-s,

$$Q = h\,\nu.$$

Because $\nu = c/\lambda$ for a wave traveling with the speed of light, we can replace $1/Q$ by λ/hc in Eq. (6.45) and get

$$\lambda - \lambda_o = \frac{h}{mc}(1 - \cos\theta).$$

In other words, when light scatters, its wavelength increases proportionally to 1 minus the cosine of its scattering angle. Setting the scale of the effect with Planck's constant, $h = 6.627 \cdot 10^{-34}$ J-s, the characteristic length in the Compton formula,

$$\frac{h}{mc} \approx 0.02426 \quad \text{Å} = 2.4 \cdot 10^{-12}\text{m},$$

which is often referred to as the Compton wavelength of the electron, a very small distance on the scale of atomic physics. If the initial photon were an X-ray with $\lambda_o = 0.7$ Å (angstrom, Å, is a convenient unit of length in atomic physics, Å $\equiv 10^{-10}$ m) and we take scattering through 45 degrees, then

$$\lambda = \lambda_o + \frac{h}{mc}(1 - \cos\theta) = 0.7 + 0.02426(0.7071) \approx 0.7018\text{Å}.$$

Scattering experiments such as this, coupled with quantum mechanics calculations of the cross sections involved, played a central role in the development of atomic physics.

PROBLEMS

6-1. Our galaxy is approximately 10^5 light-years across, and the most energetic naturally occurring particles have an energy of approximately 10^{19} eV. How long would it take a proton with this energy to travel across the galaxy as measured in the rest frame of (a) the galaxy or (b) the particle?

6-2. An electron is accelerated from rest through a voltage drop of 10^5 V and then travels at constant velocity.

a. How long does it take the electron to travel 10 m, after it has reached its final velocity?

b. What is the distance measured in the rest frame of the electron?

6-3. If all the energy used in running the accelerator at Fermilab for a full day could be collected in a box, how much heavier would the box become. (Find out about the energy requirements at Fermilab through their website, www.fnal.gov.)

6-4. Plot the total energy E versus momentum p for a particle of rest mass m_o for the cases (a) Newtonian kinematics, (b) relativistic kinematics, and (c) relativistic kinematics for a massless particle, $m_o = 0$. Note the ranges in momentum where two or more of the three curves are good approximations of one another.

6-5. In the chapter, we derived the Lorentz transformation law for energy and momentum between a frame S and a frame S' moving to the right at velocity v. Specialize to the case of the energy and momentum carried by light, $P = E/c$, and show that the transformation for radiant energy between frames S and S' for light traveling in the x direction is

$$E' = \sqrt{\frac{1 - v/c}{1 + v/c}}E.$$

Note that this result is algebraically the same as the Doppler shift for the frequencies of light observed in the two frames. In other words, the transformation laws for energy and frequency of light are identical. This is an important result for both the basic theory of electromagnetic phenomena and quantum mechanics, in which the quantum of electromagnetic energy, the photon, carries energy proportional to its frequency, $E = h\nu$, where h is Planck's constant, the cornerstone of quantum theory.

6-6. Radiant energy from the Sun is received on Earth at a rate of about 1370 J/s-m^2 on a surface perpendicular to the Sun's rays.

- **a.** What total force would be exerted on all of Earth if all the light energy were absorbed?
- **b.** Is this a large or a small force? (Compare it to the total gravitational force that the Sun exerts on the Earth.)

6-7. The average rate at which solar radiant energy reaches Earth is about 1.37×10^3 W/m^2. Assume that all this energy results from the conversion of mass to energy.

- **a.** Calculate the rate at which solar mass is being lost.
- **b.** If this rate is maintained, calculate the remaining lifetime of the Sun. (The mass of the Sun is 332,830 times the mass of the Earth, and the mass of the Earth is 5.976×10^{24} kg.)

6-8. A star of mass 10^{32} kg is surrounded by a thin, flexible spherical shell of mass 10^{25} kg. The star loses mass at the rate 10^{10} kg/s in the form of light. Suppose that all of this radiant energy is absorbed on the shell. What must the radius of the shell be so that the radiation pressure from the light balances its gravitational attraction to the star?

6-9. Light rays from the Sun hit the Earth at the rate of 1370 J/s-m^2 on a surface perpendicular to the Sun's rays.

- **a.** How much mass in the Sun is converted to energy per second to account for the radiant energy hitting Earth? (The radius of Earth is roughly $6.4 \cdot 10^6$ m and the distance from Earth to the Sun is approximately $1.5 \cdot 10^{11}$ m)
- **b.** What is the total mass converted to energy in the Sun to supply this radiant energy?
- **c.** Estimate the mass of hydrogen that must be converted to helium per second to supply this radiant energy. (Recall that most of the Sun's energy comes from fusing hydrogen into helium. The mass of a hydrogen nucleus is $1.67262 \cdot 10^{-27}$ kg, and the mass of a helium nucleus is $6.64648 \cdot 10^{-27}$ kg.)
- **d.** Estimate how long the Sun will warm Earth, accounting only for the hydrogen fusion process.

6-10. The physicist Sir Arthur Eddington, who did pioneering work in cosmology, pointed out the strength of unscreened electrostatic forces in the following dramatic fashion. Take 1 g of electrons and place them uniformly at rest in a spherical container of radius 10 cm. Calculate the mass associated with their electrostatic potential energy, and verify that it is on the order of 10^{10} kg! (Recall from electricity and magnetism that electrons repel one another through the inverse square law $F = kq^2/r^2$, where q is the electronic charge in coulombs and k is approximately $9 \cdot 10^9$ N-m^2/C^2. The electrostatic energy of a uniformly charged sphere of total charge Q and radius r is $3kQ^2/5r$. The charge of an electron is $q = 1.6 \cdot 10^{-19}$ C, and its mass is $m = 9.11 \cdot 10^{-31}$ kg.)

This alarming answer shows how important the neutrality of bulk matter really is. The negative charges of the electrons in the paper of this book, which weighs under a pound, are neutralized by the positive charges of the protons in the nuclei of the atoms that bind them, and the powerful long-range electromagnetic forces that interested Sir Arthur Eddington are canceled out.

6-11. Our derivation of $E = mc^2$ in the text using Einstein's box made two simplifications. First, we ignored the distance that the box recoils when light is in transit from one end to the other. And second, we ignored the decrease in the mass of the box when the light is in transit. Include these effects in a calculation of the same problem, and show that $E = mc^2$ can be derived perfectly.

6-12. A particle is given a relativistic kinetic energy equal to twice its rest mass. Find its resulting speed and momentum. How do these results change if the relativistic kinetic energy is five times the particle's rest mass?

6-13. Through what voltage would you have to accelerate an electron to boost its speed from rest to 99% of the speed of light? Repeat your calculation for a proton.

6-14. A relativistic electron moving in the x direction enters a region of space where there is a uniform electric field in the y direction.

a. Write down the relativistic equations of motion that describe this situation. Following Section 6.3, you will have two equations, one for the x component of the force and one for the y component.

b. Discuss the solution of the equations in part (a) qualitatively. For example, identify conserved quantities and time variable quantities. Show that the x component of the velocity of the particle decreases with time.

c. Solve for $v_x(t)$ and $v_y(t)$, assuming that initially the electron had a velocity v_o in the x direction. Denote the component of the force in the y direction F_y.

d. Suggest some practical uses for a device based on your results in an accelerator center where the electrons have v_o values very close to the speed limit c.

6-15. In science fiction stories, we find space vehicles that consist of large sails that deflect light so that the recoil of the light propels the vehicle to amazing speeds. Suppose that the Department of Energy has funded such a project, and there is a sailship in free space that feels the push of a strong, steady laser beam of light directed at it from Earth. If the sail is perfectly reflecting, calculate the mass equivalent of light required to accelerate a vehicle of rest mass m_o up to a fixed γ.

6-16. A laser with a mass of 10 kg is in free space with its beam directed toward Earth. The laser continuously emits 10^{20} photons/s of wavelength 6000 Å, as measured in its own rest frame. At $t = 0$ the laser is at rest with respect to Earth.

a. Initially how much radiant energy per second is received on Earth? (Planck's constant $h = 6.627 \cdot 10^{-31}$ kg-m^2/s, and the energy of a photon is $E = h\nu$.)

b. The radiation emitted toward Earth causes the laser to recoil away from Earth. What is the velocity of the laser relative to Earth after 10 years have elapsed on a clock at rest with respect to the laser?

c. At the time when the laser is moving with velocity v relative to Earth, how much less is the rate at which energy is received on Earth than the original rate when $v = 0$? Evaluate this for $t = 10$ years, laser time.

d. Show how an observer on Earth can explain the continually decreasing rate of reception in terms of energy considerations.

6-17. A photon rocket uses light as a propellant. If the initial and final rest masses of the rocket are M_i and M_f, show that the final velocity

v of the rocket relative to its initial rest frame is given by the equation

$$M_i/M_f = \sqrt{(1 + v/c)/(1 - v/c)}.$$

Problems on Collisions and Conservation Laws

6-18. Consider two lumps of clay each with a rest mass of m. They collide head on, one traveling to the right at speed $\frac{3}{5}c$ and the other traveling to the left at the same speed. They stick together.

a. What is the mass M of the composite lump?

b. Explain how M can be greater than $2m$.

6-19. Consider the decay of a charged pion of mass m_π initially at rest in the lab. It decays into a muon of mass m_μ and a neutrino, which we treat as massless. Find the relativistic energy, momentum, and speed of the muon in terms of m_π and m_μ.

6-20. A pion collides with a proton initially at rest in the lab and the process $\pi + p \rightarrow K + \Sigma$ occurs. What is the minimum (threshold) momentum of the pion for this process to be possible? Use these experimental masses: $m_\pi c^2 = 139$ MeV, $m_K c^2 = 494$ MeV, $m_p c^2 = 938$ MeV and $m_\Sigma c^2 = 1192$ MeV. (Hint: Consider the kinematics in the center-of-momentum and use invariants.)

6-21. When a K^o meson decays at rest into a π^+ and a π^- meson, each escapes with a speed of approximately $0.85c$. Now consider a K^o meson that is traveling at a speed of $0.9c$ relative to the lab frame. If it then decays, what is the greatest speed that one of the pions can have in the lab frame? What is the least speed?

6-22. Two identical particles, A and B, are approaching each other along a common straight line. Each particle has the same speed v, as measured in the lab. Show that the energy of particle A as measured by B is $(1 + v^2/c^2)\,(1 - v^2/c^2)^{-1}\,M_o c^2$, where M_o is A's rest mass.

6-23. A photon has energy 200 MeV and is traveling along the x axis. Suppose another photon has energy 400 MeV and is traveling along the y axis.

a. What is the total energy of this system? What is its total momentum?

b. If a single particle had the same energy and momentum, what would its mass be? What would its direction of travel, and what would its speed be?

6-24. A particle of mass M decays into two distinguishable particles of m_1 and m_2.

a. Find the energies of particles 1 and 2 in terms of M, m_1 and m_2.

A particle of energy E_1 and mass m_1 strikes a particle of mass m_2 at rest in the lab.

b. Calculate the velocity of the center-of-momentum frame in the lab.

c. Calculate the mass of the composite system of particles 1 and 2 in terms of m_1, m_2, and E_1.

6-25. A particle of rest mass m_o and relativistic kinetic energy $3m_oc^2$ strikes and sticks to a stationary particle of rest mass $2m_o$. Find the rest mass M_o of the resulting composite particle.

6-26. Kinematics of an absorption process.

a. A photon of energy E is absorbed by a stationary particle of rest mass m_o. What is the velocity and rest mass of the resulting composite particle?

b. Repeat part (a) replacing the photon by a particle of rest mass m_o and speed $0.8c$.

6-27. An atom in an excited state of energy Q_o above the ground state moves toward a scintillation counter with speed v. The atom decays to its ground state by emitting a photon of energy Q, as measured by the counter, coming completely to rest as it does so. If the rest mass of the atom is m, show that $Q = Q_o\ [1 + (Q_o/2mc^2)]$.

6-28. A neutral pion decays into two photons. The pion's rest mass is 135 MeV. Suppose it is in a secondary beam at Fermilab with a relativistic kinetic energy of 1 GeV.

a. What are the energies of the photons if they are emitted in opposite directions along the pion's original line of motion.

b. What angle is formed between the two photons if they are emitted at equal angles to the direction of the pion's motion?

6-29. An antiproton of kinetic energy 1 GeV strikes a proton at rest in the lab. (The proton and its antiparticle have identical masses, about 938 MeV/c^2.) They annihilate, and two photons emerge from the reaction, one traveling forward and one backward along the beam direction.

a. What are the energies of the two photons?

b. As measured in a reference frame of the antiproton, what energy does each photon have?

6-30. According to Newtonian mechanics, when a beam particle collides off an identical particle originally at rest, they emerge with an angle between them of exactly 90 degrees in all cases. Contrast this result to a relativistic collision:

a. If a proton of kinetic energy 500 MeV collides elastically with a proton at rest, and the protons rebound with equal energies, what is the angle between them?

b. Repeat this exercise with a proton having an initial kinetic energy of 100 GeV.

6-31. Suppose a photon has a head-on collision with an electron. What initial velocity must the electron have if the collision results in a photon recoiling straight backward with the same energy Q as it had initially?

6-32. A photon of energy E collides elastically with an electron at rest. After the collision, the photon's energy is reduced by half, and its scattering angle is 60 degrees.

a. What was its original energy? Is the frequency of this photon in the visible range?

b. A photon of energy E collides with an excited atom at rest. After the collision, the photon has the same energy, but its direction has changed by 180 degrees. If the atom is in its ground state after the collision, what was its original excitation energy?

6-33. A K meson, rest mass approximately 494 MeV/c^2, is in motion and decays into two pions, rest mass approximately 137 MeV/c^2. One of the pions emerges from the decay process at rest.

a. What is the energy of the other pion?

b. What was the energy of the original K meson?

6-34. An electron—positron pair can be produced by a gamma ray striking a stationary electron, $\gamma + e^- \rightarrow e^- + e^+ + e^-$. What is the minimum gamma ray energy that will allow this process to occur? (The positron is the electron's antiparticle. It has the same rest mass as the electron, 0.511 MeV/c^2, but the opposite charge.).

6-35. Suppose that an accelerator can give protons a kinetic energy of 300 GeV. (The rest mass of a proton is roughly 938 MeV/c^2.) Calculate the largest possible rest mass M_x of a new particle X that could be produced when the beam proton hits a stationary proton in a target, $p + p \rightarrow p + p + X$.

6-36. A positron of kinetic energy 0.511 MeV annihilates with an electron at rest, creating two photons. One photon emerges at an angle 90 degrees to the incident positron direction.

a. What are the energies of both photons? (The rest mass of an electron is 0.511 MeV/c^2. The rest mass of the positron is exactly the same.)

b. What is the direction of the second photon?

6-37. A particle of kinetic energy K collides elastically with an identical particle at rest. The two outgoing particles emerge with equal and opposite angles $\theta/2$ with respect to the incoming particle. Find the energy and momentum of each outgoing particle. Find the angle θ in terms of K and the rest mass m of one of the initial particles.

6-38. A beryllium nucleus consists of four protons and three neutrons. The mass of the beryllium nucleus is 6536 MeV/c^2, the proton is 938.28 MeV/c^2, and the neutron is 939.57 MeV/c^2.

a. Find the binding energy of beryllium. (The binding energy of a nucleus is the difference between its rest mass and the sum of the rest masses of its free constituents.)

b. Consider the reaction in which an extra neutron at rest is absorbed by a beryllium nucleus also at rest, which subsequently decays into two alpha particles. (An alpha particle consists of two protons and two neutrons bound together. Its rest mass is 3728 MeV/c^2.) What is the kinetic energy of each of the alpha particles?

6-39. What is the minimum proton energy needed in an accelerator to produce an antiproton $\overline{p}$ by the reaction $p+p\rightarrow p+p+(p+\overline{p})$. (The mass of the proton is $m_pc^2=938$ MeV, and the mass of the antiproton is the same.)

a. Do this for a colliding beam machine. (This is an accelerator, which produces two beams of equal energy that collide head on.)

b. Do this for a fixed target machine. (This is a machine, which produces one beam that is incident on a target at rest in the lab.)

c. What is the velocity of one of the final-state particles in part b?

6-40. Verify Eq. (6.32) explicitly by using the Lorentz transformation matrix written in Eq. (6.28a).

6-41. Consider a frame S'' moving in the $+x$ direction with velocity w with respect to frame S'. Let S' move in the $+x$ direction with velocity v with respect to frame S. A four-vector x''^{ν} measured in frame S'' is related to x^{σ} measured in frame S by the composition of two Lorentz boosts,

$$x''^{\nu} = \sum_{\mu} L^{\nu}{}_{\mu}(w)x'^{\mu} = \sum_{\mu\sigma} L^{\nu}{}_{\mu}(w)L^{\mu}{}_{\lambda}(v)x^{\lambda}$$

In addition, denote the relative velocity between frames S'' and S u, so

$$x''^{\nu} = \sum_{\mu} L^{\nu}{}_{\mu}(u)x^{\mu}$$

1. Show that,

$$\gamma(u) = \gamma(v)\gamma(w)(1 + vw/c^2)$$
$$u\gamma(u) = (v + w)\gamma(v)\gamma(w)$$

2. Solve for u and verify the "addition of velocities" formula,

$$u = \frac{v + w}{(1 + vw/c^2)}$$

6-42. In the text we asserted that the metric transforms as a second-rank tensor under boosts. To show this directly, write the invariant interval in terms of x and x', which are related by a boost $x'^{\mu} = \sum_{\nu} L^{\mu}{}_{\nu}x^{\nu}$,

$$ds^2 = \sum_{\mu\nu} g'_{\mu\nu}dx'^{\mu}dx'^{\nu} = \sum_{\mu\nu\rho\sigma} g'_{\mu\nu}L^{\mu}{}_{\sigma}L^{\nu}{}_{\rho}dx^{\sigma}dx^{\rho} = \sum_{\sigma\rho} g_{\sigma\rho}dx^{\sigma}dx^{\rho}$$

Verify,

$$g_{\sigma\rho} = \sum_{\mu\nu} L^{\mu}{}_{\sigma}L^{\nu}{}_{\rho}g'_{\mu\nu}$$

Invert this expression and find the equivalent statement in the text,

$$g'^{\mu\nu} = \sum_{\sigma\lambda} L^{\mu}{}_{\sigma}L^{\nu}{}_{\lambda}g^{\sigma\lambda}$$

6-43. Inertia and energy

Consider a point particle with velocity $\boldsymbol{v}$ in the x direction in frame S. Its relativistic energy is $E = \gamma(v)Mc^2$, where M is its rest mass. Suppose that the point particle is replaced by a composite particle. Take the simplest example: a collection of point particles, of masses m_i and velocities $\boldsymbol{u}_i'$, in the frame S'. As usual, the frame S' moves in the x direction at velocity $\boldsymbol{v}$. Suppose that S' is the composite particle's rest frame: the total three momentum of the point particles of masses m_i vanishes in S'. Call the total relativistic energy of the collection of point particles E' in the frame S'. Show that E, the composite particle's relativistic energy in frame S is,

$$E = \gamma(v)Mc^2$$

where $Mc^2 \equiv E' = \sum_i \gamma(u_i')m_ic^2$.

We learn that the energy (divided by c^2) of the "parts" of the composite particle in its overall rest frame is the composite particle's invariant mass (inertia).

The generalization of these ideas to a realistic situation (for example, a proton is a bound state of strongly interacting [confined] quarks) is nontrivial and requires the full machinery of field theory.

REFERENCES

[1] M. Born, Einstein's Theory of Relativity, Dover Publications, New York, 1962.
[2] A.P. French, Special Relativity, W. W. Norton, New York, 1968.

CHAPTER 7

Acceleration and Forces in Relativity: The Birth of Dynamical Fields

Contents

7.1 ACCELERATION IN RELATIVITY

Acceleration is a particularly significant and simple quantity in Newtonian mechanics: it is frame independent and establishes the differential equation that predicts the time evolution of mechanical systems, "force equals mass times acceleration." In relativity, however, the acceleration $\boldsymbol{a}$ is a complicated quantity that will play a less important role in the time evolution of relativistic systems. We need to understand it relativistically, however, on our road to Maxwell's equations.

To begin, we need to derive the rules for transforming $\boldsymbol{a}$ from one frame to another. We begin with two reference frames S and S'. S' moves along the positive x-axis with a velocity v. We have seen in Chapter 4 that if a particle has velocity $\boldsymbol{u} = (u_x, u_y, u_z)$ in S and velocity $\boldsymbol{u}' = \left(u'_x, u'_y, u'_z\right)$ in S', then

$$u_x = \frac{u'_x/v}{1 + \frac{vu'_x}{c^2}}, \quad u_y = \frac{u'_x/\gamma}{1 + \frac{vu'_x}{c^2}}, \quad u_z = \frac{u'_z/\gamma}{1 + \frac{vu'_x}{c^2}} \tag{7.1}$$

And time transforms according to the Lorentz transformation law,

$$t = \gamma(t' + vx'/c^2) \tag{7.2}$$

Special Relativity, Electrodynamics, and General Relativity
ISBN 978-0-12-813720-8
https://doi.org/10.1016/B978-0-12-813720-8.00007-6

Now consider the motion of the particle in S'. In the time interval dt', it moves a distance $dx' = u'_x dt'$ in the x-direction. According to Eq. (7.2), the corresponding time interval in S is,

$$dt = \gamma\left(dt' + \frac{vdx'}{c^2}\right) = \gamma\left(1 + \frac{vu'_x}{c^2}\right)dt' \tag{7.3}$$

Eq. (7.3) includes the phenomena of time dilation γ and the relativity of simultaneity, $1 + \frac{vu'_x}{c^2}$.

According to Eq. (7.1), a change in the velocity u_x transforms to reference frame S',

$$du_x = \frac{du'_x}{1 + vu'_x/c^2} - \left[\frac{(u'_x + v)}{(1 + vu'_x/c^2)^2}\frac{vdu'_x}{c^2}\right]$$

$$du_x = \left[\frac{1 + vu'_x/c^2}{(1 + vu'_x/c^2)^2} - \frac{(vu'_x + v^2)/c^2}{(1 + vu'_x/c^2)^2}\right]du'_x = \frac{(1 - v^2/c^2)du'_x}{(1 + vu'_x/c^2)^2}$$

$$= \frac{du'_x}{\gamma^2(1 + vu'_x/c^2)^2} \tag{7.4}$$

Dividing Eq. (7.4) by (7.3), we have the relation between a'_x and a_x,

$$\frac{du_x}{dt} = \frac{du'_x/dt'}{\gamma^3(1 + vu'_x/c^2)^3}$$

or,

$$a_x = \frac{a'_x}{\gamma^3(1 + vu'_x/c^2)^3} \tag{7.5}$$

So, in contrast to Newtonian physics where the acceleration is independent of the inertial frame in which it is calculated, here the x-components of the acceleration are proportional to one another, but the proportionality depends on v, u'_x and c^2 through algebraic factors that we have studied before.

The analogous expressions for a_x and a_y are even more complicated. We begin by differentiating u_y in Eq. (7.1),

$$du_y = \frac{du'_y}{\gamma(1 + vu'_x/c^2)} - \frac{u'_y}{\gamma(1 + vu'_x/c^2)^2}\frac{vdu'_x}{c^2} \tag{7.6}$$

Now, dividing by Eq. (7.3),

$$\frac{du_y}{dt} = \frac{du'_y/dt'}{\gamma^2\left(1 + vu'_x/c^2\right)^2} - \frac{u'_y}{\gamma^2\left(1 + vu'_x/c^2\right)^3}\frac{vdu'_x/dt'}{c^2} \tag{7.7}$$

or,

$$a_y = \frac{a'_y}{\gamma^2\left(1 + vu'_x/c^2\right)^2} - \frac{\left(vu'_y/c^2\right)a'_x}{\gamma^2\left(1 + vu'_x/c^2\right)^3} \tag{7.8}$$

The expression for a_z is obtained from Eq. (7.8) by replacing $y \rightarrow z$.

We learn from Eq. (7.8) that a_y receives contributions from both the accelerations in the x and y directions in the frame S'. For example, suppose the particle is not accelerating in the y direction in $S'\left(a'_y = 0\right)$, but it has nonzero velocity in the y direction in $S'\left(u'_y \neq 0\right)$ and is accelerating in the x direction in $S'\left(a'_x \neq 0\right)$, then it is accelerating in the y direction in S!

There is a special case of some utility in problem solving which is not complicated. Consider a particle at rest in S', $\left(u'_x = u'_y = u'_z = 0\right)$. Then its acceleration in S reads

$$a_x = a'_x/\gamma^3, \quad a_y = a'_y/\gamma^2, \quad a_z = a'_z/\gamma^2 \tag{7.9}$$

You can trace the factors of γ here to Lorentz contraction and time dilation (see the problems). Note that even in this case that the vectors $\boldsymbol{a}$ and $\boldsymbol{a}'$ are not even parallel.

7.2 TRANSFORMATION PROPERTIES OF FORCES

Now we want to understand how equations of motion and forces behave under Lorentz transformations. This discussion will be general. In other words, the results will apply to any sort of force. We will only use the force's transformation laws under boosts in this development. We shall learn some general principles and see how they apply to electromagnetism later.

Suppose the force $\boldsymbol{F}'$ acts on a particle in S',

$$\boldsymbol{F}' = \frac{d\boldsymbol{p}'}{dt'} \tag{7.10}$$

Since we know the transformation properties of $\boldsymbol{p}'$and t', we can infer the transformation properties of the force. This is our goal here.

The particle has velocity $\boldsymbol{u}'$ in frame S', and velocity $\boldsymbol{u}$ in frame S. To relate Eq. (7.10) to the same equation in S, we recall the transformation properties of momentum, and energy,

$$p_x' = \gamma(p_x - vE/c^2), \quad p_y' = p_y, \quad p_z' = p_z, \quad E' = \gamma(E - vp_x) \tag{7.11}$$

and time,

$$t' = \gamma(t - vx/c^2) \tag{7.12}$$

To write Eq. (7.10) in terms of quantities measured in frame S, we use Eqs. (7.11) and (7.12) to compute dt' and dp_x' separately. First, we relate dt' to dt, which is the time interval over which the particle travels through a distance $dx = u_x dt$, so

$$dt' = \gamma(dt - u_x v dt/c^2) = \gamma(1 - u_x v/c^2)dt \tag{7.13}$$

Now we can return to Eq. (7.11) and deduce how F_x transforms between frames,

$$F_x' = \frac{dp_x'}{dt'} = \frac{\gamma(dp_x - vdE/c^2)}{\gamma(1 - u_x v/c^2)dt} = \frac{dp_x/dt - (v/c^2)dE/dt}{(1 - u_x v/c^2)} \tag{7.14}$$

The second term in Eq. (7.14) indicates that the force in S' depends on the power dE/dt provided by the force in S. We can write the power in terms of force and velocity of the particle in S, inspired by a similar argument in Newtonian mechanics. Begin with the energy–momentum relation,

$$E^2 = m_0^2 c^4 + c^2 \boldsymbol{p}\cdot\boldsymbol{p} \tag{7.15}$$

Taking the time derivative gives

$$\frac{d}{dt}E^2 = 2E\frac{dE}{dt} = 2c^2\boldsymbol{p}\cdot\frac{d\boldsymbol{p}}{dt} \tag{7.16}$$

So,

$$\frac{dE}{dt} = \frac{c^2\boldsymbol{p}}{E}\cdot\frac{d\boldsymbol{p}}{dt} \tag{7.17}$$

The right-hand side can be simplified. $E = \gamma m_0 c^2$ and $\boldsymbol{p} = \gamma m_0 \boldsymbol{u}$, so $c^2\boldsymbol{p}/E = \boldsymbol{u}$, and Eq. (7.17) becomes,

$$\frac{dE}{dt} = \boldsymbol{F}\cdot\boldsymbol{u} \tag{7.18}$$

which is identical to the power formula in Newton's world.

Now Eq. (7.14) becomes the transformation equation for the x component for the force,

$$F'_x = \frac{F_x - (v/c^2)\boldsymbol{F}\cdot\boldsymbol{u}}{(1 - u_x v/c^2)} \tag{7.19}$$

Using the same manipulations, we can calculate F'_y and F'_z in terms of F_x, and F_y.

$$F'_y = \frac{F_y}{\gamma(1 - u_x v/c^2)}, \quad F'_z = \frac{F_z}{\gamma(1 - u_x v/c^2)} \tag{7.20}$$

Finally, since the power $\boldsymbol{F}\cdot\boldsymbol{u}$ enters Eq. (7.19), we should calculate how this quantity transforms. Since $E' = \gamma(E - vp_x)$ and $t' = \gamma(t - vx/c^2)$,

$$\frac{dE'}{dt'} = \frac{\gamma(dE - vdp_x)}{\gamma(1 - u_x v/c^2)dt} = \frac{dE/dt - vdp_x/dt}{(1 - u_x v/c^2)} \tag{7.21}$$

or,

$$\boldsymbol{F}'\cdot\boldsymbol{u}' = \frac{\boldsymbol{F}\cdot\boldsymbol{u} - vF_x}{(1 - u_x v/c^2)} \tag{7.22}$$

We learn that the four quantities (F_x, F_y, F_z) and $\boldsymbol{F}\cdot\boldsymbol{u}$ transform linearly among themselves. In general, the transformation laws are quite complicated. The transformation laws do simplify in the special case where the particle is instantaneously at rest in S, $\boldsymbol{u} = 0$. Then,

$$F'_x = F_x, \quad F'_y = \gamma F_y, \quad F'_z = \gamma F_z \tag{7.23}$$

7.3 THE DEATH OF NEWTON'S THIRD LAW, AND STATIC FORCES: THE BIRTH OF FIELDS

Let us contrast these rules with Newtonian dynamics. In Newton's world, force is invariant for point particles of fixed masses. In particular, if a force is velocity independent, such as the electrostatic force or the gravitational force, then it is velocity independent in all frames. This is clearly inconsistent with special relativity! We see from Eq. (7.19) that velocity-dependent forces are the norm. This implies that Newton's third law, that action and reaction are equal in magnitude but opposite in direction, is generally untrue! How important is this? Very! Recall that Newton's third law is the basis of momentum conservation in Newton's world. In the real world, energy—momentum conservation is a fundamental building block. How can it be that energy—momentum conservation is a property of the real world if action is not equal and opposite to reaction? We will discuss this point further in several sections below. There is much to learn from this failure.

Let us understand more simply why the failure of Newton's third law is an unavoidable aspect of relativity. Consider a force, which acts between two particles that are a considerable distance apart. The force between particles 1 and 2 will depend on the distance between particles 1 and 2 as well as their velocities, as we have learned above. (Electrodynamics will be a prime example, which we will study in detail.) Particle 1 will exert a force $\boldsymbol{F}_{12}$ on particle 2, and particle 2 will exert a force $\boldsymbol{F}_{21}$ on particle 1. Note that these forces act at *different* points, at the position of particle 2 and particle 1. So, we must compare $\boldsymbol{F}_{21}$ at $\boldsymbol{r}_1$ with $\boldsymbol{F}_{12}$ at $\boldsymbol{r}_2$ *simultaneously*. Label the two space—time points as $(\boldsymbol{r}_1, t)$ and $(\boldsymbol{r}_2, t)$. We are interested in momentum conservation at time t in this frame. But the events where we measure $\boldsymbol{F}_{21}$ and $\boldsymbol{F}_{12}$ are connected by a space-like interval. They are not causally connected! And here is the point: by viewing the events in different inertial frames, we can even change their time ordering, so they are *not* simultaneous in other frames of reference. This is an application of the "relativity of simultaneity" discussed at length in Chapters 2 and 3, and it demonstrates that the concept of action—reaction for a force that acts over a nonzero distance is inherently nonrelativistic: if one observed the validity of Newton's third law in one frame, then it need not hold in other frames. We appear to be forced to the conclusion that the total momentum of the two interacting particles is not generally conserved.

We appear to be at an impasse! How do we understand energy and momentum conservation for interacting particles in Einstein's world? Our study of collisions where the particles are well separated and noninteracting in their initial and final states were exquisite validations of the conservation of energy and momentum. We now see, however, that if we try to analyze the states of the particles when they are within each other's range of interaction, we would be at a loss. Somehow energy and momentum conservation must be true for all times in the collision process, early on when the particles are moving at constant velocities, far after the collision process when the particles are again moving at constant but different velocities, and even for those times when the forces between the particles are nonnegligible and are deflecting the particles.

Up until this point, we have been able to take Newtonian concepts such as momentum, energy, force, point particles, and time and space measurements, and carefully generalize them to a relativistic world. Now it appears that something conceptual is missing! We need a new idea! In the history of physics the missing idea was that of the electromagnetic "field." The field concept grew from the early experiments in electromagnetism and culminated in the experiments of Faraday who investigated how time varying electric and magnetic effects interplayed. When Maxwell synthesized 200 years of experiments in electromagnetism into his theory of the electromagnetic field, the ground work was laid for the theory of special relativity.

Let us summarize: In Newton's world, we can have conservation laws that are nonlocal. This works because information can travel at infinite speed. The earth attracts the moon with the same force that the moon attracts the earth. The two forces act at different points, but they always sum to zero so the momentum of the earth—moon system is conserved. There is no problem here because the forces can be compared at one instant at distant points because information can be shared instantaneously between the earth and the moon in Newton's world. The action and reaction forces are causally connected even though they are spatially separated. But, in a relativistic world this simple picture fails. Instantaneous events on the earth and the moon are not causally connected. In fact, by viewing the planetary system in two different frames, the time ordering of the events can be changed. So, the entire notion that action and reaction can be equal but opposite and the notion of action at a distance have no role in relativity.

Of course there is one situation where the action—reaction principle can apply in a relativistic world and that is where the two points of action and reaction are *coincident*. Then the conservation law becomes "local," and there is no conflict with causality. We will see as we develop the idea of the electromagnetic field that this is the way to generalize the successes of Newtonian mechanics to a relativistic world. *We are forced to turn from forces to fields. And the fields must become dynamical variables capable of carrying energy and momentum. We are forced to formulate the theory in terms of local equations.*

The concept of a "field of force" allows us to describe the interactions of particles in relativity. Instead of saying that one particle acts on another, as in Newton's world, the rules of relativity lead us to see that the particle creates a field around itself and a force acts on any particle located in this field. In Newton's world, the field just provides a static description of the interaction between particles: it is just a handy way of describing the instantaneous force between two separate particles. However, in the theory of relativity, because of the finite velocity of the propagation of interactions, the force acting on a particle at a given time is not determined by the other particle's position at that time. A change in the position of one of the particles influences other particles only after the lapse of a certain time interval. *This means that the field itself acquires physical reality—the field must become a dynamical degree of freedom with its own equation(s) of motion.* When distant particles interact, we must describe the phenomenon as a sequence of local and causal events: one particle interacts with the field at its position and a subsequent interaction occurs between that propagating field and the other particle at its distant position.

We will see how this works out in the case of electromagnetism in some detail. Maxwell's equations and the wave equation will provide the description of a fundamentally new phenomenon. All this will become clearer through concrete developments in the next few chapters.

PROBLEMS

7-1. The relationship between acceleration and force in special relativity is quite subtle:

a. Show that $F_x = \gamma^3 m_0 a_x$ in frame S.

b. Relate the acceleration a_x in frame S to that in the instantaneous rest frame ($u_{0x} = 0$) a_{0x}, $a_{0x} = \gamma^3 a_x$.

c. The results of parts a. and b. imply that $F_x = m_0 a_{0x}$, so the force in the x direction is the same in the frame in which the particle is instantaneously at rest. Is this true in the transverse directions? Show that $F_y = \frac{1}{\gamma} F_{0y}$, so the conjecture is untrue.

d. Show that $\frac{F_y}{F_x} = \frac{1}{\gamma^2} \frac{a_y}{a_x}$.

7-2. A space traveler accelerates continually at a rate of $10\ \mathrm{m/s^2}$ in her instantaneous rest frame.

a. If she starts at rest from earth, how far has she traveled at earth time t?

b. How long does it take before she attains a speed of $0.90c$?

(Hint: You may apply $a_x = \frac{a_x'}{\gamma^3 \left(1 + v u_x'/c^2\right)^3}$ with $u_x' = 0$ and realizing that $\gamma(v)$ becomes a function of time.)

CHAPTER 8

Boosting the Electrostatic Force, Electromagnetic Fields and More on Four-Vectors

Contents

8.1 THE ELECTRIC FIELD AROUND A MOVING POINT CHARGE

Now we want to take what we have learned about the transformation properties of generic forces and apply it to electrostatics [1]. This exercise will lead to our discovery of the magnetic field, the Lorentz force law of electrodynamics and eventually Maxwell's equations.

Recall some basic facts of electrostatics: If a charge is at rest, then Coulomb's law gives the electrostatic field it produces. As long as the source of the field $q_1$is at rest, then a charge q_2 experiences that field regardless of its state of motion. So, consider two frames S and S' where S' moves with velocity v in the x direction and the origins of the two frames coincide at $t = t' = 0$. If there is a point charge q_1 at rest at the origin of frame S' then the charge q_2, which is at rest at position $\boldsymbol{r}$ in the frame S and is at position $\boldsymbol{r}'_2$ in S' experiences a force at $t' = 0$ in S',

$$\boldsymbol{F}\prime = k\frac{q_1 q_2}{r'^2}\widehat{\boldsymbol{r}}' \tag{8.1}$$

where r' is the vector distance between the two particles in the frame S' and $\widehat{\boldsymbol{r}}'$ is a unit vector in that direction at $t' = 0$, as shown in Fig. 8.1A and B.

Special Relativity, Electrodynamics, and General Relativity
ISBN 978-0-12-813720-8
https://doi.org/10.1016/B978-0-12-813720-8.00008-8

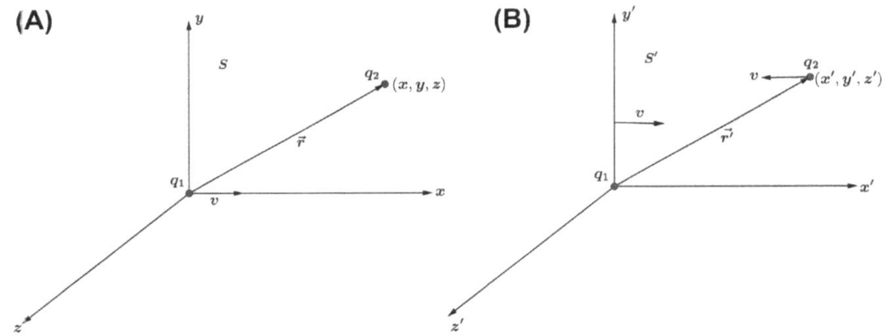

Figure 8.1 (A) The configuration of the two charges in the frame S where q_1 has velocity v at time $t = 0$ and q_2 is at rest at position (x, y, z). (B) Same situation shown in (A) but from the perspective of frame S' where the charge q_1 is at rest at the origin.

Before we continue, we should discuss the constant k and the charge q in Coulomb's law. The constant k is determined experimentally and carries the appropriate dimensions so that the left-hand side of Eq. (8.1) is a force. Later in this book we will do some numerical examples. Here we just need to note that k is defined to be a numerical constant and so it has no transformation properties. The charge q is a physical quantity and its transformation properties must be determined experimentally. In fact, it is a Lorentz invariant quantity. The experimental evidence for this fact is very precise: we are able to study the interactions of charged electrons, for example, in many environments where they have vastly different velocities. Those interactions are always given quantitatively by the hypothesis that q is velocity independent. For example, modern linear accelerators produce electrons with energies 100,000 times the electron's rest energy (0.511 MeV). These electrons are traveling at speeds approaching the speed limit, $v/c \approx 1 - \frac{1}{2} \times 10^{-10}$, and scattering experiments show no perceptible change in the electron's charge q.

Now let us return to the problem at hand and calculate the force $\boldsymbol{F}$ in frame S. At $t = 0$ q_1 is at the origin of S moving with velocity v in the x direction. The charge q_2 is at position (x, y, z) at $t = 0$, and this position maps onto (x', y', z') at $t = 0$. It helps to write Eq. (8.1) in components,

$$F'_x = kq_1q_2\frac{x'}{r'^3}, \quad F'_y = kq_1q_2\frac{y'}{r'^3}, \quad F'_z = kq_1q_2\frac{z'}{r'^3} \tag{8.2}$$

where we used $\hat{\boldsymbol{r}}' = (x'/r',\ y'/r',\ z'/r')$. It is easy to transform the force to frame S because the charge q_2 is at rest in frame S at $t = 0$, so Eq. (7.23) applies,

$$F_x = F'_x, F_y = \gamma F'_y, F_z = \gamma F'_z \tag{8.3}$$

Finally we need to relate (x, y, z) at $t = 0$ to (x', y', z'). The Lorentz transformation reads $x' = \gamma(x - vt)$, $y' = y$, $z' = z$, so at $t = 0$, $x' = \gamma x$, $y' = y$, $z' = z$. Therefore,

$$r' = \sqrt{x'^2 + y'^2 + z'^2} = \sqrt{\gamma^2 x^2 + y^2 + z^2} \tag{8.4}$$

Finally,

$$F_x = \gamma k q_1 q_2 \frac{x}{r'^3}, F_y = \gamma k q_1 q_2 \frac{y}{r'^3}, F_z = \gamma k q_1 q_2 \frac{z}{r'^3} \tag{8.5}$$

Notice a few features of the answer. First, $F_x : F_y : F_z = x : y : z$, so $\boldsymbol{F}$ points in the $\boldsymbol{r}$ direction at $t = 0$. This is the vector distance between the charge q_2 and the charge q_1 at $t = 0$. We can write,

$$\boldsymbol{F} = \gamma k q_1 q_2 \frac{\boldsymbol{r}}{r'^3} = k q_1 q_2 \frac{\gamma \boldsymbol{r}}{(\gamma^2 x^2 + y^2 + z^2)^{3/2}} = q_2 \boldsymbol{E} \tag{8.6}$$

We learn that the motion of q_1 produces a field $\boldsymbol{E} = \gamma k q_1 \boldsymbol{r}/r'^3$, which is not spherically symmetric but is radial, i.e., it points in the direction $\boldsymbol{r}$. It turns out to be a central force, and it obeys Newton's third law. Note that relativistic effects are important in the expression for $\boldsymbol{E}$. As compared with the usual Coulomb force that would act on q_2 if q_1 were also at rest in frame S, the force in this case is decreased by a factor of γ^2 if q_2 is on the x axis, weaker behind or ahead of q_1 (set $y = z = 0$ in Eq. (8.4)), and is increased by a factor of γ if q_2 is to the side of q_1 (set $x = 0$ in Eq. (8.4)). These aspects of the field $\boldsymbol{E}$ are pursued further in the problems.

8.2 THE FORCE BETWEEN TWO MOVING CHARGES

To uncover velocity-dependent forces in electrodynamics and to see the failure of Newton's third law, we need to generalize our example of Section 8.1 so that both charges are moving in frame S. To keep the algebra under control, but to illustrate the general case, let q_2 have a velocity $\boldsymbol{u}$ in the y direction at $t = 0$, as shown in Fig. 8.2A and B. In addition, choose the y axis so that q_2 lies in the x-y plane at $t = 0$ and $z' = z = 0$ at $t = 0$.

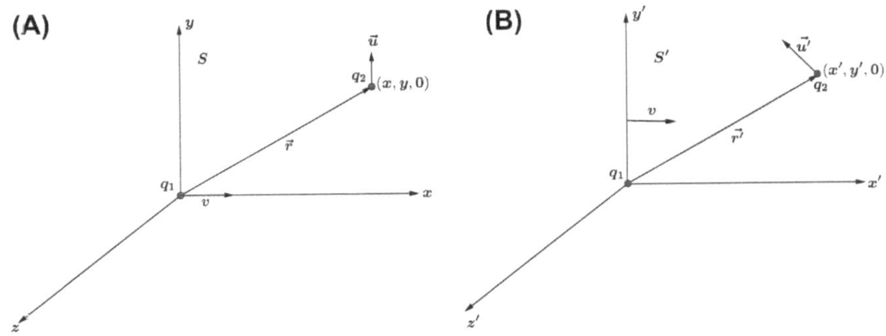

Figure 8.2 (A) Same situation as in Fig. 8.1A except charge q_2 has a velocity in the *y* direction, and it lies in the (*x*, *y*) plane. (B) Same situation shown in (A) but from the perspective of frame *S'* where the charge q_1 is at rest at the origin.

The other charge q_1 is at rest in frame S', so we know that Coulomb's law applies there, $\boldsymbol{F}' = kq_1q_2\boldsymbol{r}'/r'^3$. In components,

$$F'_x = kq_1q_2\frac{x'}{r'^3},\; F'_y = kq_1q_2\frac{y'}{r'^3},\; F'_z = 0 \tag{8.7}$$

We want to transform the force back to S where q_1 is moving with velocity v and is generating an electric current q_1v and q_2 has a velocity $\boldsymbol{u}$ in the y direction at $t = 0$. To do the transformation we need to calculate $\boldsymbol{u}\prime = \left(u'_x, u'_y, 0\right)$ in terms of $\boldsymbol{u}$ and v,

$$u'_x = \frac{u_x - v}{1 - vu_x/c^2} = -v, \quad u'_y = \frac{u_y/\gamma}{1 - vu_x/c^2} = u_y/\gamma, \quad u'_z = 0 \tag{8.8}$$

Now we have all the ingredients to calculate $\boldsymbol{F}$ using Eqs. (7.19)−(7.20),

$$F_x = \frac{F'_x + \frac{v}{c^2}\left(F'_x u'_x + F'_y u'_y\right)}{1 + vu'_x/c^2} = \frac{F'_x + \frac{v}{c^2}\left(-F'_x v + F'_y u_y/\gamma\right)}{1 - v^2/c^2}$$
$$= F'_x + \gamma^2\frac{vu_y}{\gamma c^2}F'_y$$

Now use Eq. (8.7) and $x' = \gamma x$, $y' = y$ and find,

$$F_x = kq_1q_2\frac{x'}{r'^3} + \gamma\frac{vu_y}{c^2}kq_1q_2\frac{y'}{r'^3} = \gamma kq_1q_2\frac{x}{r'^3} + \gamma kq_1q_2\frac{vu_y}{c^2}\frac{y}{r'^3}$$

where $r' = \sqrt{\gamma^2x^2 + y^2 + z^2}$ and we can collect terms into an "expected" electrostatic piece and a new piece, which is velocity dependent,

$$F_x = \frac{\gamma k q_1 q_2}{r'^3}\left(x + \frac{v u_y}{c^2} y\right) \tag{8.9}$$

Next,

$$F_y = \frac{F'_y/\gamma}{1 + v u'_x/c^2} = \frac{k q_1 q_2 y'/\gamma r'^3}{1 - v^2/c^2} = \gamma k q_1 q_2 \frac{y}{r'^3} \tag{8.10}$$

Finally,

$$F_z = 0 \tag{8.11}$$

Let us focus on the piece of the force Eq. (8.9), which depends on the velocity of the second particle,

$$F_{x,mag} = \frac{\gamma k q_1 q_2}{r'^3} \frac{v u_y}{c^2} y \tag{8.12}$$

where we have put a subscript "mag" on this component of the force because it will be interpreted in terms of a magnetic field produced by the moving charge q_1. In fact, Eq. (8.12) is one example of the Lorentz force law, which we are developing step by step. Let us anticipate its final form: the Lorentz force law for a particle of charge q_2 and velocity $\boldsymbol{u}$ propagating through an electric field $\boldsymbol{E}$ and a magnetic field $\boldsymbol{B}$ is,

$$\boldsymbol{F} = q_2(\boldsymbol{E} + \boldsymbol{u} \times \boldsymbol{B}) \tag{8.13}$$

Note the cross product of two vectors, $\boldsymbol{u} \times \boldsymbol{B}$, which appears here. Its definition and properties are probably familiar, but they are reviewed in Appendix D for completeness. Recall that the direction of the cross product of two vectors is given by the "right-hand rule." Since Eq. (8.12) states that the magnetic force is in the x direction and the velocity $\boldsymbol{u}$ is in the y direction, the magnetic field $\boldsymbol{B}$ must point in the z direction so that the cross product $\boldsymbol{u} \times \boldsymbol{B}$ is in the x direction. We can read off the magnitude of B from Eq. (8.12),

$$B = \frac{\gamma k q_1}{r'^3} \frac{v y}{c^2} \tag{8.14}$$

Now we can write $\boldsymbol{F}$ in terms of $\boldsymbol{E}$ and $\boldsymbol{B}$,

$$F_x = q_2 E_x + q_2 u_y B, \quad F_y = q_2 E_y, \quad F_z = 0 \tag{8.15}$$

Now let us get the vector character of $\boldsymbol{F}$, $\boldsymbol{E}$, and $\boldsymbol{B}$ in one equation. Since $\boldsymbol{B}$ points in the z direction and is made up of $\boldsymbol{v}$ and $\boldsymbol{r}$ (see Eq. (8.14)), and since $\boldsymbol{v}$ points in the x direction and $\boldsymbol{r}$ lies in the x-y plane, the cross product of $\boldsymbol{v}$ and $\boldsymbol{r}$ lies in the z direction. So, we infer that,

$$\boldsymbol{B} = \frac{\gamma k q_1}{r'^3}\frac{\boldsymbol{v}\times\boldsymbol{r}}{c^2} \tag{8.16}$$

where we observed that $\boldsymbol{v}\times\boldsymbol{r} = v_x y\widehat{\boldsymbol{z}}$ by writing out the cross product. It is even more instructive to identify $\boldsymbol{E}$ in Eq. (8.16),

$$\boldsymbol{B} = \frac{1}{c^2}\boldsymbol{v}\times\left(\frac{\gamma k q_1}{r'^3}\boldsymbol{r}\right) = \frac{1}{c^2}(\boldsymbol{v}\times\boldsymbol{E}) \tag{8.17}$$

where we identified $\boldsymbol{E}$ from Eq. (8.6). The virtue of Eq. (8.17) is that it shows that $\boldsymbol{B}$ is generated by the motion (velocity $\boldsymbol{v}$) of the source q_1 of the electric field.

$\boldsymbol{B}$ is not an independent, new concept. *It is a consequence of special relativity and the existence of Coulomb's law for static charges.* Note that since the electric field is radial, the magnetic field circulates around the moving charge.

Let us discuss this discovery. In the frame S' where q_1 is at rest, we have Coulomb's law for the static electric field. The physics in the moving frame S where q_1 has a velocity $\boldsymbol{v}$ is obtained from the physics in the frame S' by applying a Lorentz transformation. There is nothing really "new" here—just consequences of special relativity. We find a force in the x direction, which depends on $\boldsymbol{v}$, the velocity of the Lorentz transformation, and $\boldsymbol{u}$, the velocity of the charge q_2. When we take the force apart into the terms $q_2\boldsymbol{u}\times\boldsymbol{B}$, we see that $\boldsymbol{B}$ is generated by the current $q_1\boldsymbol{v}$ of the source charge q_1 in the frame S and that the $\boldsymbol{B}$ field exerts a force on q_2 proportional to its current $q_2\boldsymbol{u}$. In the problem set we fill out the derivation of the Lorentz force law by considering additional kinematic arrangements of q_1 and q_2.

Finally let us discuss the failure of Newton's third law in the context of Eq. (8.15). We see that q_2 experiences a magnetic force in the x direction due to the motion of charge q_1 in the x direction. Similarly, we could change our perspective and calculate the magnetic force on charge q_1 due to the motion of q_2. Similar steps to those above show that the magnetic force on q_1 points in the y direction and has the same magnitude. So, the two magnetic forces cannot sum to zero!

This observation leaves us with the puzzle discussed briefly in Chapter 7: energy—momentum conservation is a basic principle of all theories of physics, but how can this principle apply to electromagnetism where Newton's third law is violated? Some new concepts must appear in the relativistic theory to insure energy—momentum conservation. We had a short discussion of these new concepts (fields and new dynamical degrees of freedom) in Chapter 7, but there is more to come in later portions of this and the following chapter.

8.3 TRANSFORMING E AND B BETWEEN FRAMES

We will see that a consequence of relativity is that the electric and magnetic fields develop dynamics given by Maxwell's equations. Our next step toward this goal is to obtain their properties under boosts between inertial frames. One way to do this is to study devices such as capacitors and solenoids and describe them in frames related by boosts. We can calculate how charges, charge densities, and currents transform, and their transformation laws will determine those of the fields. This traditional approach is illustrated in the problems. The approach we take here is to exploit the Lorentz force law. We have derived the Lorentz force law for a particle of charge q moving with velocity $\boldsymbol{u}$ in frame S,

$$\boldsymbol{F} = q(\boldsymbol{E} + \boldsymbol{u} \times \boldsymbol{B}) \tag{8.18}$$

This law should hold in all inertial frames such as S',

$$\boldsymbol{F}' = q(\boldsymbol{E}' + \boldsymbol{u}' \times \boldsymbol{B}') \tag{8.19}$$

Since we know how a generic force $\boldsymbol{F}$ transforms, we can infer from Eq. (7.19)—(7.22) how $\boldsymbol{E}$ and $\boldsymbol{B}$ transform. Let us do the details.

To begin, recall the transformation laws for the components of a force,

$$F'_x = \frac{F_x - (v/c^2)\boldsymbol{F}\cdot\boldsymbol{u}}{(1 - u_x v/c^2)}, \quad F'_y = \frac{F_y}{\gamma(1 - u_x v/c^2)}, \quad F'_z = \frac{F_z}{\gamma(1 - u_x v/c^2)} \tag{8.20}$$

$$\boldsymbol{F}'\cdot\boldsymbol{u}' = \frac{\boldsymbol{F}\cdot\boldsymbol{u} - vF_x}{(1 - u_x v/c^2)} \tag{8.21}$$

We use the first equation of Eq. (8.20) to transform the x component of the Lorentz force,

$$q\left(E'_x + (\boldsymbol{u}' \times \boldsymbol{B}')_x\right) = \frac{q\left(E_x + (\boldsymbol{u} \times \boldsymbol{B})_x - \frac{v}{c^2}\boldsymbol{F}\cdot\boldsymbol{u}\right)}{(1 - u_x v/c^2)} \tag{8.22}$$

Now we need to write out the cross products in Eq. (8.22) in components, and we need to evaluate $\boldsymbol{F}\cdot\boldsymbol{u}$ for the Lorentz force,

$$\boldsymbol{F}\cdot\boldsymbol{u} = q(\boldsymbol{E} + \boldsymbol{u}\times\boldsymbol{B})\cdot\boldsymbol{u} = q\boldsymbol{E}\cdot\boldsymbol{u} \tag{8.23}$$

Now Eq. (8.22) becomes,

$$E'_x + u'_y B'_z - u'_z B'_y = \frac{E_x - \frac{vu_x}{c^2}E_x + u_y B_z - u_z B_y - \frac{vu_y}{c^2}E_y - \frac{vu_z}{c^2}E_z}{(1 - u_x v/c^2)}$$

Now we can use our transformation laws for velocities to replace primed quantities by unprimed ones on the left-hand side and on the right-hand side we can isolate E_x,

$$E'_x + \frac{u_y B'_z - u_z B'_y}{\gamma(1 - u_x v/c^2)} = E_x + \frac{u_y B_z - u_z B_y - \frac{vu_y}{c^2}E_y - \frac{vu_z}{c^2}E_z}{(1 - u_x v/c^2)} \tag{8.24}$$

This equation is hiding three independent equations in it. The point is that u_y and u_z are independent variables, so the only way Eq. (8.24) can be true in general is if the coefficients of u_y and u_z are separately zero. We learn three transformation laws,

$$E'_x = E_x, \quad B'_y = \gamma\left(B_y + \frac{v}{c^2}E_z\right), \quad B'_z = \gamma\left(B_z - \frac{v}{c^2}E_y\right) \tag{8.25}$$

We see that the components of the electric and magnetic fields mix among themselves under boosts! This indicates that electric effects in one frame can be interpreted as magnetic effects in another.

But we are not done. There are three more components in the two vectors $\boldsymbol{E}$ and $\boldsymbol{B}$. We turn to the second equation in Eq. (8.20). We specialize to the Lorentz force again and have,

$$q\left(E'_y + (\boldsymbol{u}\prime \times \boldsymbol{B}\prime)_y\right) = \frac{q\left(E_y + (\boldsymbol{u}\times\boldsymbol{B})_y\right)}{\gamma(1 - u_x v/c^2)} \tag{8.26}$$

Now we write the cross product in components, and we use the transformation law for velocities to eliminate $\boldsymbol{u}'$,

$$E'_y - u'_x B'_z + u'_z B'_x = E'_y - \frac{(u_x - v)B'_z + \frac{u_z}{\gamma}B'_x}{(1 - u_x v/c^2)} = \frac{E_y - u_x B_z + u_z B_x}{\gamma(1 - u_x v/c^2)}$$

Matching the coefficients of u_z, we learn that,

$$B'_x = B_x \tag{8.27}$$

Now we multiply through by $(1 - u_x v/c^2)$ and find,

$$E'_y - \frac{vu_x}{c^2}E'_y + vB'_z - u_x B'_z = \frac{1}{\gamma}E_y - \frac{u_x}{\gamma}B_z + \frac{u_z}{\gamma}B_x \tag{8.28}$$

But we know from Eq. (8.25) how B'_z transforms, $B'_z = \gamma\left(B_z - \frac{v}{c^2}E_y\right)$. Substituting into Eq. (8.28) gives,

$$E'_y - \frac{vu_x}{c^2}E'_y + \gamma v B_z - \gamma\frac{v^2}{c^2}E_y - u_x\gamma B_z + u_x\gamma\frac{v}{c^2}E_y = \frac{1}{\gamma}E_y - \frac{u_x}{\gamma}B_z + \frac{u_z}{\gamma}B_x$$

The coefficient of u_x must be zero, so

$$-\frac{v}{c^2}E'_y - \gamma B_z + \gamma\frac{v}{c^2}E_y = -\frac{1}{\gamma}B_z$$

Finally, combining the two terms proportional to B_z, we have $\left(\gamma - \frac{1}{\gamma}\right) = \gamma\frac{v^2}{c^2}$, giving,

$$-\frac{v}{c^2}E'_y = -\gamma\frac{v}{c^2}E_y + \gamma\frac{v^2}{c^2}B_z$$

which becomes the transformation law for E'_y,

$$E'_y = \gamma(E_y - vB_z) \tag{8.29}$$

And finally, we could do the same analysis on the z component of Eq. (8.26), or argue by symmetry and find our last transformation law,

$$E'_z = \gamma(E_z + vB_y) \tag{8.30}$$

The virtue of these derivations is their fundamental character: the requirement of the "covariance" of the Lorentz force law dictates how the components of the electric and magnetic fields mix with each other under boosts. Various exercises in the problem set below present physical illustrations of these general, rather abstract results.

8.4 MORE ON INVARIANTS AND FOUR-VECTORS

At this point in our studies we have met many quantities with different transformation properties under boosts. The most important were four-vectors. Recall the four coordinates,

$$x^\mu = (x^0, x^1, x^2, x^3) = (ct, x, y, z)$$

that transforms under boosts according to the Lorentz transformation rules,

$$x'^0 = \gamma\left(x^0 - \frac{v}{c}x^1\right), \quad x'^1 = \gamma\left(x^1 - \frac{v}{c}x^0\right), \quad x'^2 = x^2, \quad x'^3 = x^3$$

We have met other collections of four variables that transform in this fashion. The energy—momentum four-vector is a particularly significant one. If we can express physical laws in terms of four-vectors, we can formulate Postulate 1 of relativity particularly elegantly.

Let us review some of the properties of four-vectors and let us discover more. We write four-vectors $A^\mu = (A^0, A^1, A^2, A^3)$ and $B^\mu = (B^0, B^1, B^2, B^3)$. The "square" or "length" of a four-vector is

$$A^2 = A\cdot A = A_0^2 - \boldsymbol{A}^2 \tag{8.31}$$

We can also define the "inner product" of two four-vectors,

$$A\cdot B = A^0B^0 - A^1B^1 - A^2B^2 - A^3B^3 \tag{8.32}$$

and can show that the inner product of two four-vectors is a Lorentz invariant.

We can construct additional four-vectors that are related to the three vectors we have already studied. Recall the differential element,

$$dx^\mu = (dx^0, dx^1, dx^2, dx^3) \tag{8.33}$$

It has an invariant length,

$$ds^2 = dx\cdot dx = c^2dt^2 - d\boldsymbol{x}^2 \tag{8.34}$$

The proper time $d\tau$ is related to the invariant interval ds^2, if it is time-like, $ds^2 > 0$,

$$d\tau^2 = ds^2/c^2 = dt^2 - \frac{1}{c^2}d\boldsymbol{x}^2 \tag{8.35}$$

For a particle moving with velocity $\boldsymbol{u}$,

$$d\tau^2 = dt^2 - \frac{1}{c^2}\boldsymbol{u}^2 dt^2 = dt^2/\gamma^2 \tag{8.36}$$

which is an expression of time dilation,

$$dt/d\tau = \gamma(u) \tag{8.37}$$

Using $d\tau$ we can construct other four-vectors. The "four-velocity" is,

$$u^\mu = \frac{dx^\mu}{d\tau} = \left(c\frac{dt}{d\tau}, \frac{d\boldsymbol{x}}{d\tau}\right) \tag{8.38}$$

Using Eq. (8.37) we have,

$$u^\mu = \left(c\frac{dt}{d\tau}, \frac{d\boldsymbol{x}}{dt}\frac{dt}{d\tau}\right) = \gamma(u)(c, \boldsymbol{u}) \tag{8.39}$$

u_μ provides a natural way to generalize the three velocity to the four-dimensional world of Minkowski. The invariant length of u_μ is,

$$u^2 = \gamma^2 c^2 - \gamma^2 \boldsymbol{u}^2 = \frac{1}{(1 - \boldsymbol{u}^2/c^2)}(c^2 - \boldsymbol{u}^2) = c^2 \tag{8.40}$$

An easier way to evaluate the invariant u^2 is to choose a convenient frame. This is the rest frame of the particle where $u_\mu = (c, 0)$ and the result $u^2 = c^2$ is immediate.

We can similarly generalize the three-dimensional acceleration to four dimensions with the four-acceleration,

$$a^\mu = \frac{du^\mu}{d\tau} = \frac{d^2x^\mu}{d\tau^2} \tag{8.41}$$

a_μ has a more complicated relation to $\boldsymbol{a}$ than u_μ has to $\boldsymbol{u}$,

$$a^\mu = \gamma\frac{du^\mu}{dt} = \gamma\left(\frac{d\gamma}{dt}c, \frac{d\gamma}{dt}\boldsymbol{u} + \gamma\boldsymbol{a}\right) \tag{8.42}$$

Note that in the instantaneous rest frame of the particle where $u = 0$ and $\gamma = 1$ and $d\gamma/dt = 0$,

$$a_\mu = (0, \boldsymbol{a}).$$

We have already introduced the relativistic momentum, $\boldsymbol{p} = \gamma m\boldsymbol{u}$, and energy $E = \gamma mc^2$. We have formed the four-momentum, $p^\mu = (p^0, p^1, p^2, p^3) = (E/c, \boldsymbol{p}) = \gamma m(c, \boldsymbol{u})$. We observed earlier that,

$$p^\mu = m\frac{dx^\mu}{d\tau} = mu^\mu$$

The fact that $u^2 = u\cdot u = c^2$ gives the energy—momentum relation of a free particle,

$$p\cdot p = E^2/c^2 - \boldsymbol{p}^2 = m^2 u\cdot u = m^2c^2$$
$$E^2 = \boldsymbol{p}^2c^2 + m^2c^4$$

The usefulness of p_μ suggests that we introduce a "four-force,"

$$f^\mu = \frac{dp^\mu}{d\tau} = m\frac{du^\mu}{d\tau} = m\frac{d^2x^\mu}{d\tau^2} \tag{8.43}$$

Recall that the three-dimensional force $\boldsymbol{f} = d\boldsymbol{p}/dt$ and the power $dE/dt = \boldsymbol{u}\cdot\boldsymbol{f}$ mix under boosts. We can relate these quantities to the four-force,

$$f^\mu = \frac{dp^\mu}{d\tau} = \gamma(\boldsymbol{u})\frac{d}{dt}(E/c, \boldsymbol{p}) = \gamma(\boldsymbol{u})(\boldsymbol{f}\cdot\boldsymbol{u}/c, \boldsymbol{f}) \tag{8.44}$$

The four-force will be useful when we write the Lorentz force law in a covariant fashion. It will also be important when we develop the equation of motion of particles in general relativity.

The virtue in defining and identifying four-vectors is that four-vectors all share the same simple transformation law, the Lorentz transformation law of x^μ. Under boosts the components of each four-vector mix among themselves in a universal fashion. This is much simpler than our particular transformation laws for the various three-vector quantities that initiated this development. Relations between four-vectors are "physical" in the sense that they have the same form in every reference frame. The covariance of the laws of physics is well expressed in this language.

PROBLEMS

8-1. Let us consider the boosted Coulomb field in more generality by generalizing the discussion in Section 8.1.

Calculate the electric and magnetic fields of a point charge moving with velocity $\boldsymbol{v}$. The particle starts at the origin at $t = 0$. We measure the electric field $\boldsymbol{E}$ at point $\boldsymbol{r}$. Express your answers in terms of the vector $\boldsymbol{R} = \boldsymbol{r} - \boldsymbol{v}t$, which is the vector between the point where you measure the electric field and the position of the charge at time t.

8-2. Let us consider more configurations of the two particles in Section 8.2 to develop a full derivation of the Lorentz force law from electrostatics and relativity.

In the frame S at $t=0$ let q_1 be at the origin and moving along the x axis at velocity v, and let q_2 be at point y on the y axis and also moving along the x axis at velocity v. So, both particles are at rest in frame S' at $t'=0$. In S' we have,

$$F'_x = 0, \quad F'_y = \frac{kq_1q_2}{y'^2}, \quad F'_z = 0$$

a. Show that,

$$F_y = \frac{1}{\gamma}\frac{kq_1q_2}{y^2}$$

b. If q_2 had been stationary in S, show that we would have had,

$$F_y^{elec} = \gamma\frac{kq_1q_2}{y^2}$$

c. Show that the difference of these two forces in part a and b is,

$$F_y^{mag} = -\frac{v^2}{c^2}F_y^{elec}$$

and can be interpreted as a magnetic force exerted on the moving charge q_2 by the moving source q_1.

8-3. Redo the analysis in problem Section 8.2 in the case where q_2 moves with the velocity u_x in the x direction in frame S and $q_1$is moving along the x axis at velocity v and at $t=0$ is at the origin in frame S. Show that,

$$F_y^{mag} = -\frac{vu_x}{c^2}F_y^{elec}$$

8-4. Let us obtain the Lorentz transformations of electric and magnetic fields in some special cases using the construction of capacitors and solenoids.

Consider a capacitor at rest in frame S. Its square plates of side l are parallel to the x-z plane. The plates have a charge/area of σ and $-\sigma$ in frame S.

a. What is the electric field strength E inside the capacitor in terms of σ?

b. Now measure the electric field E' in a frame S', which moves in the x direction at velocity v. Show that $E'_y = \gamma E_y$.

c. Now orient the capacitor so that its plates are parallel to the y-z plane. Show that $E'_x = E_x$.

Now consider a solenoid oriented along the x axis. It has n turns of wire per length, and its length l is much longer than its radius r. The wire carries a current I.

d. Find the magnetic field inside the solenoid in terms of n and I.

e. Now measure the properties of the solenoid in frame S'. How are n', l', and B', related to n, l, and B?

8-5. Show that the Lorentz force law is unchanged under time reversal: $t \to -t$, $\boldsymbol{E} \to \boldsymbol{E}$, and $\boldsymbol{B} \to -\boldsymbol{B}$. Explain the change in sign of magnetic field under time reversal.

8-6. Using the transformation rules for the electric and magnetic fields between frames, show that these two quantities are invariant under boosts,

a. $\boldsymbol{E}^2 - c^2\boldsymbol{B}^2$ and

b. $\boldsymbol{E}\cdot\boldsymbol{B}$

We learn that 1. If the magnitudes of E and cB are equal in any frame, then they are equal in all frames, and 2. If $\boldsymbol{E}$ and $\boldsymbol{B}$ are perpendicular in any reference frame, then they are perpendicular in all frames.

8-7. Let us consider the motion of a relativistic particle in a uniform electric field in more detail.

The particle has charge e, and the uniform electric field $\boldsymbol{E}$ points in the x direction. The motion will be confined to a plane, which we label x-y. Initially $p_{x,0} = 0$, $p_{y,0} = p_0$, $x_0 = 0$, and $y_0 = 0$.

a. Show that the particle's relativistic energy E_{kin} at time $t > 0$ is,

$$E_{kin} = \sqrt{m^2c^4 + c^2p_0^2 + (ceEt)^2}$$

b. Use $v = pc^2/E_{kin}$ to find,

$$\frac{dx}{dt} = \frac{c^2eEt}{\sqrt{m^2c^4 + c^2p_0^2 + (ceEt)^2}}$$

which we can integrate and find,

$$x(t) = \frac{1}{eE}\sqrt{m^2c^4 + c^2p_0^2 + (ceEt)^2}$$

c. Similarly, show that,

$$\frac{dy}{dt} = \frac{c^2p_0}{\sqrt{m^2c^4 + c^2p_0^2 + (ceEt)^2}}$$

which you can integrate and find,

$$y(t) = \frac{cp_0}{eE}\sinh^{-1}\left(\frac{ceEt}{\sqrt{m^2c^4 + c^2p_0^2}}\right)$$

$\left(\text{Hint: To do the integral, note } \frac{d\sinh^{-1}z}{dz} = \frac{1}{\sqrt{1+z^2}}\right)$

d. Solve the result of part b. for t, substitute into the result of part c. and find the curve along which the particle moves,

$$x = \frac{\sqrt{m^2c^4 + c^2p_0^2}}{eE}\cosh\left(\frac{eEy}{cp_0}\right)$$

(Hint: Use the formula $\cosh(x) = \sqrt{1 + \sinh^2 x}$)

e. This curve is called a "catenary" in classical differential geometry. Draw it using a graphics program for various choices of parameters. Suppose that $v \ll c$. Show that the catenary reduces to a parabola, a familiar result in classical mechanics,

$$x = \frac{eE}{2mv_0^2}y^2 + const$$

f. In this relativistic mechanics problem, the x and y motions do *not* decouple as they do in Newtonian mechanics. Consider the equation for dy/dt and explain the new ingredient. Compare the large t behavior of $y(t)$ in part c. with the Newtonian result.

8-8. Let us solve the problem of a charged particle moving in a uniform magnetic field.

The Lorentz force law of a charged particle in a constant uniform $\boldsymbol{B}$ chosen in the z direction,

$$\frac{d}{dt}\boldsymbol{p} = e\boldsymbol{v} \times \boldsymbol{B}$$

Write $p = Ev/c^2$ and realize that the relativistic energy E is constant in time because the magnetic field does no work. The equation of motion simplifies,

$$\frac{E}{c^2}\frac{d\boldsymbol{v}}{dt} = e\boldsymbol{v} \times \boldsymbol{B}$$

Show that this vector equation can be written in Cartesian coordinates,

$$\frac{d}{dt}v_x = \omega v_y, \frac{d}{dt}v_y = -\omega v_x, \frac{d}{dt}v_z = 0$$

where $\omega = eBc^2/E$. Identify two coupled oscillators with the solution,

$$v_x(t) = \sqrt{v_x^2 + v_y^2}\cos(\omega t + \alpha), v_y(t) = -\sqrt{v_x^2 + v_y^2}\sin(\omega t + \alpha)$$

where α is a constant and $\sqrt{v_x^2 + v_y^2}$ is also time independent. Now show that,

$$x(t) = x_0 + r\sin(\omega t + \alpha), \quad y(t) = y_0 + r\cos(\omega t + \alpha)$$

where the radius of the circular motion is $r = \sqrt{v_x^2 + v_y^2}/\omega$. Also,

$$z(t) = z_0 + v_{0z}t$$

So the three-dimensional motion is a helix.

8-9. Consider the relativistic motion of a charged particle in parallel and uniform electric and magnetic fields, following a discussion in Landau and Lifshitz, *The Classical Theory of Fields* [2].

Let $\boldsymbol{E}$ and $\boldsymbol{B}$ point in the z direction.

The magnetic field has no influence on the z direction so we know from previous problems that,

$$z(t) = \frac{1}{eE}\sqrt{m^2c^4 + c^2p_0^2 + (ceEt)^2} = E_{kin}/eE$$

For the motion in the x-y plane, start with the Lorentz force law and then substitute $\boldsymbol{v} = \boldsymbol{p}c^2/E_{kin}$,

$$\frac{d}{dt}p_x = -eBv_y, \frac{d}{dt}p_y = -eBv_x$$

$$\frac{d}{dt}p_x = -\frac{eBc^2}{E_{kin}}p_y, \frac{d}{dt}p_y = -\frac{eBc^2}{E_{kin}}p_x,$$

So, we have coupled oscillators, but the effective angular frequency has its own t dependence. So define,

$$\frac{d}{dt}\varphi = eBc^2/E_{kin}$$

a. Show that the coupled equations are solved by,

$$p_x = \sqrt{p_x^2 + p_y^2}\cos(\varphi),\ p_y = -\sqrt{p_x^2 + p_y^2}\sin(\varphi)$$

where $\sqrt{p_x^2 + p_y^2}$ is constant in time.

b. Show that

$$ct = \frac{\sqrt{m^2c^4 + c^2p_0^2}}{eE}\sinh\left(\frac{E}{cB}\varphi\right)$$

by integrating the differential equation for φ and using the integral

$$\int \frac{dx}{\sqrt{a^2 + x^2}} = \sinh^{-1}\left(\frac{x}{a}\right)$$

c. The equation for p_x can be written as an equation for $dx/d\varphi$ using $\frac{d}{dt}\varphi = eBc^2/E_{kin}$. Use this strategy and show that,

$$\frac{dx}{d\varphi} = \frac{\sqrt{p_x^2 + p_y^2}}{eB}\cos(\varphi),\ \frac{dy}{d\varphi} = -\frac{\sqrt{p_x^2 + p_y^2}}{eB}\sin(\varphi)$$

Therefore,

$$x(t) = \frac{\sqrt{p_x^2 + p_y^2}}{eB}\sin(\varphi),\ y(t) = \frac{\sqrt{p_x^2 + p_y^2}}{eB}\cos(\varphi)$$

d. Combine the formulas above for $z(t)$ with the result from part b. and find,

$$z(\varphi) = \frac{\sqrt{m^2c^4 + c^2p_0^2}}{eE}\cosh\left(\frac{E}{cB}\varphi\right)$$

Now we can describe the resulting motion: the trajectory is a helix with fixed radius $\frac{\sqrt{p_x^2+p_y^2}}{eB}$, but an increasing stretch in the z

direction, along which the particle moves with a decreasing angular velocity $d\varphi/dt = eB/E_{kin}$ and a velocity in the z direction, which increases toward the limit c.

8-10. Consider the transformation equations relating $\boldsymbol{E}$ and $\boldsymbol{B}$ in frame S to $\boldsymbol{E}'$ and $\boldsymbol{B}'$ in frame S'.

a. Show that if $\boldsymbol{B}' = 0$ in S', then in S, $\mathbf{B} = \boldsymbol{v} \times \mathbf{E}/c^2$.

b. Show that if $\boldsymbol{E}' = 0$ in S', then in S, $\mathbf{E} = -\boldsymbol{v} \times \mathbf{B}$.

c. Prove that if $\boldsymbol{E}$ and $\boldsymbol{B}$ are perpendicular in some frame S, then there exists a frame S' in which the field is purely electric or purely magnetic. The velocity $\boldsymbol{v}$ of S' relative to S is perpendicular to $\boldsymbol{E}$ and $\boldsymbol{B}$ and equal to $\mathbf{c}^2 B/E$ in the first case and E/B in the second.

REFERENCES

[1] A.P. French, Special Relativity, W.W. Norton, New York, 1968.
[2] L.D. Landau, E.M. Lifshitz, The Classical Theory of Fields, Pergamon Press, Oxford, 1962.

CHAPTER 9

Maxwell's Equations of Electrodynamics and the Wave Equation

Contents

9.1 BOOSTING GAUSS' LAW AND DISCOVERING MAXWELL'S EQUATIONS

Now we are almost ready to obtain Maxwell's equations from electrostatics and special relativity, but we need to write Coulomb's law in a more fundamental fashion. We have been writing,

$$\boldsymbol{F} = k\frac{q_1 q_2}{r^2}\widehat{\boldsymbol{r}} = q_2 \boldsymbol{E}(\boldsymbol{r}) \tag{9.1}$$

for the force between a charge q_1 at rest at the origin of frame S and a charge q_2 at position $\boldsymbol{r}$. This formula applies when these are the only charges in the system. If there are other charges or boundary conditions in the system then they will effect the electric field at position $\boldsymbol{r}$. This will not do. We need a precise statement of the fact that the charge at the origin produces an electric field in its immediate vicinity. This will have to be a differential equation. When that equation is solved under appropriate conditions, Coulomb's law Eq. (9.1) should follow.

To do this, we note that the electric field in Eq. (9.1) falls off as the square of the distance from the charge. Following the original ideas of the great mathematician Carl Friedrich Gauss, this means that we can detect the presence of the charge at the origin by measuring the electric flux through a distant closed surface surrounding the origin. Let $d\boldsymbol{a}$ be an infinitesimal element of surface area. $d\boldsymbol{a}$ has a magnitude equal to the area of the surface element, and it has a direction perpendicular to the infinitesimal

Special Relativity, Electrodynamics, and General Relativity
ISBN 978-0-12-813720-8
https://doi.org/10.1016/B978-0-12-813720-8.00009-X

surface element. Consider a sphere of radius R around the origin and calculate the total electric flux through it from Eq. (9.1),

$$\oint \boldsymbol{E}\cdot d\boldsymbol{a} = \oint k\frac{q_1}{R^2}\widehat{\boldsymbol{r}}\cdot d\boldsymbol{a} = \oint k\frac{q_1}{R^2}R^2 d\Omega = 4\pi k q_1 \tag{9.2}$$

Here $d\Omega$ is the infinitesimal solid angle on the sphere of radius R. $R^2 d\Omega$ is the magnitude of the infinitesimal surface element $d\boldsymbol{a}$. Note that the factors of R cancel in the integrand—the R^{-2} fall off of the electric field is canceled by the R^2 growth of the surface area of the sphere. The solid angle integrates to 4π for a sphere that encloses the origin. The reader can consult Appendix D for a summary of these and other useful results from vector calculus.

There are two more observations about this equation that are useful. First, recall Gauss' theorem that relates the integral of the flux of any vector field $\boldsymbol{V}(\boldsymbol{r})$ through a closed surface to the volume integral of the divergence of that vector field inside the surface,

$$\oint \boldsymbol{V}\cdot d\boldsymbol{a} = \oiiint \nabla\cdot\boldsymbol{V}\, d^3r \tag{9.3}$$

This wonderful theorem, which is also reviewed in Appendix D, frees our discussion Eq. (9.2) from the specific spherical surface used there. Here ∇ is the gradient operator, which in Cartesian coordinates reads $\nabla = \left(\frac{\partial}{\partial x}, \frac{\partial}{\partial y}, \frac{\partial}{\partial z}\right)$ and the "divergence" of $\boldsymbol{V}$ is $\nabla\cdot V = \left(\frac{\partial V_x}{\partial x}, \frac{\partial V_y}{\partial y}, \frac{\partial V_z}{\partial z}\right)$. The second observation is that q_1 in Eq. (9.2) might be a macroscopic charge, which is concentrated around the origin and can be expressed as a volume integral over a charge density ρ,

$$q_1 = \oiiint \rho\, d^3r \tag{9.4}$$

Taking $\boldsymbol{V} = \boldsymbol{E}$, and substituting Eqs. (9.3) and (9.4) into Eq. (9.2) we learn that,

$$\oiiint \nabla\cdot\boldsymbol{E}\, d^3r = 4\pi k \oiiint \rho\, d^3r \tag{9.5}$$

But this equality must hold for *any* volume surrounding the origin, which includes the charge. This can only be true if the integrands themselves are equal point-by-point,

$$\nabla\cdot\boldsymbol{E} = 4\pi k\rho \tag{9.6}$$

This is the local differential equation we sought. It expresses the fact that charge densities generate electric fields locally. It is fundamental.

Another critical property of Eq. (9.6) is that it is a "linear system." For example, if we have a problem where ρ_1 is the source of an electric field $\boldsymbol{E}_1$, and we have another problem where ρ_2 is the source of an electric field $\boldsymbol{E}_2$,

$$\nabla \cdot \boldsymbol{E}_1 = 4\pi k \rho_1, \quad \nabla \cdot \boldsymbol{E}_2 = 4\pi k \rho_2 \tag{9.7}$$

then if we add the two equations, we see that $\rho_1 + \rho_2$ is the source of an electric field $\boldsymbol{E}_1 + \boldsymbol{E}_2$,

$$\nabla \cdot (\boldsymbol{E}_1 + \boldsymbol{E}_2) = 4\pi k(\rho_1 + \rho_2) \tag{9.8}$$

This has the important consequence that Eq. (9.6) is true for a general charge distribution ρ.

Eq. (9.6) is so general and useful, because it is a local differential equation, that it is generally given the name "Gauss' law," named after its inventor.

In the problem set at the end of this chapter, we show that if ρ is concentrated at the origin and there are no other charges in the system, then Coulomb's law Eq. (9.1) follows from Gauss' law. But Eq. (9.6) is particularly useful because it is a local differential equation that expresses the idea that charges are the sources of electric fields point-by-point in space—time. It also applies to the case where ρ has explicit time dependence. Eq. (9.6) works in this case because ρ and $\boldsymbol{E}$ enter the equation at the *same* space—time point. On the other hand, Coulomb's law Eq. (9.1) is *not* true for a time-dependent ρ because information travels at the speed limit and the resulting electric field must contain this fact in its spatial dependence far from the origin. If ρ has time dependence then currents must be flowing, as we will discuss below, and these currents will generate time-dependent magnetic fields, which will generate additional time-dependent electric fields. In those cases Gauss' law will not be sufficient to calculate the total electric field. When we derive Maxwell's equations below, we will see how things work out in this general case.

Now let us study the charge density, in more detail. Suppose that all the charges are at rest in the frame S, and label the charge density there ρ_0. When we boost a charge density to a velocity $\boldsymbol{u}$, we generate a new charge density ρ and we create a current density $\boldsymbol{J} = \rho \boldsymbol{u}$. Since ρ is a ratio of charge per volume, and the charge is invariant under boosts while volumes contract in the direction of the boost by a factor $\gamma(\boldsymbol{u})$, we learn that,

$$\rho = \frac{q}{V} = \frac{q_0}{V_0/\gamma} = \gamma \frac{q_0}{V_0} = \gamma \rho_0 \tag{9.9}$$

and the current density is $\boldsymbol{J} = \rho \boldsymbol{u} = \gamma \rho_0 \boldsymbol{u}$.

Compare these results to the energy–momentum carried by a particle, $E = \gamma m_0 c^2$ and $\boldsymbol{p} = \gamma m_0 \boldsymbol{u}$. We see that ρ and $\boldsymbol{J}$ share the same kinematics factors with E and $\boldsymbol{p}$, and thus form a four-vector,

$$J^\mu = (c\rho, J_x, J_y, J_z) \tag{9.10}$$

And this implies that they obey the Lorentz transformation laws between frames S and S',

$$\rho' = \gamma\left(\rho - \frac{v}{c^2}J_x\right), \quad J'_x = \gamma(J_x - v\rho), \quad J'_y = J_y, \quad J'_z = J_z \tag{9.11}$$

We see that ρ and J_x mix under boosts such as E and p_x or such as t and x! Since ρ is the source of electric fields and J_x is the source of magnetic fields, this mixing can also be viewed as a consequence of the mixing of electric and magnetic fields that we derived earlier from requiring that the Lorentz force law applies in all inertial frames, Section 8.3. We will see that Eq. (9.11) is an essential ingredient in our derivation of Maxwell's equations below.

Another crucial property of the charge density and the current density is conservation of charge. As we discussed earlier when considering Newton's third law, conservation laws should be expressed through local equations in relativity. Gauss' theorem will be essential here. Conservation of charge means that charge is neither created nor destroyed in physical processes. However, currents flow and change the local charge density when they do so. Consider the flow of current through an infinitesimal surface area, $\boldsymbol{J} \cdot d\boldsymbol{a}$. This quantity counts the amount of charge passing through the surface element per second. So, if one sums up this quantity over a closed surface enclosing a volume V, one counts up the amount of charge leaving that volume per second,

$$\oiint \boldsymbol{J} \cdot d\boldsymbol{a} = -\frac{d}{dt}\iiint \rho \, d^3 r \tag{9.12}$$

Using Gauss' theorem, we can write for a stationary volume,

$$\iiint \nabla \cdot \boldsymbol{J} d^3 r = -\iiint \frac{\partial \rho}{\partial t} d^3 r \tag{9.13}$$

Since this equality holds for any volume, it must be true as a local differential equation,

$$\nabla \cdot \boldsymbol{J} + \frac{\partial \rho}{\partial t} = 0 \tag{9.14}$$

This is an elegant expression of charge conservation. In fact this equation is fully relativistic! It can be written as the inner product of two four-vectors, following the notation introduced in Chapter 6, Section 6.5. There we saw that $\partial_\mu = \partial/\partial x^\mu = (\partial/\partial ct, \partial/\partial x, \partial/\partial y, \partial/\partial z) = (\partial/\partial ct, \nabla)$ is a covariant four-vector, so

$$\partial \cdot J = \sum_\mu \partial_\mu J^\mu = \frac{\partial}{\partial ct} \cdot c\rho + \nabla \cdot \mathbf{J} = \partial\rho/\partial t + \nabla \cdot \mathbf{J} = 0 \tag{9.15}$$

is a Lorentz invariant conservation law,

$$\begin{aligned} \partial \cdot J = \partial' \cdot J' = \sum_\mu \partial_\mu J^\mu = \sum_\mu \partial'_\mu J'^\mu = \partial\rho/\partial t + \nabla \cdot \mathbf{J} \\ = \partial\rho'/\partial t' + \nabla' \cdot \mathbf{J}' = 0 \end{aligned} \tag{9.16}$$

Let us check explicitly that it is true in any inertial frame, if it is true in one of them. In other words, consider the expression Eq. (9.14) in frame S', $\nabla' \cdot J' + \frac{\partial \rho'}{\partial t'} = 0$. We know how each variable in this equation transforms between frames, so we can rewrite it in terms of variables referred to frame S. Since $\partial_\mu = \partial/\partial x^\mu = (\partial/\partial ct, \nabla)$ is a covariant four-vector, it transforms as $x_\mu = \sum_\sigma g_{\mu\sigma} x^\sigma = (ct, -x, -y, -z)$,

$$\frac{\partial}{\partial t'} = \gamma\left(\frac{\partial}{\partial t} + v\frac{\partial}{\partial x}\right), \quad \frac{\partial}{\partial x'} = \gamma\left(\frac{\partial}{\partial x} + \frac{v}{c^2}\frac{\partial}{\partial t}\right), \quad \frac{\partial}{\partial y'} = \frac{\partial}{\partial y}, \quad \frac{\partial}{\partial z'} = \frac{\partial}{\partial z} \tag{9.17}$$

So, we can check Eq. (9.16) explicitly using Eq. (9.11), the transformation law for the four-vector current J^μ, and Eq. (9.17),

$$\begin{aligned} \nabla' \cdot J' + \frac{\partial \rho'}{\partial t'} &= \gamma\left(\frac{\partial}{\partial x} + \frac{v}{c^2}\frac{\partial}{\partial t}\right)\gamma(J_x - v\rho) + \frac{\partial J_y}{\partial y} + \frac{\partial J_z}{\partial z} \\ &\quad + \gamma\left(\frac{\partial}{\partial t} + v\frac{\partial}{\partial x}\right)\gamma\left(\rho - \frac{v}{c^2}J_x\right) \\ \nabla' \cdot J' + \frac{\partial \rho'}{\partial t'} &= \gamma^2\left(\frac{\partial J_x}{\partial x} - v\frac{\partial \rho}{\partial x} + \frac{v}{c^2}\frac{\partial J_x}{\partial t} - \frac{v^2}{c^2}\frac{\partial \rho}{\partial t}\right) + \frac{\partial J_y}{\partial y} + \frac{\partial J_z}{\partial z} \\ &\quad + \gamma^2\left(\frac{\partial \rho}{\partial t} - \frac{v}{c^2}\frac{\partial J_x}{\partial t} + v\frac{\partial \rho}{\partial x} - \frac{v^2}{c^2}\frac{\partial J_x}{\partial x}\right) \\ \nabla' \cdot J' + \frac{\partial \rho'}{\partial t'} &= \gamma^2\left(1 - \frac{v^2}{c^2}\right)\frac{\partial J_x}{\partial x} + \gamma^2\left(1 - \frac{v^2}{c^2}\right)\frac{\partial \rho}{\partial t} + \frac{\partial J_y}{\partial y} + \frac{\partial J_z}{\partial z} \\ \nabla' \cdot J' + \frac{\partial \rho'}{\partial t'} &= \nabla \cdot J + \frac{\partial \rho}{\partial t} = 0 \end{aligned} \tag{9.18}$$

So, after a flurry of algebra all is well: we have the same conservation law (called the "continuity" equation) in all frames.

This work emphasizes that local differential equations allow us to capture conservation laws, equations of motion and dynamics in a fashion that makes the relativistic character, and causality of the physics manifest. The covariant four-vector $\partial_\mu = (\partial_0, \partial_1, \partial_2, \partial_3) = \left(\frac{\partial}{\partial x_0}, \nabla\right)$ will prove useful in other applications and theoretical developments.

Now we are in position to turn to the central result of this chapter: the derivative of Maxwell's equation from electrostatics and special relativity. We have discussed at length why such a derivation should be possible, and now we have the machinery to carry it out. We begin with Gauss' law Eq. (9.6), our expression of the fact that charge densities are the sources of electric fields as a local differential equation. This equation must be true in all frames, and yet it is puzzling because it involves only derivatives in the spatial variables and none in time! It is not like an equation of motion, which has a first or second time derivative in it and predicts how a physical system evolves in time. In "old-fashioned" language it is called an "equation of constraint." If we assume the validity of this equation in frame S' and then boost to frame S, we will generate an equation with time derivatives. How does this work?

Let us do the exercise. We begin with,

$$\nabla' \cdot \boldsymbol{E}' = 4\pi k \rho'$$

And we use Eq. (9.17) to write the gradient in terms of derivatives in S, we use Eqs. (8.25), (8.29) and (8.30) to transform the fields and Eq. (9.11) to transform the charge density,

$$\gamma\left(\frac{\partial}{\partial x} + \frac{v}{c^2}\frac{\partial}{\partial t}\right)E_x + \frac{\partial}{\partial y}\gamma(E_y - vB_z) + \frac{\partial}{\partial z}\gamma(E_z + vB_y) = 4\pi k\gamma\left(\rho - \frac{v}{c^2}J_x\right)$$

$$\gamma(\nabla \cdot E - 4\pi k\rho) - v\gamma\left(-\frac{\partial E_x}{c^2 \partial t} - \frac{\partial B_z}{\partial y} + \frac{\partial B_y}{\partial z} - \frac{4\pi k}{c^2}J_x\right) = 0 \qquad (9.19)$$

The overall factor of γ in Eq. (9.19) can be divided out. Now we have two expressions in the two parentheses, one is multiplied by the arbitrary velocity v of the boost and the other is not. How can their sum be zero? *It must be that each term is individually zero*! The first expression in parentheses in Eq. (9.18) is just Gauss' law in frame S. Since this expression must vanish, we learn that Gauss' law must hold in all frames if it holds in one. This leaves the second parentheses, which also must vanish. This is clearly one component of a vector equation. If we did the boost in the y direction, we would get the second component and if we did the boost in the z direction, we would get the third component. Identifying a "curl" of the magnetic field, $\nabla \times \boldsymbol{B}$, we have,

$$\nabla \times \boldsymbol{B} = \frac{4\pi k}{c^2}\boldsymbol{J} + \frac{1}{c^2}\frac{\partial \boldsymbol{E}}{\partial t} \tag{9.20}$$

which we recognize as the Ampere–Maxwell equation with the "displacement" current, $\frac{1}{c^2}\frac{\partial \boldsymbol{E}}{\partial t}$, predicted from first principles! This equation makes precise and in full generality that moving charges, currents, create magnetic fields, and time-varying electric fields.

This equation was first discovered in lab experiments, which involved quasistationary currents and so the last term in the equation, the displacement current, was missed. When Maxwell put the equations of electric and magnetic fields into one system, he found that they were inconsistent with current conservation. He invented the displacement current to patch up that flaw. Experiments with strongly time-dependent currents, executed by H. Hertz two decades after Maxwell invented his equations, validated the discovery. Those experiments discovered electromagnetic radiation. We will see below that the displacement current is an essential ingredient in the propagation of electromagnetic fields. In our approach, current conservation is incorporated from the beginning so we evaded history's error! (Ah, hindsight!) To see the necessity of the displacement current, take the divergence of Eq. (9.20),

$$\nabla \cdot (\nabla \times \boldsymbol{B}) = \frac{4\pi k}{c^2}\nabla \cdot \boldsymbol{J} + \frac{1}{c^2}\frac{\partial \nabla \cdot \boldsymbol{E}}{\partial t} \tag{9.21}$$

Now use the identity "the divergence of a curl of any vector field is zero" to see that the left-hand side of this equation vanishes identically. (The proof of this identity can be done by simply writing out the expression in Cartesian coordinates, as reviewed in Appendix D, and noting that all the terms cancel.) Now, on the right-hand side replace the divergence of the electric field with $4\pi k\rho$ using Gauss' law. Then Eq. (9.21) becomes our equation for current conservation, Eq. (9.14), and we are done.

Before we discuss the physical content of the Ampere–Maxwell equation, let us derive the remaining Maxwell equation. Our discussion here started with Gauss' law, a constraint equation for the electric field. We have a similar equation for the magnetic field. The absence of magnetic charges implies that,

$$\nabla \cdot \boldsymbol{B} = 0 \tag{9.22}$$

In analogy to Gauss' law for the electric field, this equation means that there are no magnetic monopoles, the analogue of electric charge. We know from experience in the lab that magnetic fields are generated by electric currents. We also know that permanent magnets have individual North and South poles. We do not find localized magnetic poles ("monopoles") in our world, so the right-hand side of Eq. (9.22) is zero. Eq. (9.22) is usually taken for granted in the world of physics, so it does not even have a special name! If magnetic monopoles are eventually discovered, there will be regions of space where the right-hand side of Eq. (9.22) is nonzero. But as of today, there are no magnetic monopoles so Eq. (9.22) stands and the only source of magnetic fields in classical physics are electric currents.

Eq. (9.22) is a "constraint" equation similar to Gauss' law for electrostatics. Let us see what we learn by viewing it from a different inertial frame. We start with Eq. (9.22) in frame S',

$$\nabla' \cdot \boldsymbol{B}\prime = 0 \tag{9.23}$$

Now substitute the variables in frame S' with variables in frame S using Eq. (9.17) and Section 8.3,

$$\gamma\left(\frac{\partial}{\partial x} + \frac{v}{c^2}\frac{\partial}{\partial t}\right)B_x + \frac{\partial}{\partial y}\gamma\left(B_y + \frac{v}{c^2}E_z\right) + \frac{\partial}{\partial z}\gamma\left(B_z - \frac{v}{c^2}E_y\right) = 0$$

$$\gamma\left(\frac{\partial}{\partial x}B_x + \frac{\partial}{\partial y}B_y + \frac{\partial}{\partial z}B_z\right) - v\gamma\left(-\frac{1}{c^2}\frac{\partial B_x}{\partial t} - \frac{1}{c^2}\frac{\partial E_z}{\partial y} + \frac{1}{c^2}\frac{\partial E_y}{\partial z}\right) = 0$$

$$\gamma(\nabla \cdot \boldsymbol{B}) - v\gamma\left(-\frac{1}{c^2}\frac{\partial B_x}{\partial t} - \frac{1}{c^2}(\nabla \times \boldsymbol{E})_x\right) = 0$$

By the same argument we applied to understand and extract the content from Eq. (9.19), we learn that $\nabla \cdot \boldsymbol{B} = 0$, and that,

$$\nabla \times \boldsymbol{E} = -\partial \boldsymbol{B}/\partial t \tag{9.24}$$

where, as in our derivation of the Ampere–Maxwell equation, we have anticipated that boosts in the y and z directions will provide the full vector character of the result, which we recognize as Faraday's law! Recall that Faraday's law was discovered in lab experiments by observing that time-varying magnetic fields induced "electromotive" forces in closed loops. This phenomenon was the basis of myriad inventions in electromagnetic motors, generators, and other devices.

To obtain Maxwell's equations in their full generality, we must accommodate general currents and charge densities. We already argued in the case of Gauss' law of electrostatics that we can obtain the law for a general charge density ρ from the special case of a point charge by using the principle of linear superposition. Now we need to do the same for the current density $\boldsymbol{J}$ in the Ampere–Maxwell law. In Eq. (9.6), we could consider infinitesimal pointlike charge densities and generate infinitesimal pointlike currents in a particular direction by performing the appropriate boost. But then we could consider boosts of any velocity in any direction in our x-y-z space and by linearly superposing the resulting Ampere–Maxwell equations, we could obtain the general case. For example, we could even generate the case of a closed loop of electric current and could predict the magnetic field everywhere in that environment. The only constraint that our resulting distribution of current density $\boldsymbol{J}$ and charge density ρ satisfy is that of current conservation, $\nabla \cdot \boldsymbol{J} + \frac{\partial \rho}{\partial t} = 0$, which is built into the construction as we have already discussed.

Now we have all of Maxwell's equations,

$$\nabla \cdot \boldsymbol{E} = 4\pi k\rho \quad \text{(Gauss' Law)}$$

$$\nabla \times \boldsymbol{B} = \frac{4\pi k}{c^2}\boldsymbol{J} + \frac{1}{c^2}\frac{\partial \boldsymbol{E}}{\partial t} \quad \text{(Ampere − Maxwell Law)}$$

$$\nabla \cdot \boldsymbol{B} = 0 \quad \text{(No Magnetic Monoples)}$$

$$\nabla \times \boldsymbol{E} = -\partial \boldsymbol{B}/\partial t \quad \text{(Faraday's Law)}$$

We should reflect on what has been accomplished here. We started with Coulomb's law of electrostatics, wrote it as a local differential equation, Gauss' law for a point particle, and then generalized it to an arbitrary charge distribution using the principle of linear superposition. Then we boosted Gauss' law to a moving frame where we calculated the transformed charge density and current density. Since we had inferred how electric and magnetic fields, coordinates, and time and their derivatives transformed under boosts, we could compute all the pieces of the equation from first principles. We did the same for Gauss' law for the magnetic field. The result was Maxwell's equations. Finally, since the equations are linear we could invoke the principle of linear superposition again and infer that they applied to a general distribution of charges and currents that satisfy current conservation.

We have learned that for Gauss' law, $\nabla \cdot \boldsymbol{E} = 4\pi k\rho$, to make sense relativistically, it *must* be accompanied by three other equations, the Maxwell–Ampere equations, $\nabla \times \boldsymbol{B} = \frac{4\pi k}{c^2}\boldsymbol{J} + \frac{1}{c^2}\frac{\partial \boldsymbol{E}}{\partial t}$. These *four* equations are in fact intertwined inexorably by relativity—you cannot have one without the others because they mix under boosts. The physical reality is the group of four of them. Similarly, for $\nabla \cdot \boldsymbol{B} = 0$ and Faraday's law.

The right-hand side of Gauss' law is the charge density, which is the zeroth component of the four-vector J^μ, the electric current. The Maxwell–Ampere law has the currents J^i, $i = 1, 2, 3$, on their right-hand sides. The fact that all four components of J^μ mix under boosts is the reason why Gauss' law becomes four equations (a quartet) in relativity. This suggests that our work is not done until we learn to write the left-hand side of Gauss' law and Maxwell–Ampere's law in terms of four-vectors and tensors. Only then will we see the true physical law residing there. We shall accomplish this goal in the next chapter. We shall also make similar progress on $\nabla \cdot \boldsymbol{B} = 0$ and Faraday's law.

Now we can apply Maxwell's equations to all the electromagnetic devices invented over the years. When we choose convenient frames of reference to solve particular problems, we can use our knowledge of relativity to provide the needed transformations.

Let us put Ampere's and Faraday's laws into a more physical and familiar context. Consider a closed loop l, which is the boundary of a surface S. Denote an infinitesimal surface area on S as $\boldsymbol{da}$, and take the surface integral of Ampere–Maxwell's law,

$$\oiint \nabla \times \boldsymbol{B} \cdot \boldsymbol{da} = \frac{4\pi k}{c^2} \oiint \boldsymbol{J} \cdot \boldsymbol{da} + \frac{1}{c^2} \oiint \frac{\partial \boldsymbol{E}}{\partial t} \cdot \boldsymbol{da} \tag{9.25}$$

We need to make two observations here to write this result in standard form. First, there is Stoke's theorem from vector calculus (Appendix D): Consider a vector field $\boldsymbol{V}$. The flux of the curl of the vector field $\boldsymbol{V}$ through a surface S equals the line integral of $\boldsymbol{V}$ around the boundary l of the surface,

$$\oiint \nabla \times \boldsymbol{V} \cdot \boldsymbol{da} = \oint \boldsymbol{V} \cdot \boldsymbol{dl} \tag{9.26}$$

We apply this famous theorem to the left-hand side of Eq. (9.25), which becomes $\oint \boldsymbol{B} \cdot \boldsymbol{dl}$, the line integral of the magnetic field around the closed loop l that forms the boundary of the surface S. Our second observation is that on the right-hand side of Eq. (9.25) we have the quantity $\oiint \boldsymbol{J} \cdot \boldsymbol{da}$ which is the amount of charge that passes through the surface S per second. It is traditional to call this quantity I. If the electric current were confined to a wire at rest in the lab, then I is the amount of charge that passes a point on the wire per second. Now Eq. (9.26) becomes,

$$\oint \boldsymbol{B}\cdot d\boldsymbol{l} = \frac{4\pi k}{c^2} I + \frac{1}{c^2} \oiint \frac{\partial \boldsymbol{E}}{\partial t}\cdot d\boldsymbol{a} \tag{9.27}$$

This expression reduces to Ampere's law for stationary currents because the displacement current vanishes in that case. In problems with simple geometries, Ampere's law is a convenient way to calculate the magnetic field generated by time-independent currents. Some applications are included in the problems that follow this chapter, but traditional textbooks on electromagnetism have a wealth of challenging examples.

Finally, let us apply Stoke's theorem to Faraday's law,

$$\oiint \nabla \times \boldsymbol{E}\cdot d\boldsymbol{a} = -\oiint \frac{\partial \boldsymbol{B}}{\partial t}\cdot d\boldsymbol{a}$$

$$\oint \boldsymbol{E}\cdot d\boldsymbol{l} = -\frac{d}{dt} \oiint \boldsymbol{B}\cdot d\boldsymbol{a} \tag{9.28}$$

which states that the "time rate of change of the magnetic flux through a fixed surface induces an electromotive force $\oint \boldsymbol{E}\cdot d\boldsymbol{l}$ around its boundary." In traditional textbooks, this form of Faraday's law is applied to devices with simple geometries where both sides are easily evaluated and used to make quantitative predictions. We also include some examples in the problems at the end of the chapter. The minus sign in Eq. (9.28) is the source of "Lenz's law," also discussed in the problems.

9.2 THE WAVE EQUATION FOR LIGHT AND THE DERIVATION THAT LIGHT TRAVELS AT THE SPEED LIMIT

We have been assuming throughout our discussions that light provides an example of the speed limit c of special relativity. Now that we have Maxwell's equations it is a "simple" exercise to demonstrate this fact, which plays such an important role in physics. When Maxwell first derived the wave equation, physicists were not comfortable with the idea of waves propagating though empty space, and they held onto Newton's vision of space—time. It took the work of many physicists to find the path forward, and it was Einstein's synthesis and emphasis on fundamentals that brought us to the modern view that we present here.

Maxwell's equations state that electric and magnetic fields are generated by charges and currents. To make a wave of electric and magnetic fields, we must impart some energy into these fields. We can do this by accelerating some charges somewhere in space. It is analogous to shaking the end of a taut long rope and creating waves traveling down it. As the waves travel away from the shaking end, they take on dynamics of their own and travel with a particular amplitude and speed and carry energy and momentum with them. A wave traveling in a direction $\boldsymbol{k}$ at velocity $\boldsymbol{v}$ can be described by a function,

$$A(\boldsymbol{k}\cdot(\boldsymbol{r}-\boldsymbol{v}t)) \tag{9.29}$$

Choosing the "wave vector" $\boldsymbol{k}$ in the x direction, the argument of the function A becomes $k_x(x - v_x t)$. Note that if t increases by Δt, then x must increase by $v_x \Delta t$ to keep the argument (phase) of the function A unchanged. This observation establishes that $\boldsymbol{v}$ is the velocity of the wave. Since the argument of the function A should be dimensionless, the "wave vector" $\boldsymbol{k}$ has dimensions length^{-1}. In applications with periodic motion, the wave number $|\boldsymbol{k}|$ is $2\pi/\lambda$, where λ is the wavelength. More on this below.

We learn from these preliminaries that for a traveling electromagnetic wave, the electric and magnetic fields have the functional forms, $\boldsymbol{E}(\boldsymbol{k}\cdot(\boldsymbol{r}-\boldsymbol{v}t))$ and $\boldsymbol{B}(\boldsymbol{k}\cdot(\boldsymbol{r}-\boldsymbol{v}t))$. In regions of space far from their sources, they must satisfy Maxwell's equations without sources,

$$\begin{aligned}
\nabla\cdot\boldsymbol{E} &= 0\\
\nabla\cdot\boldsymbol{B} &= 0\\
\nabla\times\boldsymbol{B} &= \frac{1}{c^2}\frac{\partial \boldsymbol{E}}{\partial t}\\
\nabla\times\boldsymbol{E} &= -\frac{\partial \boldsymbol{B}}{\partial t}
\end{aligned}$$

Substituting the traveling wave functional form into the first equation, Gauss' law in empty space, gives, defining the phase $\alpha = \boldsymbol{k}\cdot(\boldsymbol{r}-\boldsymbol{v}t)$,

$$\nabla\cdot\boldsymbol{E}(\boldsymbol{k}\cdot(\boldsymbol{r}-\boldsymbol{v}t)) = \boldsymbol{k}\cdot\frac{\partial \boldsymbol{E}(\alpha)}{\partial\alpha} = \frac{\partial}{\partial\alpha}\boldsymbol{k}\cdot\boldsymbol{E}(\alpha) = 0$$

where we used the fact that $\boldsymbol{k}$ is a constant. We learn that $\boldsymbol{k}\cdot\boldsymbol{E} = 0$. This is called the "transversality condition." It means that the electric field is perpendicular to the propagation direction of the wave. If $\boldsymbol{k}$ points in the x direction and the wave is linearly polarized, then $\boldsymbol{E}$ could point in the

y direction. Let us analyze this case first where the phase simplifies to $\alpha = k_x(x - v_x t)$ and $\boldsymbol{E} = (0, E_y, 0)$.

Similarly $\nabla \cdot \boldsymbol{B} = 0$ implies that $\boldsymbol{k} \cdot \boldsymbol{B} = 0$, another transversality condition.

Now we need to determine the relative orientation of $\boldsymbol{B}$ and $\boldsymbol{E}$. Since $\boldsymbol{E}$ is in the y direction, $\nabla \times \boldsymbol{E} = \hat{\boldsymbol{z}} \frac{\partial E_y}{\partial x}$ where we used the fact that $\boldsymbol{E}$ depends only on x, not y or z. From Faraday's law, this implies that $\boldsymbol{B}$ lies in the z direction.

In summary, we see that both $\boldsymbol{E}$ and $\boldsymbol{B}$ are perpendicular to the direction of propagation of the wave, and they are perpendicular to each other with $\boldsymbol{E} \times \boldsymbol{B}$ pointing in the direction of propagation of the wave, $\boldsymbol{k}$. Note that these aspects of the traveling wave are true in every frame. This follows from the fact that Maxwell's equations hold in any inertial frame, or from the fact that light travels at the speed limit and therefore its state must be Lorentz invariant, as we shall now discuss in more detail.

Let us find the wave equation that applies here. Ampere's and Faraday's laws reduce to,

$$\frac{\partial E_y}{\partial x} = -\frac{\partial B_z}{\partial t}, \quad \frac{\partial B_z}{\partial x} = -\frac{\partial E_y}{c^2 \partial t} \tag{9.30}$$

Therefore, the time dependence in the electric field feeds into the magnetic field (the displacement current in Maxwell's equation is essential here), and the time dependence of the magnetic field feeds into the electric field. This is analogous to the equations for coupled oscillators familiar from classical mechanics. We can decouple these equations by applying $\frac{\partial}{c^2 \partial t}$ to the first of these equations and $-\frac{\partial}{\partial x}$ to the second equation,

$$\frac{\partial^2 E_y}{c^2 \partial t \partial x} = -\frac{\partial^2 B_z}{c^2 \partial t^2}, \quad -\frac{\partial^2 B_z}{\partial x^2} = \frac{\partial^2 E_y}{c^2 \partial x \partial t}$$

Now add these two equations to produce a second-order differential equation for B_z,

$$\left(\frac{\partial^2}{\partial x^2} - \frac{\partial^2}{c^2 \partial t^2} \right) B_z = 0 \tag{9.31}$$

Analogous manipulations produce a decoupled differential equation for E_y,

$$\left(\frac{\partial^2}{\partial x^2} - \frac{\partial^2}{c^2 \partial t^2} \right) E_y = 0 \tag{9.32}$$

Each of these equations is solved with a traveling wave form, $\boldsymbol{E}(k_x(x - v_x t))$ and $\boldsymbol{B}(k_x(x - v_x t))$. Now substitute this wave form into Eq. (9.31) and use the chain rule to find,

$$\left(\frac{\partial^2}{\partial x^2} - \frac{\partial^2}{c^2 \partial t^2}\right) B_z(\boldsymbol{k} \cdot (\boldsymbol{r} - \boldsymbol{v}t)) = k_x^2 \left(1 - \frac{v^2}{c^2}\right) \frac{d^2 B_z}{d^2 \alpha} = 0$$

We learn that the traveling wave form is indeed a solution, but Maxwell's equations imply that *the wave must travel at the speed limit*, $v = c$, as promised in Chapter 1. A similar argument applies as well to E_y.

It is easy to generalize this discussion to wave propagation in a general direction $\boldsymbol{k}$. We would then find the wave equations,

$$\left(\frac{\partial^2}{\partial x^2} + \frac{\partial^2}{\partial y^2} + \frac{\partial^2}{\partial z^2} - \frac{\partial^2}{c^2 \partial t^2}\right) \boldsymbol{B} = 0, \quad \left(\frac{\partial^2}{\partial x^2} + \frac{\partial^2}{\partial y^2} + \frac{\partial^2}{\partial z^2} - \frac{\partial^2}{c^2 \partial t^2}\right) \boldsymbol{E} = 0 \tag{9.33}$$

The transversality conditions read $\boldsymbol{k} \cdot \boldsymbol{B} = 0$ and $\boldsymbol{k} \cdot \boldsymbol{E} = 0$. In addition, Faraday's and Ampere's laws imply that $\boldsymbol{B}$ and $\boldsymbol{E}$ are perpendicular, and the ratio of their magnitudes is,

$$\frac{E}{B} = c \tag{9.34}$$

The "wave operator" $\left(\frac{\partial^2}{\partial x^2} + \frac{\partial^2}{\partial y^2} + \frac{\partial^2}{\partial z^2} - \frac{\partial^2}{c^2 \partial t^2}\right) = \left(\nabla \cdot \nabla - \frac{\partial^2}{c^2 \partial t^2}\right)$, is very important in electrodynamics and field theory in general. It is an invariant operator, and this fact implies that electromagnetic waves travel at the speed limit c independently of the velocity of the wave's source. Let us check these points. The claim is that $\left(\nabla \cdot \nabla - \frac{\partial^2}{c^2 \partial t^2}\right)$ is invariant under boosts,

$$\left(\nabla \cdot \nabla - \frac{\partial^2}{c^2 \partial t^2}\right) = \left(\nabla' \cdot \nabla' - \frac{\partial^2}{c^2 \partial t'^2}\right) \tag{9.35}$$

To prove this, we apply the Lorentz transformation equations for the derivatives, Eq. (9.17), and carry through some algebra, which should be quite familiar by now. Or we can note that the wave operator is the length squared of a four-vector operator and therefore is guaranteed to be an invariant,

$$\partial \cdot \partial = \partial_0^2 - \partial_1^2 - \partial_2^2 - \partial_3^2 = \frac{\partial^2}{c^2 \partial t^2} - \nabla \cdot \nabla$$

$$\partial \cdot \partial = \partial' \cdot \partial' = \partial_0'^2 - \partial_1'^2 - \partial_2'^2 - \partial_3'^2 = \frac{\partial^2}{c^2 \partial t'^2} - \nabla' \cdot \nabla'$$

This invariance proves that if an electromagnetic wave travels at a speed c in one frame, then it does so in all inertial frames. This provides an example of Postulate 2 of special relativity. This completes an important logical loop in this development of the theory.

A familiar and useful wave form for traveling waves is a sinusoidal function such as $\sin(\boldsymbol{k} \cdot (\boldsymbol{r} - \boldsymbol{v}t))$ or $\cos(\boldsymbol{k} \cdot (\boldsymbol{r} - \boldsymbol{v}t))$. Substituting these expressions into Maxwell's equations without sources and into the wave equation, we verify that these are solutions as long as $|\boldsymbol{v}| = c$. The reader should verify that $k = |\boldsymbol{k}| = 2\pi/\lambda$ and $c/\lambda = \nu$ where ν is the frequency of the wave in cycles per second and λ is the wavelength of the regular periodic oscillation.

Since these waves satisfy a Lorentz invariant wave equation, they can be written in invariant form. All observers in different inertial frames must measure the same phase of a wave of light since the wave propagates at the speed limit with respect to all of them. This suggests that we can introduce a four-vector, which has the wave vector $\boldsymbol{k}$ as its spatial components,

$$k^\mu = (2\pi/\lambda, \boldsymbol{k})$$

Let us justify this construction. First, k^μ is lightlike, $k \cdot k = \left(\frac{2\pi}{\lambda}\right)^2 - \boldsymbol{k}^2 = 0$, and second, the inner product of k^μ and x^μ gives the phase of the wave, $\alpha = k \cdot x = \frac{2\pi}{\lambda} \cdot ct - \boldsymbol{k} \cdot \boldsymbol{x}$, which equals $\frac{2\pi}{\lambda}(ct - x)$ if $\boldsymbol{k}$ points along the x axis. Since $k^\mu = \partial^\mu \alpha$ and α is Lorentz invariant, k^μ must be a four-vector because ∂^μ is one. So, in the discussion above, one could write $A(k \cdot x)$ for the traveling electromagnetic wave and write $\sin(k \cdot x)$ or $\cos(k \cdot x)$ in the sinusoidal special cases.

In this discussion, we have emphasized linearly polarized waves for simplicity, but we can also make circularly polarized waves by taking linear superpositions. For example, to make a right-handed circularly polarized wave propagating in the z direction, we can take the sum of waves linearly polarized in the x and y directions,

$$(E_x, E_y, E_z) \sim (\cos k(ct - z), \sin k(ct - z), 0)$$

Maxwell's equations predict a magnetic field $(B_x, B_y, B_z) \sim \frac{1}{c}(-E_y, E_x, 0)$, which is perpendicular to the electric field with $\boldsymbol{E} \times \boldsymbol{B}$ pointing along the $+z$ axis. Note that $(E_x, E_y) \sim (\cos \mathrm{k}(ct - z), \sin \mathrm{k}(ct - z))$ describes a vector of unit length that at fixed $z = 0$ rotates around the origin in the x-y plane in a counterclockwise direction with a period $\Delta T = 2\pi/kc = 1/\nu$. This is a right-handed circularly polarized wave. To make a left-handed circularly polarized wave, replace E_y with $-E_y$ in the expression for the electric field. In the quantum description of radiation, right-handed (left-handed) circularly polarized waves describe a photon with its spin 1 aligned (antialigned) with the propagation direction of the wave. The spin of the photon will be discussed further in Chapter 11.

PROBLEMS

9-1. Let us do some introductory problems in Gauss' law in integral form, Eqs. (9.2)−(9.5).

- **a.** Find the electric field $\boldsymbol{E}$ outside a uniformly charged solid sphere of radius R and total charge Q.
- **b.** Repeat part a. for a very long cylinder, which has a charge density $\rho = \rho_0$. If the cylinder has a radius R, get the $\boldsymbol{E}$ field both inside and outside the cylinder.
- **c.** Repeat part b. for an infinite plane that carries a uniform surface charge/area $\sigma = \sigma_0$.

9-2. Let us do some introductory problems on Ampere's law in its integral form, Eq. (9.27).

- **a.** Find the magnetic field $\boldsymbol{B}$ outside an infinite uniform surface current, $\boldsymbol{S} = S\widehat{\boldsymbol{x}}$, flowing over the x-y plane as shown in Fig. 9.1. Here S has the units of charge/length.

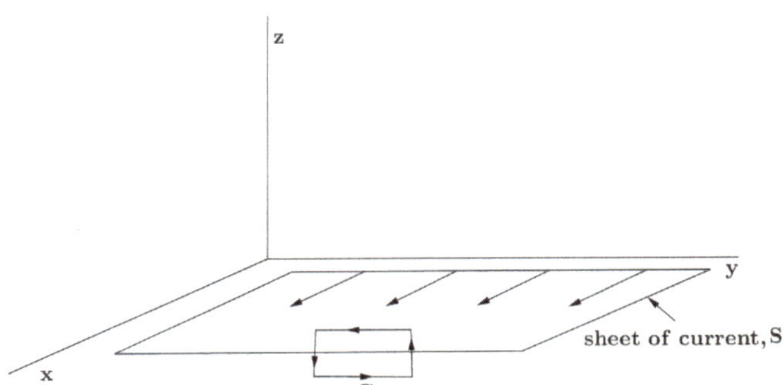

Figure 9.1 There is a surface current in the (x, y) plane and a loop in the (y, z) plane. Applying Ampere's law to the loop yields a formula for the magnetic field generated by the uniform, stationary surface current.

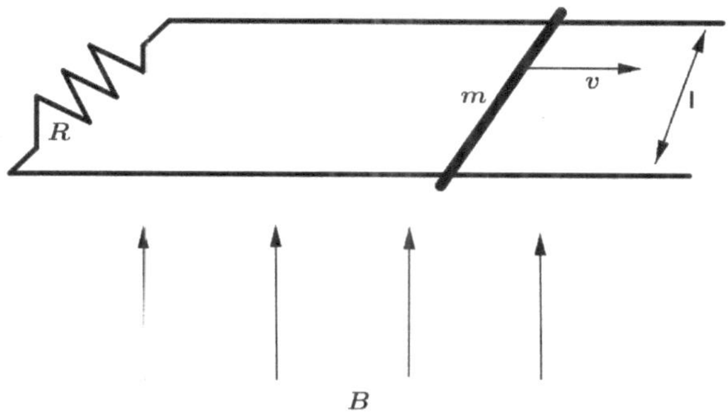

Figure 9.2 The bar travels at velocity v and Faraday's law implies that the variable magnetic flux through the circuit formed by the bar, the wires, and the resistor causes a current to flow.

b. Find the $\boldsymbol{B}$ field inside a long solenoid consisting of n closely wound turns per unit length on a cylinder of radius R and carrying a steady current I.

9-3. Let us analyze an experiment on Faraday's law that you may have done in a physics lab.

A metal bar of mass m slides without friction on two parallel conducting rails a distance l apart as shown in Fig. 9.2. A resistor R is connected between the rails and a uniform magnetic field $\boldsymbol{B}$ pointing upward fills the entire region.

a. If the bar moves at velocity v as shown, what is the current in the resistor? In what direction does it flow? (Recall Ohm's law, $V = RI$, where V is the voltage drop across the resistor, $V = \int \boldsymbol{E} \cdot d\boldsymbol{l}$, and the line integral is taken from the beginning to the end of the resistor. R is the resistance.)

b. What is the magnetic force on the bar? What is its direction?

c. Suppose the initial velocity of the bar was v_0. Its initial kinetic energy was $\frac{1}{2}mv_0^2$ ($v_0^2 \ll c^2$, so Newtonian kinematics work very accurately.). Show that all this energy shows up in the resistor by the time the bar comes to rest.

9-4. Faraday's law plays a central role in many alternating current (AC) devices. Here is a generator.

A square loop of side a is mounted on a vertical shaft and rotated at an angular velocity ω. A uniform magnetic field $\boldsymbol{B}$ points to the right in Fig. 9.3. Find a formula for the electromotive force in the wire.

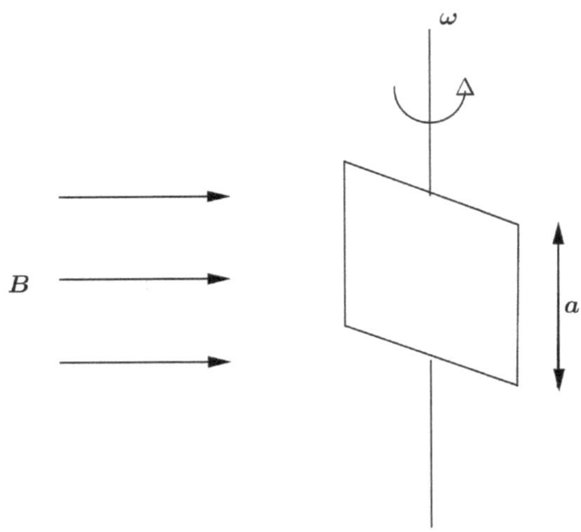

Figure 9.3 The rotating square loop experiences a variable magnetic flux that causes an electromotive force to develop in the loop in accord with Faraday's law.

9-5. Here is a puzzling lab demonstration illustrative of Faraday's law.

A line of charge per unit length λ is glued onto the rim of a wheel of radius b, which is suspended horizontally so that it is free to rotate. In the central region, out to radius a, there is a uniform magnetic field $\boldsymbol{B}$ pointing up. Now the field is turned off. What happens?

Student A says that nothing happens because the magnetic field does not touch the rim.

Student B says that the wheel rotates and she calculates the angular frequency ω of the rotation using Faraday's law.

a. Which student is right? Explain.

b. If you voted for student B, calculate ω assuming all the mass of the wheel is on its rim. Show that ω does not depend on the rate at which the magnetic field was turned off.

Now for a puzzle. If student B were right, then we have generated a final state with nonzero angular momentum. But conservation of angular momentum is a sacred principle of physics. Where might the angular momentum have been in the initial setup of the demonstration? (We will see later in the text that electromagnetic fields carry energy, momentum, and angular momentum. This is a challenging subject!)

9-6. In the text, we saw many examples where electromagnetic fields carry energy and momentum. Let us see how energy flow and energy conservation follows directly from Maxwell's equations.

We begin with Faraday's law and Ampere–Maxwell's law,

$$\nabla \times \boldsymbol{B} = \frac{1}{c^2}\frac{\partial \boldsymbol{E}}{\partial t} + \frac{4\pi k}{c^2}\boldsymbol{J}$$

$$\nabla \times \boldsymbol{E} = -\frac{\partial \boldsymbol{B}}{\partial t}$$

a. Take the dot product of the first equation with $\boldsymbol{E}$, and take the dot product of the second equation with $\boldsymbol{B}$, add them and simplify and verify,

$$\frac{1}{2}\frac{\partial}{\partial t}\left(\frac{1}{c^2}\boldsymbol{E}^2 + \boldsymbol{B}^2\right) = -\frac{4\pi k}{c^2}\boldsymbol{E}\cdot\boldsymbol{J} - \nabla\cdot(\boldsymbol{E}\times\boldsymbol{B})$$

(Hint: Use the vector identity $\nabla\cdot(\boldsymbol{F}\times\boldsymbol{G}) = \boldsymbol{G}\cdot(\nabla\times\boldsymbol{F}) - \boldsymbol{F}\cdot(\nabla\times\boldsymbol{G})$, which holds for any $\boldsymbol{F}$ and $\boldsymbol{G}$.)

b. If the current is generated by an ensemble of point particles show that,

$$\iiint \boldsymbol{J}\cdot\boldsymbol{E}d^3r = \sum e\boldsymbol{v}\cdot\boldsymbol{E} = \frac{\partial}{\partial t}E_{kin}$$

where E_{kin} is the sum of the relativistic energies of the ensemble of particles. (Hint: The last equality was illustrated in the text for a single particle. It is a consequence of the Lorentz force law and the fact that electric fields do work on charged particles.)

c. Integrate the result of part a., integrate it over a spatial volume, use part b. and Gauss' theorem to derive,

$$\frac{\partial}{\partial t}\left(\oiiint\left(\frac{\boldsymbol{E}^2 + c^2\boldsymbol{B}^2}{8\pi k}\right)d^3r + E_{kin}\right) = -\oiint\left(\frac{c^2}{4\pi k}\boldsymbol{E}\times\boldsymbol{B}\right)\cdot d\boldsymbol{a}$$

where E_{kin} includes only those particles inside the volume.

This equation is an expression of energy conservation: the rate of change of the total energy inside a volume is accounted for by the energy per unit time that flows into the volume through its boundary surface. Further work on this topic shows that $W = \frac{\boldsymbol{E}^2+c^2\boldsymbol{B}^2}{8\pi k}$ is the energy density stored in the electromagnetic fields, and $\boldsymbol{S} = \frac{c^2}{4\pi k}\boldsymbol{E}\times\boldsymbol{B}$ is the energy flux carried by the fields (Poynting vector, famous in the subject).

The general case of momentum involves more mathematical technology, and you will learn about it in more advanced courses. Conservation laws that supplant Newton's third law follow.

9-7. Let us illustrate the concepts of energy density and energy flux in some simple problems. The goal of this problem is to motivate some of the work and comments in Problem **9-6**. Refer to that problem for various definitions.

Consider a capacitor with large isolated square plates $L \times L$, which are at distance d apart. Suppose that $d \ll L$. The capacitor is initially uncharged. Charge it up slowly by taking positive charge from one plate directly to the other until an electric field E_0 exists between the plates.

a. Show that the energy required is

$$E_{tot} = \left(\frac{1}{8\pi k}E_0^2\right)L^2 d$$

Next, consider a very long solenoid of n closely wound turns of wire per length on a cylinder of radius R and length L. Use Faraday's law to calculate the energy required to slowly bring the magnetic field in the solenoid to a value B_0.

b. Show that the energy required is

$$E_{tot} = \left(\frac{c^2}{8\pi k}B_0^2\right)(\pi R^2 L)$$

Consider current flowing down a wire of length L and radius R. There is a potential difference of V between the ends of the wire so the electric field along it is $E = \frac{V}{L}$. A current I flows down the wire. The current produces a magnetic field around the wire.

c. Calculate the magnetic field at the surface of the wire from Ampere's law,

$$B = \frac{I}{2\pi R}$$

d. Calculate the Poynting vector on the surface of the wire, and verify that its magnitude is,

$$S = \frac{V}{L}\frac{I}{2\pi R} = \frac{VI}{2\pi RL}$$

Verify that it points radially inward.

Note that the power developed by the circuit is VI. (A charge Q is being transported through a potential difference of V, so the energy required is VQ and the power is the time derivative of this, $V\frac{d}{dt}Q = VI$.) So the Poynting vector is indeed the power per area.

CHAPTER 10

Magnetism in the Lab, the Discovery of Relativity, and the Way Forward

Contents

10.1 HOW CAN MAGNETIC FORCES BE LARGE IF THEY ARE RELATIVISTIC EFFECTS?

Let us consider the puzzling fact that magnetic forces are large in the lab although magnetic effects are relativistic and are suppressed relative to electric effects by powers of $c \approx 3.0 \times 10^8$ m/s. This experimental fact is a counterexample to the widely held belief that relativistic effects are only sizeable for particles traveling near the speed limit c. In this example, we shall see that the electrons that move through a typical conducting wire travel at speeds of a fraction of a millimeter per second and are the sources of the powerful magnetic fields in typical electromagnets.

To begin, consider an electron moving parallel to an uncharged conducting wire, which carries a current I as shown in Fig. 10.1. Label the

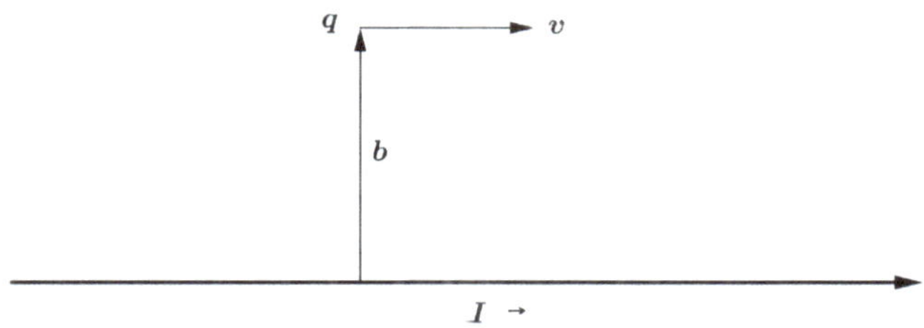

Figure 10.1 A charge q moves parallel to an uncharged conducting wire, which carries a current I. The charge's velocity v matches the velocities of the electrons in the wire.

Special Relativity, Electrodynamics, and General Relativity
ISBN 978-0-12-813720-8
https://doi.org/10.1016/B978-0-12-813720-8.00010-6

charge of the electron q, its velocity v, and its transverse distance from the wire b. Choose v to match the velocities of the electrons in the wire. To calculate the force on the electron, we could calculate the magnetic field generated by the current I using Ampere's law in the lab frame. Then the Lorentz force law would give us the force on the electron. But it will teach us more to do the calculation in the rest frame of the electron.

Before doing any quantitative calculations, let us argue that there is an attractive force between the electron and the wire in the electron's rest frame. In Fig. 10.2 we show the single electron q moving parallel to the wire with velocity v and distance b from it. Inside the wire we show the heavy Cu^+ ions that are at rest in the lab frame, and we show the electrons moving with velocity v in the wire. We have noted carefully that in the lab frame the distance between neighboring *stationary* Cu^+ ions is identical to the distance between the *moving* electrons, l_0. This equality must be true because we can measure to very high precision that the wire is locally electrically neutral in the lab frame. In our model, think of the electrons as moving through the wire in a "musical chairs" fashion, canceling the positive charges of the Cu^+ ions as they go.

We display these ideas in Fig. 10.3, a Minkowski diagram of the world lines of the electrons and the Cu^+ ions. The world lines of the Cu^+ ions are vertical because they are at rest in the lab frame S, and the world lines of the electrons tilt to the right with $\tan\theta = v/c$. The distances between neighboring Cu^+ ions is l_0 and the distance between neighboring electrons is also, l_0. This is shown in the figure and is true on any line of given t, any horizontal cut through the figure.

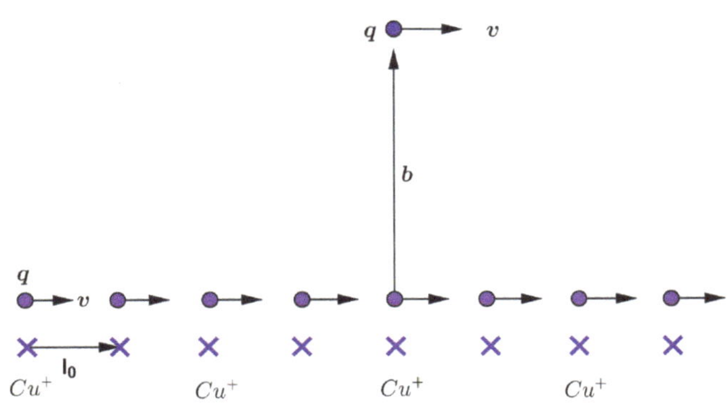

Figure 10.2 Same as Fig. 10.1 but showing the moving electrons and the stationary copper ions in the wire in the lab frame.

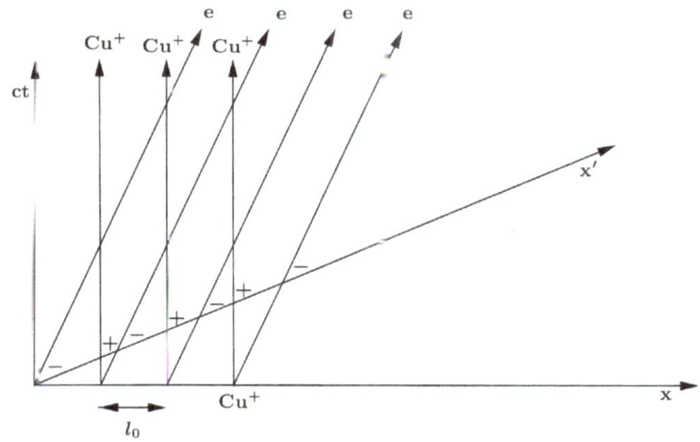

Figure 10.3 Minkowski diagram of Fig. 10.2. The x' axis is the line of constant $ct' = 0$ in the electron's rest frame S'.

Now for the interesting part. Let us check that the electron q at distance b from the wire measures a positive charge per length on the wire. In other words, there is an electric field present in the frame of the electron and since it corresponds to a positive charge per unit length on the wire, it produces an attraction. When we do a short calculation of the Lorentz contraction, we shall see that this electric field gives the correct attractive force.

To understand this we take the Minkowski diagram we have just drawn and put the x' axis on it, as shown in Fig. 10.3. The x' axis is the line of constant $ct' = 0$ in the electron's rest frame S'. Note that the world lines of any of the electrons are parallel to the ct' axis, and the x' axis tilts up, above the x axis, by $\tan\theta = v/c$. In this frame the distance between neighboring Cu^+ ions is l_+, and the distance between neighboring electrons in the wire is labeled l_-. These distances are measured at constant ct', along any line parallel to the x' axis. Because of the tilt of the x' axis, we immediately see that $l_- > l_-$. But this means that the negative charge per unit length due to the electrons is smaller than that due to the Cu^+ ions. Therefore, the electron at distance b from the wire is attracted to it by an electric force in its rest frame.

Now let us solve the problem quantitatively. In Fig. 10.4, we show the electron's rest frame S', and we use the same labeling as in the Minkowski diagram. To begin, we need to compare the two figures showing the frames S and S', so we can relate the distances l_0, l_+, and l_-. Since the electrons in the wire are at rest in S', l_- is the proper distance between them. Since they

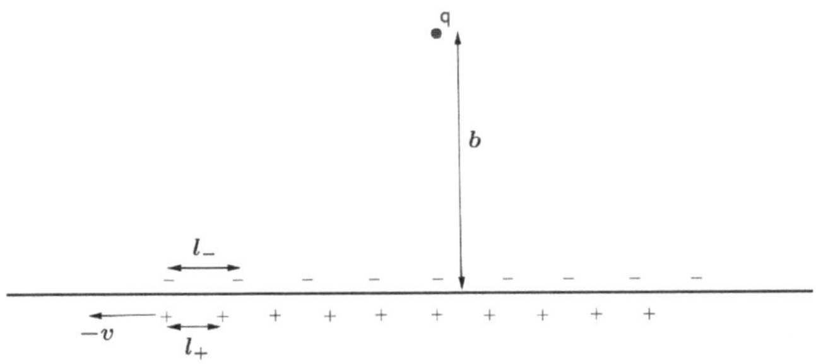

Figure 10.4 View of the wire from the perspective of the electron's rest frame S'.

are moving with velocity v in the lab frame S where the distance between them is called l_0, the lengths are related by Lorentz contraction,

$$l_0 = l_-/\gamma \tag{10.1}$$

If we define λ_0 to be the charge per unit length of the Cu^+ ions in the lab, then $-\lambda_0$ is the charge per unit length of the electrons in the lab since the wire is neutral there. Since the distance between the electrons is larger in S' by a factor of γ, $l_- = \gamma l_0$, the charge per unit length of the electrons in S', λ_-, is reduced accordingly,

$$\lambda_- = -\lambda_0/\gamma \tag{10.2}$$

Similarly, since the Cu^+ ions have velocity $-v$ in S', the distance between them in S' contracts,

$$l_+ = l_0/\gamma \tag{10.3}$$

and their charge density increases,

$$\lambda_+ = \gamma\lambda_0 \tag{10.4}$$

Collecting these results, the net charge density of the wire in the frame S', is

$$\lambda' = \lambda_+ + \lambda_- = \lambda_0\left(\gamma - \frac{1}{\gamma}\right) \tag{10.5}$$

This charge density generates an electric field in S', which we calculate using Gauss' law,

$$\frac{1}{4\pi k}\oint \boldsymbol{E}'\cdot d\boldsymbol{a}' = Q \tag{10.6}$$

In this application, we choose the closed surface to be a cylinder concentric with the wire and having radius b and height h' as shown in Fig. 10.5. Since

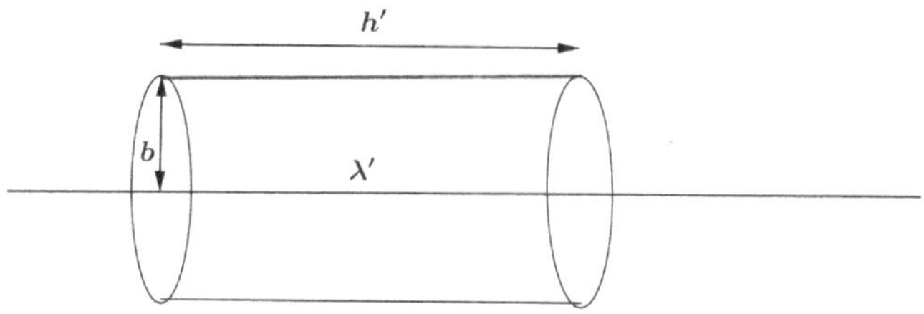

Figure 10.5 Gaussian cylinder used to calculate the electric field E' in the frame S' of the moving electrons.

the symmetry of the figure indicates that the electric field $\boldsymbol{E}'$ points radially outward, we can evaluate both sides of Eq. (10.6) and learn that,

$$\frac{1}{4\pi k}2\pi b h' E' = \lambda' h'$$

So,

$$E' = 2k\frac{\lambda'}{b} \tag{10.7}$$

Now we can write down the force on the electron of charge q in its rest frame S',

$$F'_y = 2k\frac{\lambda'}{b}q = \frac{2k\lambda_0}{b}\left(\gamma - \frac{1}{\gamma}\right)q = \gamma\frac{2k\lambda_0}{b}\left(1 - \frac{1}{\gamma^2}\right)q = \gamma\frac{v^2}{c^2}\frac{2k\lambda_0}{b}q \tag{10.8}$$

where we observed that the force is transverse to the wire, and we called this direction y.

Our last task consists of relating F'_y to F_y, the force that the electron experiences in the lab. Since $F'_y = dp'_y/dt'$, we need only recall the Lorentz transformations to complete this task. First, p_y is invariant under boosts in the x direction, so $dp'_y = dp_y$. Next, dt' is the proper time of the electron because it is at rest in S'. Therefore, $dt = \gamma dt'$. We conclude that,

$$F'_y = \frac{dp'_y}{dt'} = \frac{dp_y}{dt/\gamma} = \gamma\frac{dp_y}{dt} = \gamma F_y \tag{10.9}$$

Now we have our final result. The force that the moving electron feels in the lab frame is,

$$F_y = \frac{v^2}{c^2}\frac{2k\lambda_0}{b}q \tag{10.10}$$

Since q is negative for an electron, the force is attractive, as observed above. In addition, it appears to be tiny because it is of order $\frac{v^2}{c^2}$. However, as a lecture demonstration with ordinary parallel wires shows, the force is appreciable for typical conditions. Unscreened electric forces such as $2k\frac{\lambda_0}{b}$ are gigantic and are typically not seen in applications because bulk matter is usually neutral. (If you studied Problem 6.9, you learned this earlier.) We will estimate the forces between parallel wires in more detail later in this section.

Let us check that this approach to our problem gets the conventional answer.

First, in the conventional calculation we obtain the magnetic field in the lab from Ampere's law, as illustrated in Fig. 10.6. Ampere's law states that the line integral of the magnetic field around a closed loop C is proportional to the current that pierces the surface spanned by C,

$$\oint \boldsymbol{B}\cdot d\boldsymbol{l} = \frac{4\pi k}{c^2} I \tag{10.11}$$

One can argue on the basis of the symmetry of the straight wire in the figure that the magnetic field circulates round the wire. Choosing the closed loop in Ampere's law to be a circle concentric with the wire and of radius b, we see that $\boldsymbol{B}$ is parallel to $d\boldsymbol{l}$ and the integral becomes $2\pi bB$, so

$$B = \frac{2kI}{c^2 b} \tag{10.12}$$

where the current I, the amount of charge passing a fixed point in the wire per second, is $-\lambda_0 v$.

Therefore, the electron at distance b from the wire feels a magnetic field,

$$B = -\frac{v}{c^2}\frac{2k\lambda_0}{b} \tag{10.13}$$

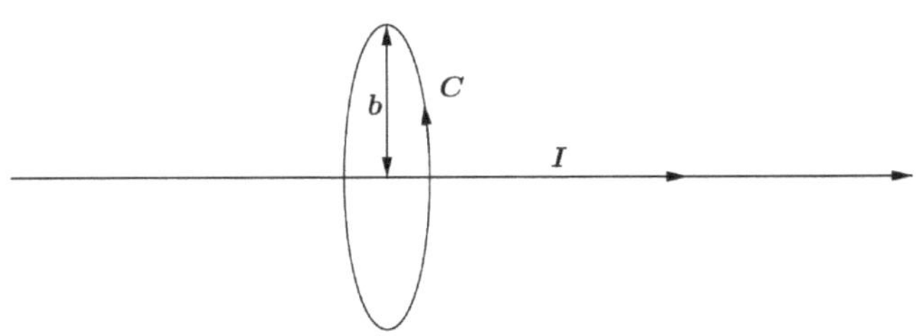

Figure 10.6 An imaginery loop C around the wire used to apply Ampere's law for the calculation of the magnetic field generated by the current I in the lab frame.

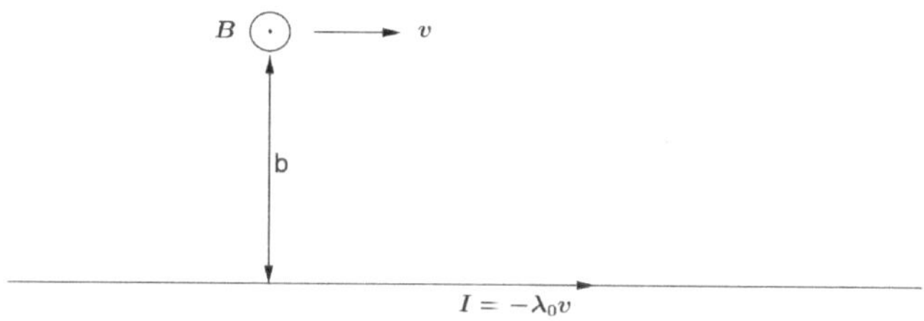

Figure 10.7 The magnetic field generated by the current in the wire at the location of the moving electron at a transverse distance b from the wire.

where the minus sign indicates that B comes out of the paper as shown in Fig. 10.7. The Lorentz force law, $\boldsymbol{F} = q\boldsymbol{v} \times \boldsymbol{B}$, produces an attractive force between the negatively charged electron and the wire of magnitude,

$$vBq = \frac{v^2}{c^2}\frac{2k\lambda_0}{b}q \tag{10.14}$$

which agrees perfectly with Eq. (10.10).

Now let us consider two parallel wires at distance b apart each carrying a current I in the same direction. There is the wire of the previous discussion and the second wire at the distance b from the first wire where the electron had been. So, the magnetic field due to the first wire at the position of the second wire is given by Eq. (10.12) and is perpendicular to the plane containing the wires. We can easily calculate the force that this field produces on all the electrons in the length l of the second wire. The amount of charge in the length l of wire is $\lambda_0 l$ and its velocity is v, so the Lorentz force law predicts the force on this length of wire to be,

$$F = v(\lambda_0 l)B = v(\lambda_0 l)\frac{2kI}{c^2 b} = \frac{2k}{c^2}\frac{I^2}{b}l \tag{10.15}$$

We can substitute typical parameters into this formula to compare with lab demonstrations of magnetic forces. First, the constant k, which establishes the scale of forces in Coulomb's law is roughly 8.99×10^9 Nm^2/C^2. Let the wires be 1 cm apart and choose currents of 10 A. Then the arithmetic yields $F = 2 \cdot 10^{-3}$newtons per meter, which is a sufficient force to bend typical copper wires as they attract one another. (Recall that 1 pound-force = 4.45 N) In fact, if you have ever constructed electromagnets from typical wire coils, you know how effective they are at picking up nails, deflecting compass needles, etc.

So far we have not calculated the electron's velocity v in the wire. It is instructive to do this, following in the foot steps of A. P. French [1]. We will start with the experimental current $I = 10$ A and calculate v, the velocity of the electrons producing that current. 10 A means that 10 C pass any fixed point on the wire per second. Since each electron carries a charge of roughly 1.6×10^{-19} C, $10/1.6 \times 10^{-19} \approx 6 \times 10^{19}$ electrons pass any fixed point on the wire per second. Suppose the cross-sectional area of the wire is 1 mm$^2 = 10^{-2}$ cm^2. Then in 1 s all the electrons in a volume of size $v \times 10^{-2}$ cm^2 pass any fixed point on the wire. How many electrons are in this volume? We need the properties of Cu to calculate this. The molar mass of Cu is 63.54, so 63.54 g of Cu contains Avogadro's number (6.02×10^{23}) of Cu atoms. The density of Cu is 8.96 g/cm^3. So, 1 cm^3 contains $8.96/63.54 \approx 0.14$ M of Cu, which contains $0.14 \times 6.02 \times 10^{23} \approx 0.85 \times 10^{23}$ Cu atoms. Since one Cu atom contributes one mobile electron to the wire, this is also the number of mobile electrons in a cubic centimeter of wire. Combining these facts, the number of electrons that pass a fixed point on the wire per second is,

$$0.85 \times 10^{23}(v \cdot 10^{-2}) \approx 6 \cdot 10^{19} \tag{10.16}$$

Solving this equation for v we learn that the velocity of the electrons in the wire is really tiny,

$$v \approx 0.07 \ \text{cm/s} \tag{10.17}$$

We come to the astonishing conclusion that v is a tiny velocity on the scale of the speed limit,

$$\frac{v^2}{c^2} \approx (2.3 \times 10^{-12})^2 \approx 5.3 \times 10^{-24} \tag{10.18}$$

The reader should reflect on the fact that this number is responsible for the discovery of special relativity! The experiments of Ampere, Faraday, and others who discovered that magnetism was produced by currents were actually measuring a Lorentz contraction factor that deviated from unity by 2.65×10^{-24}.

Curious.

10.2 ELECTRODYNAMICS FORMULATED IN COVARIANT NOTATION

Now that we have derived the Lorentz force law and Maxwell's equations from electrostatics and special relativity, it is time to write these equations in a language that embodies the principles of relativity. Recall that when we found the transformation laws for the electric and magnetic fields, we left the subject in an unfinished state. Under boosts the electric and magnetic fields satisfy the rules,

$$E'_x = E_x, \quad E'_y = \gamma(E_y - vB_z), \quad E'_z = \gamma(E_z + vB_y)$$

$$B'_x = B_x, \quad B'_y = \gamma\left(B_y + \frac{v}{c^2}E_z\right), \quad B'_z = \gamma\left(B_z - \frac{v}{c^2}E_y\right) \tag{10.19}$$

So, all the components of the two three-vectors mix under boosts. This teaches us that one must think of the collection of two three-vectors as the "electromagnetic field." This is the quantity that has true physical existence. It is not E or B alone since their components mix. As you discovered in the problem sets, under various physical circumstances one can transform to a frame where "magnetic effects" become "electric effects" and vice versa. The problem we face here is how to express these results in a relativistic language that remains true in all inertial frames. We know how to proceed with the four-vector position, velocity, energy—momentum, etc. But what about the six components of these two three-vectors?

We need a relativistic object that can accommodate six three-vector components and can incorporate the transformation law Eq. (10.19). A second rank tensor $T^{\mu\nu}$ has $4 \times 4 = 16$ components, so this does not fit. The metric tensor $g^{\mu\nu}$ is symmetric, $g^{\mu\nu}=g^{\nu\mu}$, so it has 10 independent components. We need a quantity with six independent components. As we noted above, the six components are two three-vectors. So, if some second rank tensor is required, its zero—zero component, F^{00}, say, must vanish. To be relativistic, its $0-0$ component must then vanish in every frame to accommodate Eq. (10.19), $F'^{00} = F^{00} = 0$. This requirement puts restrictions on the tensor $F^{\mu\nu}$ and tells us how to construct it. Being a second rank tensor, it must transform under boosts as,

$$F'^{\mu\nu} = \sum_{\sigma\lambda} L^{\mu}{}_{\sigma} L^{\nu}{}_{\lambda} F^{\sigma\lambda} \tag{10.20}$$

where L^{μ}_{σ} are the coefficients of the Lorentz transformation,

$$x'^{\mu} = \sum_{\nu} L^{\mu}{}_{\nu} x^{\nu} \tag{10.21}$$

with $L^0{}_0 = \gamma,\ L^0{}_1 = -\frac{v}{c}\gamma,\ L^1{}_0 = -\frac{v}{c}\gamma,\ L^1{}_1 = \gamma,\ L^2{}_2 = 1,\ L^3{}_3 = 1$ and all other elements in the Lorentz transformation vanishing. If we write out the transformation for F^{00} we find,

$$\begin{aligned} F'^{00} &= \sum_{\sigma\lambda} L^0{}_{\sigma} L^0{}_{\lambda} F^{\sigma\lambda} \\ &= L^0{}_0 L^0{}_0 F^{00} + L^0{}_1 L^0{}_0 F^{10} + L^0{}_0 L^0{}_1 F^{01} + L^0{}_1 L^0{}_1 F^{11} \end{aligned}$$

For $F'^{00} = 0$ if $F^{00} = 0$, we see that F^{11} must vanish and $F^{10} = -F^{01}$. If we consider boosts in the y direction and in the z direction, we learn $F^{20} = -F^{02}$, $F^{30} = -F^{03}$ and $F^{22} = F^{33} = 0$. Progress!

We can dig deeper into $F^{\mu\nu}$ now that we know how the first row is related to the first column, $F^{0\mu} = -F^{\mu 0}$. Consider the transformation law for F^{02} and F^{20},

$$F'^{02} = \sum_{\sigma\lambda} L^0{}_\sigma L^2{}_\lambda F^{\sigma\lambda} = L^0{}_0 L^2{}_2 F^{02} + L^0{}_1 L^2{}_2 F^{12}$$

$$F'^{20} = \sum_{\sigma\lambda} L^2{}_\sigma L^0{}_\lambda F^{\sigma\lambda} = L^2{}_2 L^0{}_0 F^{20} + L^2{}_2 L^0{}_1 F^{21}$$

Since $F^{02} = -F^{20}$ we learn that $F^{12} = -F^{21}$ as well. More considerations of this type show that $F^{ij} = -F^{ji}$ for i and j ranging from 1 to 3.

Collecting everything, we learn that $F^{\mu\nu}$ must be *antisymmetric*, $F^{\mu\nu} = -F^{\nu\mu}$. This symmetry reduces the number of independent components of $F^{\mu\nu}$ to six, so the electric and magnetic fields fit perfectly! But we still must check that Eq. (10.20) reproduces Eq. (10.19) for an antisymmetric field tensor $F^{\mu\nu}$. Let us write out Eq. (10.20) using the L^μ_ν coefficients in Eq. (10.21). Some algebra yields,

$$F'^{01} = F^{01}, \quad F'^{02} = \gamma\left(F^{02} - \frac{v}{c}F^{12}\right), \quad F'^{03} = \gamma\left(F^{03} + \frac{v}{c}F^{31}\right)$$

$$F'^{23} = F^{23}, \quad F'^{31} = \gamma\left(F^{31} + \frac{v}{c}F^{03}\right), \quad F'^{12} = \gamma\left(F^{12} - \frac{v}{c}F^{02}\right) \tag{10.22}$$

We see a perfect match with either,

$$F^{0i} = E_i/c, \quad F^{12} = B_3, \quad F^{31} = B_2, \quad F^{23} = B_1$$

or,

$$\widetilde{F}^{0i} = B_i, \quad \widetilde{F}^{12} = -E_3/c, \quad \widetilde{F}^{31} = -E_2/c, \quad \widetilde{F}^{23} = -E_1/c$$

Note that $\widetilde{F}^{\mu\nu}$ is obtained from $F^{\mu\nu}$ using the substitution $E_i/c \rightarrow B_i$ and $B_i \rightarrow -E_i/c$. This is the symmetry, called "electromagnetic duality," we noticed before in the problem sets that interchanges Faraday's law with Ampere–Maxwell's law in regions of space–time where the current density vanishes, $J^\mu = 0$.

As matrices $F^{\mu\nu}$ and $\widetilde{F}^{\mu\nu}$ read,

$$F^{\mu\nu} = \begin{pmatrix} 0 & E_x/c & E_y/c & E_z/c \\ -E_x/c & 0 & B_z & -B_y \\ -E_y/c & -B_z & 0 & B_x \\ -E_z/c & B_y & -B_x & 0 \end{pmatrix} \tag{10.23a}$$

and,

$$\widetilde{F}^{\mu\nu} = \begin{pmatrix} 0 & B_x & B_y & B_z \\ -B_x & 0 & -E_z/c & E_y/c \\ -B_y & E_z/c & 0 & -E_x/c \\ -B_z & -E_y/c & E_x/c & 0 \end{pmatrix} \tag{10.23b}$$

When physicists talk about "the electromagnetic field" they are referring to $F^{\mu\nu}$. They are not referring to either $\boldsymbol{E}$ or $\boldsymbol{B}$. We have seen in the problem sets how electric effects become magnetic effects under boosts, and we have learned a lot about electricity and magnetism through these exercises. However, everything electromagnetic resides in $F^{\mu\nu}$. Its components transform under boosts, and some elements can vanish in certain frames, and other components made more prominent. $F^{\mu\nu}$ cannot be boosted to zero, because if $F^{\mu\nu} = 0$ in one frame it would be zero in *all* frames according to Eq. (10.20).

Our next task is to write the Lorentz force law and Maxwell's equations in terms of $F^{\mu\nu}$ and its dual. First, consider the two Maxwell equations with sources,

$$\nabla \cdot \boldsymbol{E} = 4\pi k\rho \tag{10.24a}$$

$$\nabla \times \boldsymbol{B} - \frac{1}{c^2}\frac{\partial \boldsymbol{E}}{\partial t} = \frac{4\pi k}{c^2}\boldsymbol{J} \tag{10.24b}$$

We have ∂^μ, $F^{\mu\nu}$, and J^μ to work with. The gradient ∂^μ should apply to $F^{\mu\nu}$ because Maxwell's equations have the gradients of field strengths on the left-hand side of Eq. (10.24). J^μ should appear on the right-hand side. Note that $\sum_\nu \partial_\nu F^{\mu\nu}$ transforms as a four-vector V^μ because the other index ν is fully contracted. One can check using the transformation laws for ∂^μ, and $F^{\mu\nu}$ that this is so (see the problem set). So, let us try,

$$\sum_\nu \partial_\nu F^{\mu\nu} = \frac{4\pi k}{c^2} J^\mu \tag{10.25}$$

Let us check the $\mu = 0$ component of this equation first,

$$\sum_\nu \partial_\nu F^{0\nu} = \partial_0 F^{00} + \partial_1 F^{01} + \partial_2 F^{02} + \partial_3 F^{03}$$

$$\sum_\nu \partial_\nu F^{0\nu} = \frac{1}{c}\left(\frac{\partial E_x}{\partial x} + \frac{\partial E_y}{\partial y} + \frac{\partial E_z}{\partial z}\right) = \frac{1}{c}(\nabla \cdot \boldsymbol{E}) = \frac{4\pi k}{c^2} J^0 = \frac{4\pi k}{c^2} c\rho$$

So, we have Gauss' law, Eq. (10.24a)! We leave it to the problem set to check that the $\sum_\nu \partial_\nu F^{i\nu}$ components of this equation produce the Maxwell–Ampere equation Eq. (10.24b).

This gives two of the Maxwell equations. The homogeneous equations are given by the dual of $F^{\mu\nu}$,

$$\sum_\nu \partial_\nu \widetilde{F}^{\mu\nu} = 0 \tag{10.26}$$

Writing out the $\mu = 0$ component gives,

$$\sum_\nu \partial_\nu \widetilde{F}^{0\nu} = \partial_0 \widetilde{F}^{00} + \partial_1 \widetilde{F}^{01} + \partial_2 \widetilde{F}^{02} + \partial_3 \widetilde{F}^{03}$$

$$\sum_\nu \partial_\nu \widetilde{F}^{0\nu} = \frac{\partial B_x}{\partial x} + \frac{\partial B_y}{\partial y} + \frac{\partial B_z}{\partial z} = \nabla \cdot \boldsymbol{B} = 0$$

which reproduces the constraint equation, the absence of magnetic monopoles. The equations $\sum_\nu \partial_\nu \widetilde{F}^{i\nu} = 0$ for $i = 1, 2, 3$ produce Faraday's equation, $\nabla \times \boldsymbol{E} = -\partial \boldsymbol{B}/\partial t$, as discussed in the problem set.

What have we achieved? Now the four Maxwell equations are expressed as two covariant equations. This makes explicit that they are true in all inertial frames. Expressing the equations in terms of four-vectors and tensors, objects which have a frame-independent physical significance, shows their compatibility with the two postulates of special relativity. Writing the equations in terms of components (indices) allows us to manipulate them in conventional terms, but their physical significance transcends these technicalities. Modern mathematical notation, which we cannot present here, emphasizes the frame independence of these equations ab initio.

Next we must write the Lorentz force law in this language. The force law reads,

$$\frac{d\boldsymbol{p}}{dt} = \boldsymbol{F} = q(\boldsymbol{E} + \boldsymbol{v} \times \boldsymbol{B}) \tag{10.27}$$

Here $\boldsymbol{p} = \gamma m \boldsymbol{v}$ is the relativistic three-momentum, which comprises three components of the relativistic four-momentum, $p^\mu = m u^\mu$, and u^μ is the four-vector velocity, $u^\mu = dx^\mu/d\tau$, and $d\tau = dt/\gamma$ is the proper time differential. Let us write the right hand side of Eq. (10.27) in covariant form. We have the four-vector u^μ and tensor $F^{\mu\nu}$ at our disposal and have to generate a four-vector $dp^\mu/d\tau$. The first construction to try is,

$$dp^\mu/d\tau = q\sum_\nu u_\nu F^{\nu\mu} \tag{10.28}$$

Let us write out the $\mu = 1$ component of this conjecture,

$$dp^1/d\tau = q\sum_\nu u_\nu F^{\nu 1} = q(u^0F^0 \; - u^1F^{11} - u^2F^{21} - u^3F^{31})$$

$$dp^1/d\tau = \gamma dp^1/dt = q(\gamma cE_x/c + \gamma v_y B_z - \gamma v_z B_y) = \gamma q\big(E_x + (v \times B)_x\big)$$

And, we see that the γ factors cancel to reproduce the x component of Eq. (10.27)!

This completes our discussion of the relativistic character of Maxwell's equations and the Lorentz force law. These equations make the compatibility of these equations with the postulates of relativity explicit. The "old-fashioned" way of writing these equations obscured their relativistic nature. However, for practical applications and physical insight and intuition, the "old" expression of the physics is often much more immediate. However, a true master of this subject can jump between formulations as needed.

10.3 NEXT STEPS

Let us review some of the lessons we have learned.

Special relativity has shown us that Newton's concept of a force, which acts as "action at a distance" does not generalize to Einstein's world. Newton's third law fails as a consequence. These concepts are replaced by local conservation laws, local differential equations, and fields.

The electrostatics of Newton's world produces the Lorentz force law when special relativity is applied. Velocity-dependent forces that violate Newton's third law are unavoidable. In addition, when Coulomb's law is expressed as a local differential equation, Gauss' law, special relativity elevates the electric field to a dynamical variable and Ampere—Maxwell's equation is a consequence. Similarly, the absence of magnetic monopoles expressed as a local condition implies Faraday's law, and the magnetic field becomes a dynamical variable. As a consequence we no longer think of separate electric and magnetic effects: under Lorentz transformations, electric effects in one frame become magnetic effects in another. In fact we saw that the electromagnetic field satisfies a Lorentz invariant wave equation, which implies that light travels at the speed limit in all inertial frames.

In Newton's world the dynamical variable in these developments was just the charged particle. It was the source of electrostatic forces. The charged particle consisted of one degree of freedom. The electric field was given by a constraint equation and had no dynamics of its own. In Einstein's world the validity of the constraint equations, Gauss' law for the electric field and the divergence-free condition for the magnetic field (no monopoles), in all inertial frames imply that the electric and magnetic fields become dynamical variables, and Maxwell's equations were found. In fact the electromagnetic field brings two new degrees of freedom into the theory. There are two degrees of freedom because a freely propagating electromagnetic field is described by transverse electric and magnetic fields. The electric field could be chosen in either of two directions perpendicular to the direction of propagation of the wave. In both cases the magnetic field was determined in terms of the electric field. The two independent polarization states of the electric field are the two independent degrees of freedom of the new dynamical field.

So relativity has profoundly changed our understanding of electricity and magnetism. But what of the charged particle? We have not made much progress here. We went from Newtonian kinematics to relativistic kinematics. However, our formulation has not embodied the principles and phenomena of relativity that we have discussed for the electromagnetic field. When we discussed collisions and used the conservation of energy—momentum to answer kinematic questions, we referred to the fact that special relativity allows for the conversion of mass to energy and energy to mass to motivate the creation of elementary particles in relativistic collisions. Those collisions were constrained by conservation laws of the underlying theories of elementary particles, but the details were not discussed. Nonetheless, this phenomenology displays that in relativistic theories we can expect particle number conservation to be violated, collisions starting with two particles but ending with 10, 20, or more are commonplace at high-energy physics labs. In Newton's world the integrity of a particle and its mass were paramount, but in relativity the creation and destruction of particles is fundamental. Therefore, our formulation of particle mechanics in this book is incomplete.

How could we formulate particle creation and destruction? *We should copy the success we have had with the electromagnetic field and look for a field description of matter!* This would have to be done within the context of quantum physics. Such a development is beyond the scope of this book, but it is good to see where we are headed. It turns out that particle creation and annihilation are required processes, just as photon creation and absorption are in a quantum formulation of electrodynamics. We saw how Maxwell's equations led to the wave equation describing freely propagating electromagnetic waves. In a field theory of particles, the creation and destruction of particles are equally inevitable. The theory would be inconsistent without these effects just as electromagnetism would be inconsistent if radiation were somehow suppressed.

Let me tell you a little about how things work out. In a quantum field theory of particles you need to formulate locality following our development of electromagnetism. When you do that you can build in energy–momentum conservation for interacting particles. You also learn the wonderful fact that to preserve locality in the interactions, the theory *predicts* the existence of antiparticles and associated processes such as electron–positron production in variable electromagnetic fields! The mathematics of local field theory causes the proliferation of degrees of freedom again. Attempts to describe radiation and absorption of photons within a single charged particle framework will necessarily fail. Another element in Newton's world must be abandoned for a richer alternative.

All of this constitutes another developing, modern story that you can look forward to participating in.

PROBLEMS

10-1. Generalize the problem of two parallel wires carrying identical currents I to the case where wire 1 carries current I_1 and wire 2 carries current I_2. These different currents come about because in the lab frame the mobile electrons in wire 1 move in the x direction with velocity v_1, and the mobile electrons in wire 2 move with velocity v_2. Show that the force on a length l of wire is,

$$F = \frac{2k}{c^2}\frac{I_1 I_2}{b} l$$

where b is the distance between the wires and the force is attractive if the currents are in the same direction and repulsive otherwise.

10-2. Consider an electron e at distance b from the wire again. But in this case suppose that the electron e is at rest in the lab frame S. Recall that the wire is electrically neutral and carries a current I in the lab. We model the current as before by supposing that the mobile electrons have a velocity v to the right, and the Cu^+ ions are at rest in the lab. As before, call the frame in which the mobile electrons are at rest S'. Since the wire is neutral in the lab and the electron e is not moving, it experiences no force in the lab.

a. Argue that since the electron e experiences no force in the lab frame, it experiences no force in any inertial frame.

b. Consider the wire and electron e in the rest frame of the mobile electrons S'. Show that the net charge per unit length of the wire in the frame S' is $\frac{v^2}{c^2}\gamma\lambda_+$ where λ_+ is the charge per unit length of the Cu^+ ions in the lab frame S.

c. Calculate the electric field at the position of the electron in the frame S' using Gauss' law and the result of part b.

d. Calculate the magnetic field at the position of the electron in the frame S' using the Ampere–Maxwell law and the result of part b.

e. Calculate the electric force on the electron e in frame S', and calculate the magnetic force on the electron e in frame S' using the Lorentz force law. Show that these two forces cancel and give the required result that the force on the electron e is zero in frame S'.

10-3. Relativistic notation for Maxwell's equations and the Lorentz force law.

Let us fill out some of the discussion and derivations in Section 10.2.

a. Using the transformation laws for ∂_ν and $F^{\mu\nu}$, show that $\sum_\nu \partial_\nu F^{\mu\nu}$ transforms as a four-vector V^μ. In other words, when upper and lower indices, such as ν, are fully contracted (summed over) they effectively produce a scalar.

b. Show that $\sum_\mu \partial_\mu F^{\mu i} = \frac{4\pi k}{c^2} J^i$ for $i = 1, 2, 3$ produce the Maxwell–Ampere equation.

c. Show that $\sum_\mu \partial_\mu \widetilde{F}^{\mu i} = 0$ for $i = 1, 2, 3$ produce Faraday's equation.

d. Write out the $\mu = 0$ component of $dp^\mu/d\tau = q\sum_\nu u_\nu F^{\nu\mu}$ and compare it to the equations in our discussion of forces, work, and energy.

10-4. The four-vector potential and Maxwell's equations.

In Appendix F we introduced the four-vector potential for the electromagnetic field,

$$A^\mu = (V/c, \boldsymbol{A})$$

which could be used to generate the electric and magnetic fields,

$$\boldsymbol{E} = -\nabla V - \partial \boldsymbol{A}/\partial t, \quad \boldsymbol{B} = \nabla \times \boldsymbol{A} \tag{10.29}$$

Let us write the field tensor $F^{\mu\nu}$ in terms of ∂^ν and A^μ. Since $F^{\mu\nu}$ is antisymmetric, a candidate formula that has the correct transformation laws is,

$$F^{\mu\nu} = \partial^\mu A^\nu - \partial^\nu A^\mu \tag{10.30}$$

a. Show that Eq. (10.30) produces Eq. (10.29) when you write out the Cartesian coordinates. Note that this is an elegant discovery because it shows us that you can make

$F^{\mu\nu}$ out of two four-vectors! This is much less clumsy than the discussion in the text, but it relies on using the four-vector A^μ, which is not directly measureable in the lab (only $\boldsymbol{E}$ and $\boldsymbol{B}$ are physical).

Recall that we chose the Lorenz gauge in Appendix F to find the wave equation for A^μ. The Lorenz gauge condition was,

$$\nabla \cdot \boldsymbol{A} = -\frac{1}{c^2}\frac{\partial V}{\partial t} \tag{10.31}$$

b. Show that the Lorenz gauge condition can be written,

$$\sum_\mu \partial_\mu A^\mu = 0$$

which shows that it is fully relativistic, i.e., if it is chosen in one frame, it is true in all frames.

c. Show that the potential formulation of $F^{\mu\nu} = \partial^\mu A^\nu - \partial^\nu A^\mu$, automatically solves the two homogeneous Maxwell equations, $\sum_\nu \partial_\nu \widetilde{F}^{\mu\nu} = 0$.

d. Write out the Maxwell equations $\sum \partial_\nu F^{\mu\nu} = \frac{4\pi k}{c^2} J^\mu$ in terms of A^μ using Eq. (10.30), and show "that in the Lorenz gauge $\sum_\mu \partial_\mu A^\mu = 0$, they become the wave equation,

$$\sum_\nu \partial_\nu \partial^\nu A^\mu = \left(\frac{1}{c^2}\frac{\partial^2}{\partial t^2} - \nabla \cdot \nabla\right) A^\mu = \frac{4\pi k}{c^2} J^\mu$$

10-5. Electrodynamic invariants. Maxwell energy–momentum stress tensor.

The field strength tensor $F^{\mu\nu}$ and its dual $\widetilde{F}^{\mu\nu}$ can be used to make electrodynamic invariants. Consider the "contractions,"

$$\sum_{\mu\nu} F^{\mu\nu}F_{\mu\nu}\,, \quad \sum_{\mu\nu} F^{\mu\nu}\widetilde{F}_{\mu\nu}, \quad \sum_{\mu\nu} \widetilde{F}^{\mu\nu}\widetilde{F}_{\mu\nu}$$

a. Evaluate each of these expressions in terms of $\boldsymbol{E}$ and $\boldsymbol{B}$, and relate them to the invariants discussed in the problem set on electrodynamic invariants in Chapter 8.

We can make another second rank tensor, which is quadratic in the electric and magnetic fields $\sum_{\sigma} F^{\mu\sigma}F^{\nu}_{\sigma}$. This becomes the Maxwell stress tensor if we make it traceless,

$$T^{\mu\nu} = \sum_{\sigma} F^{\mu\sigma}F^{\nu}_{\sigma} - \frac{1}{4}g^{\mu\nu}\left(\sum_{\lambda\sigma} F^{\lambda\sigma}F_{\lambda\sigma}\right)$$

b. Show that $T^{\mu\nu}$ is symmetric, $T^{\mu\nu} = T^{\nu\mu}$, and traceless, $\sum_{\mu} T^{\mu}_{\mu} = 0$.

c. Write out the T^{0i}, $i = 1, 2, 3$, components and compare with the Poynting vector introduced in Problem 8.9. Recall that the Poynting vector gives the energy per unit area passing through a surface per unit time ("energy flux"). Interpret T^{00} as well.

d. Show that $T^{\mu\nu}$ is conserved, $\partial_{\mu}T^{\mu\nu} = 0$, in a source-free ($J^{\mu} = 0$) region of space–time. You will learn in your first field theory course that the energy–momentum carried by charged matter fields can also be expressed as a second rank tensor, $T^{\mu\nu}_{M}$, and the sum, $T^{\mu\nu} + T^{\mu\nu}_{M}$, is conserved, $\partial_{\mu}\left(T^{\mu\nu} + T^{\mu\nu}_{M}\right) = 0$. This implies that energy and momentum can flow between the matter and the electrodynamic fields, but it does so in a locally conserved fashion.

REFERENCE

[1] A.P. French, Special Relativity, W. W. Norton, New York, 1968.

CHAPTER 11

Introduction to General Relativity

Contents

11.1 THE EQUIVALENCE PRINCIPLE, GRAVITY, AND APPARENT FORCES

Everyone knows the story of Isaac Newton and the falling apple. There was a plague in Great Britain, so the students were sent out of the cities to reduce their chances of catching the contagion. Supposedly Newton relaxed under an apple tree in his childhood home, Woolsthrope Manor, contemplating the current ideas of mechanics. When an apple fell on his head, he invented the idea of the gravitational force—Earth's enormous mass exerted a force on the apple, breaking its stem and causing a collision with Newton's precious head. This event led Newton, over the course of later months back at the university, to the force law of gravity,

Special Relativity, Electrodynamics, and General Relativity
ISBN 978-0-12-813720-8
https://doi.org/10.1016/B978-0-12-813720-8.00011-8

$$\mathbf{F}_{12} = -G\frac{m_1 m_2}{r_{12}^2}\widehat{\mathbf{r}}_{12}, \tag{11.1}$$

where G is Newton's constant ($G \approx 6.67 \cdot 10^{-11}$ Nm^2/kg^2), which sets the scale for gravitational forces; m_1 and m_2 are the masses, which are a distance r_{12} apart; $\mathbf{F}_{12}$ is the force that m_1 exerts on m_2; and $\widehat{\mathbf{r}}_{12}$ is the unit vector pointing from m_1 to m_2. Newton arrived at the inverse square character of the force law, Eq. (11.1), to explain the extensive planetary data accumulated by Kepler and others. In fact, as recorded in the Principia Mathematica, Newton *derived* the force law Eq. (11.1) from Kepler's three laws of planetary motion using the foundations of the differential geometry of curves which he also invented. Choosing m_1 to be the mass M_s of the Sun and m_2 to be the mass of Earth M_e, the equation of motion of the Earth around the Sun is given by Newton's second law, force equals mass times acceleration (see also Fig. 11.1).

$$M_e\ddot{\mathbf{r}}_e = -G\frac{M_e M_s}{|\mathbf{r}_e - \mathbf{r}_s|^2}\widehat{\mathbf{r}}_{es}. \tag{11.2}$$

A crucial element of Eq. (11.2) is the fact that the mass of the accelerating body, Earth in this case, cancels out of the equation of motion. We describe this by saying that the inertial mass, the mass on the left-hand side of Newton's second law (mass times acceleration equals force), equals the gravitational mass, the mass in the gravitational force expression. It is said that Galileo was the first physicist to investigate this point, before Newton codified classical dynamics, and establish the equality of these two masses

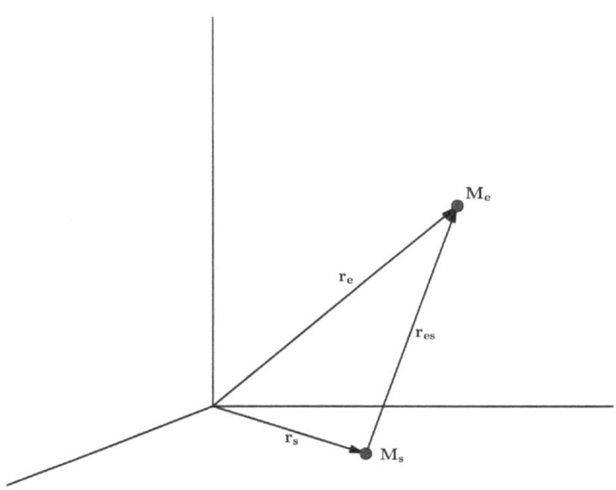

Figure 11.1 Coordinates of the Earth–Sun system used in Newton's gravitational force law.

experimentally. Galileo, through his assistants, dropped masses off the Leaning Tower of Pisa and observed that they accelerated identically in the gravitational field provided by Earth. Modern experiments pioneered by Lörand Eötvös in the early days of the 20th century and many others in modern times have established the equality of the gravitational and inertial masses to high precision, better than one part in 10^{13}! The axiom that the two masses are strictly identical evolved into a central ingredient in the soon-to-be famous equivalence principle. Under Einstein, the equivalence principle developed into the statement that there is no way to distinguish the local effects of a gravitational field from those in an accelerating reference frame free of external forces. The equivalence principle allows us to understand accelerating reference frames in terms of gravity and gravity in terms of accelerating reference frames. This principle is explained and discussed in much greater detail as we journey forth.

The equivalence principle and the inverse square law of gravity are both under constant experimental scrutiny by high-precision experiments. We accept both ideas as exact throughout our discussions. However, if one or both should fail ever so slightly, many topics of theoretical physics would need fundamental changes.

Classical physicists understood that the principle of equivalence made gravity a very special phenomenon. Other forces (such as electrostatics) or mechanical devices (such as springs) produce accelerations that are inversely proportional to the mass of the body. However, there are forces besides gravity which are familiar from day-to-day experiences that produce accelerations that are independent of the mass of the body. These are called "apparent" forces and are strictly geometrical in origin. For example, when you drive a car and accelerate from a stop sign or decelerate at a red traffic light, you experience forces of this type. Another example consists of centripetal and Coriolis forces. These are the forces that occur when you measure acceleration in a rotating coordinate system, called a noninertial frame of reference. Recall that an inertial frame is one in which an isolated body moves in a straight line at constant velocity. To change the body's velocity, a force must be applied. The velocity of the body changes according to Newton's second law, $\mathbf{f} = m\mathbf{a}$, where $\mathbf{f}$ is the applied force and $\mathbf{a}$ is the body's acceleration, $\mathbf{a} = d^2\mathbf{r}/dt^2$. A simple example of a rotating, noninertial frame is afforded by a turntable spinning at a constant angular velocity ω in an otherwise inertial environment. From the perspective of a coordinate system fixed in the rotating turntable, a body moving at a constant velocity in the inertial frame is accelerating. Clearly this acceleration is independent of the body's mass and is purely a result of the coordinate transformation between the inertial and the rotating noninertial frames.

Let us work out the details of the centrifugal and Coriolis accelerations from scratch in the context of Newton's world. Later we will revisit rotating reference frames in the context of general relativity.

If a mass m is at rest on a turntable that is rotating in an otherwise inertial frame, the mass experiences a centripetal force; and if it is moving with respect to the turntable, it experiences a Coriolis force as well. To begin, pin the mass at distance r from the axis of the turntable. If the angular velocity of the turntable is ω, then the transverse velocity v_t of the mass is ωr, as shown in Fig. 11.2.

Because the direction of v_t is changing, the mass is subject to an inward acceleration, called the centripetal acceleration. We can calculate it by viewing Fig. 11.3, which shows the velocity at time t and at time $t+\Delta t$. The angle between the two velocities in the figure is $\omega\Delta t$, so the change in the velocity is $v_t\omega\Delta t$ and its direction is inward. So,

$$dv_t/dt = v_t\omega = r\omega^2 = v_t^2/r,$$

which is the familiar expression for the centripetal acceleration. If we release the mass abruptly, it accelerates outward, as viewed in the rotating frame, with an acceleration equal and opposite to the centripetal acceleration.

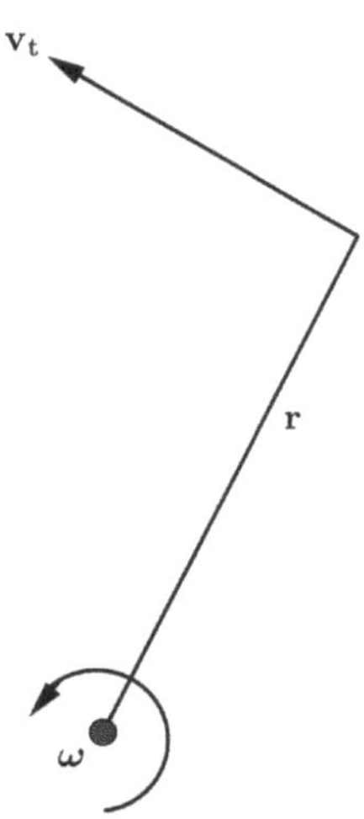

Figure 11.2 Transverse velocity of a rotating particle a distance r from the center of a turntable.

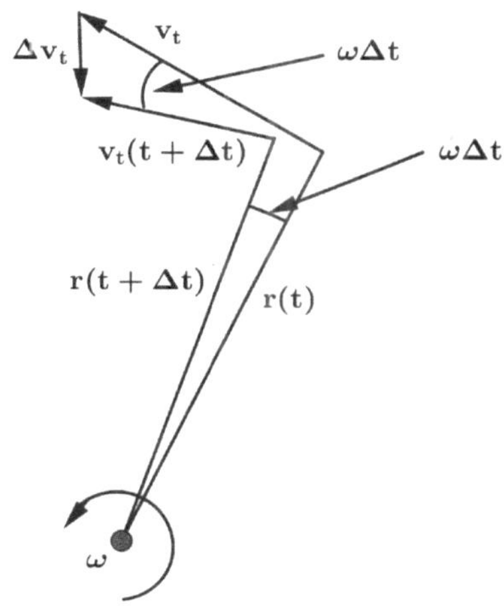

Figure 11.3 Calculation of the centripetal acceleration of a rotating mass point. The particle's transverse velocity is shown at time t and at time $t + \Delta t$ and the difference Δv_t is taken.

This acceleration is called the centrifugal acceleration. When you swing a mass on a string, you feel the centrifugal acceleration pulling the string taut.

Next, recall the Coriolis acceleration. This acceleration acts perpendicular to the direction of motion of the mass relative to the turntable (to the right of the particle's velocity if ω is positive) and has a magnitude of $2v'\omega$, where v' is the velocity of m relative to the turntable. Let us derive this result from scratch. First, we consider transverse motion, motion at a constant r around a circle, and then we look at radial motion.

So, consider a particle with a given transverse velocity v. Decompose v into two pieces: the velocity of the turntable at r, $v_t = \omega r$; and the velocity difference, $v' = v - v_t$, or the velocity of the mass m relative to the turntable at that point. The full centripetal acceleration is then

$$\frac{v^2}{r} = \frac{(v_t + v')^2}{r} = \frac{v_t^2}{r} + 2v'\omega + \frac{v'^2}{r}.$$

The second term on the right-hand side is the Coriolis acceleration. For the particle to remain on a circle of fixed r, its contact to the turntable must provide this additional inward acceleration, $2v'\omega$. The particle itself pushes on the turntable with a force $-2v'\omega m$, which is called the Coriolis force. The third term in the equation, v'^2/r, is the additional centripetal acceleration produced by the increased speed.

Next, let m have a radial velocity v' along an axis of the turntable. As the particle travels outward, its transverse velocity increases and its direction of motion changes, so there are two contributions to its acceleration. Because the angular velocity ω of the turntable is fixed, its transverse velocity is $v_t = \omega r$. But r changes because the mass has a radial velocity v' with respect to the turntable, $\Delta r = v' \Delta t$. So, the magnitude of the transverse velocity changes $\Delta v_t = \omega \Delta r = v' \omega \Delta t$, and we see that the first contribution to the acceleration, which is clearly perpendicular to its direction of motion, is $v' \omega$. The second contribution to the acceleration comes from the change in direction of v' as the mass m moves radially on the rotating turntable. Because the turntable rotates through the angle $\omega \Delta t$ in the time interval Δt, the mass develops an additional transverse velocity, $v' \omega \Delta t$. We see that both contributions to the particle's transverse velocity act in the same direction and sum to $2v' \omega$, as shown in Fig. 11.4.

If we speak about forces, we see that the turntable pushes on the mass with a transverse force of $2v' \omega m$, so that it continues in a radial direction.

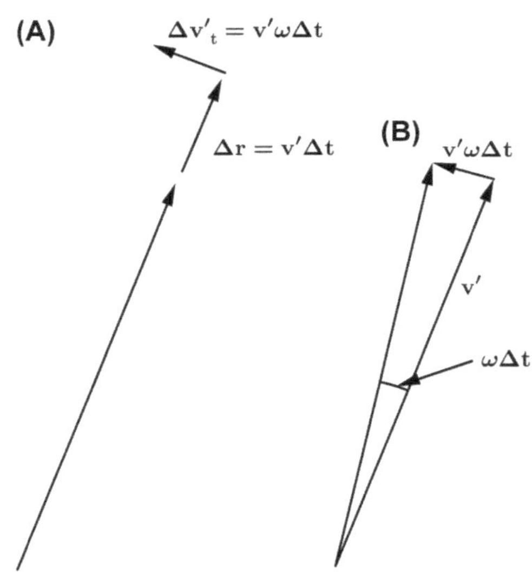

Figure 11.4 Calculation of the Coriolis force on a particle moving radially on a turntable. Part (A) shows the component due to the particle's change in its radial position, and part (B) shows the component due to the change in the direction of its radial velocity.

So, if the particle is not attached to the turntable, it accelerates to the right of its direction of motion, as measured in the turntable frame, at a rate

$$a_{\text{Coriolis}} = -2v'\omega,$$

which is the well-known Coriolis acceleration result we sought.

In summary, for both radial and transverse velocities relative to the turntable, we have the same expression for the Coriolis acceleration. Because any velocity can be decomposed into a radial component and a transverse component, our result for the Coriolis acceleration is perfectly general.

Let us collect these results in two formulas expressing Newton's second law in a rotating frame described by plane polar curvilinear coordinates (r, θ). The relation between the Cartesian coordinates in the inertial frame and the rotating plane polar coordinates is,

$$x = r\cos(\theta + \omega t) \quad y = r\sin(\theta + \omega t) \tag{11.3}$$

where ω is the rate of rotation of the turntable. Our exercise in kinematics above produced the equations of motion for the radial and angular variables complete with centripetal forces, Coriolis forces and external forces (f_r, f_θ),

$$\frac{d^2r}{dt^2} - r\left(\omega + \frac{d\theta}{dt}\right)^2 = f_r \tag{11.4a}$$

$$\frac{d^2\theta}{dt^2} + \frac{2}{r}\frac{dr}{dt}\left(\omega + \frac{d\theta}{dt}\right) = f_\theta \tag{11.4b}$$

It is easy to check in special cases that these equations capture all the ingredients in the intuitive discussion of centripetal and Coriolis effects above. It is instructive, but somewhat tedious, to begin with Newton's second law written in Cartesian coordinates,

$$\frac{d^2x}{dt^2} = f_x \quad \frac{d^2y}{dt^2} = f_y \tag{11.5}$$

apply the transformation Eq. (11.3) and arrive at Eq. (11.4). We will develop systematic ways to do such calculations in Chapter 12. (see Problems 12.8 and 12.25)

This exercise teaches us that there are *two* sources of "apparent" forces in the equations of motion Eq. (11.4): a. the use of curvilinear coordinates, the case $\omega = 0$ in Eq. (11.4), and b. the noninertial character of the rotating frames, $\omega \neq 0$. There are clearly two distinct geometrical effects and are usually discussed separately. We will not follow tradition here because they enter the equations of motion in the same fashion. When we develop the equations of general relativity, we will see that sources of energy–momentum produce curved space–time and noninertial frames, and all these effects will be rederived in a new context. We will see that gravitational "forces" enter equations such as Eq. (11.4) on the left-hand side with "apparent" forces and *not* on the right-hand side where the Lorentz force of electromagnetism, for example, will appear!

What have we learned so far? We see that force-free straight-line motion in an inertial frame is interpreted as accelerated motion in a noninertial frame. If we introduce forces to describe the accelerated motion in the noninertial frame, we should use the term "apparent" for this reason. These accelerations are determined purely by the relation of the inertial and the noninertial coordinate systems and can be derived without any knowledge of mechanics. All we need is geometry.

Now we come to the good stuff. Einstein argued that gravity is also an apparent force! Consider first a body moving in a constant gravitational field where it experiences a constant acceleration $-g$ independent of its mass. This problem is indistinguishable from the motion of a body in a frame free of gravity but accelerating upward at the rate g. These facts were well appreciated by classical physicists in Newton's era, but this statement of the equivalence principle was popularized and pursued in the context of the relativistic theory of space–time by Einstein. Einstein posed the equivalence—the impossibility of distinguishing the physics in a constant gravitational field from that in an accelerating frame—by imagining a physicist doing experiments in an elevator. The elevator is accelerating upward at a rate g, and, Einstein claimed, if the elevator has no windows so the physicist cannot see the tricks being played on him, there is no *local* experiment he can run that can distinguish this environment from one at rest on the surface of a planet where gravity generates the approximately uniform acceleration g. (On the surface of Earth, Newton's gravitational force law gives $g = GM_e/R_e^2 \approx 9.8\ m/s^2$, where R_e is the radius of Earth, neglecting all other planetary masses.)

We come to the conclusion that uniform gravity is an apparent force much like centripetal and Coriolis forces, familiar from our experiences with turntables. As with any apparent force, we can transform to another coordinate system where the apparent force vanishes identically. In the case of a constant gravitational field, we can just consider an observer in free fall in that environment. According to the equivalence principle, this free-falling frame of reference is force-free and is an inertial frame where isolated masses travel along straight lines.

The idea that gravity can be transformed away by passing to a free-falling, *inertial* frame of reference is very useful. Since we know that masses move along straight lines in inertial frames, we can use the equivalence principle to solve any mechanics problem in a small enough region of space−time where a given gravitational field can be treated as uniform—just consider the motion from the perspective of a freely falling frame where special relativity holds, solve the problem, and finally map it back to the coordinates an observer would use at rest in the gravitational field. (It is important here to check that there are no external nongravitational forces, such as electrostatics, in the environment. These forces are not "apparent" and cannot be transformed away by a slick choice of reference frame.)

But, as emphasized by Einstein, one can also go beyond ordinary mechanics problems because the equivalence principle applies to *any* process. An interesting application concerns the deflection of light by a gravitational field. Because light moves along straight lines in inertial frames, the equivalence principle implies that it experiences an acceleration when it moves transverse to a gravitational field, as shown in Fig. 11.5. Consider an initially horizontal beam of light in the Earth's gravitational field, labeling x

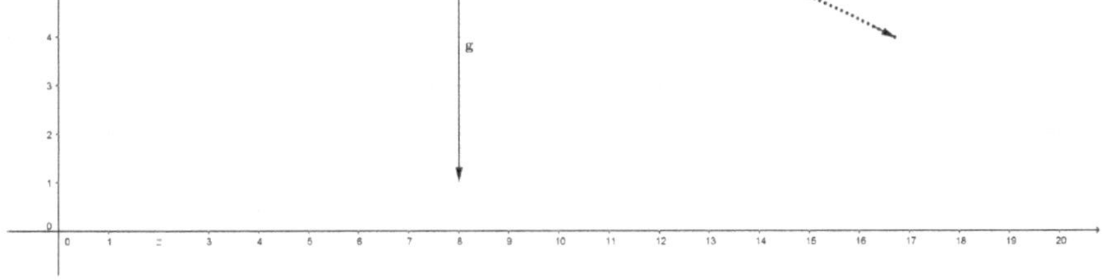

Figure 11.5 A light ray moving in a uniform transverse gravitational field.

as the horizontal position and y the vertical position. Then the equivalence principle predicts the local bending of light,

$$\frac{d^2y}{dx^2} \cong \frac{d^2y}{c^2dt^2} \cong -\frac{g}{c^2}$$

As emphasized by Rindler [1], this is a remarkable result, which has far reaching implications. It states that since light travels at a finite speed, it has "weight"! No other assumptions about light need be invoked in this argument. Light, and all other physical phenomena, must travel on locally curved paths in gravity. This point strongly motivates the idea that gravity is an aspect of geometry and does not belong in the menagerie of forces! The argument above just gives the *local* bending of light in a uniform gravitational field. To compare to experiments we need to consider the path light traverses past a stellar object such as the Sun. The full machinery of general relativity is required here, as we will illustrate later in this chapter. Historically the deflection of light by the gravitational field of the Sun was computed using Newtonian mechanics in the early days of the 19th century. Einstein repeated the calculation in the context of relativistic, curved space—time in 1919 and found that the Newtonian prediction is too small by a factor of two.

The experimental observation of the deflection of light in a gravitational field comes from a lensing effect, as illustrated in Fig. 11.6. When light passes by a large astronomical object, it is "attracted," as implied by the equivalence principle, and it is deflected as shown in the figure. The effect was observed originally by carefully measuring the background stars around the halo of the Sun during an eclipse and comparing those measurements to the positions of the background stars when the Sun is in a different part of the sky. These measurements are difficult; only after data were accumulated over many years was it possible to rule in favor of the relativistic prediction. In fact, modern radio astronomy techniques using

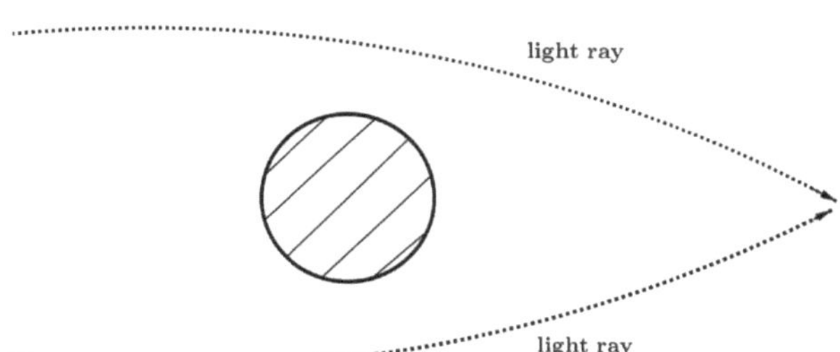

Figure 11.6 Light rays are attracted to a massive body. A lensing effect.

interferometry were the first to give decisive results with small enough errors to really distinguish between Newton and Einstein.

Finally, let us reconsider the scattering of light in the presence of a large star somewhat more critically. Because the direction of the gravitational acceleration varies as we move around the star, the use of the equivalence principle must be stated more carefully. As we pass around the Sun, the equivalence principle can be applied only *locally*; that is, only over a region of space—time where the gravitational field is essentially uniform can we find a freely falling inertial frame in which the field is essentially eliminated. The theory speaks of "local inertial frames" to accommodate spatially varying gravitational fields. We certainly cannot eliminate the effects of gravity from large space—time regions! The spatial dependence of gravitational fields means that the mathematical details of the theory change from point to point. For example, a body falling in a spatially varying gravitational field executes straight-line motion in each local inertial frame approximating the varying gravitational field. The actual trajectory of the body is obtained by patching together its simple trajectories in contiguous inertial frames. This sounds awfully complicated. Mathematically, the language for this motion is neatly given by differential geometry. It will take several sections and developments to make these ideas precise and to obtain some classic predictions of general relativity, the gravitational redshift, the resolution of the twin paradox, the deflection of light in a gravitational field, and other related phenomena.

In later sections of this chapter and in Chapter 12, we will introduce and illustrate the concepts and equations of differential geometry and general coordinate transformations so that we can consider the motion of particles and light in strong gravitational fields, such as those near a Schwarzschild black hole. We shall see that there is a distance from the black hole, called the Schwarzschild radius, which acts as a point of no return: although a particle can pass through this distance from the outside, nothing from the inside can ever return to the outside. There is no equivalent to such a phenomenon in Newtonian physics because it is a consequence of the geometry of space—time inside the Schwarzschild radius. The geodesic equations for particle motion in a curved space—time will be developed as well and will be illustrated in classic differential geometry where visualization is possible and helpful. Tidal forces will be analyzed in the context of the geometry of space—time and non-Euclidean geometry, and intrinsic curvature of space—time will emerge. Gravitational waves will be considered and the realistic situation where gravitational waves are observed as small variations on a flat Minkowski space—time of special relativity will be developed in some detail. The first detection of gravitational waves in 2015 by the LIGO collaboration will be presented and discussed. We shall see that gravitational waves are emitted from oscillating quadrupole moments of binary systems, and we will make estimates of the physical parameters involved. Along the way, we will introduce the Einstein field equations. Our discussions will be limited because this is a prefield theory course, so that only simple examples of the energy—momentum tensor are accessible to us. We will see in several cases that the equivalence principle will serve us well both for physical arguments as well as in mathematical developments, but we shall travel beyond it and discuss the general relativistic version of tidal forces, space—time curvature, the intrinsic nonlinearity of general relativity, and the physics of the event horizon around a black hole. We will compare and contrast electromagnetism, special relativity, and general relativity. In general relativity we will see that the Newtonian gravitational potential is replaced by the metric, which will become a dynamical field that will be the origin of dynamical space—time geometry. This development will parallel our earlier discussion where special relativity forced the electrostatic potential to be replaced by the dynamic electromagnetic field. Many of these concepts will play out in a discussion of the generation and observation of gravitational waves.

11.2 MOTION IN A ROTATING, RELATIVISTIC REFERENCE FRAME: SPATIAL CURVATURE AND THOMAS PRECESSION

In special relativity, we plot the world line of a particle's motion on a Minkowski diagram (Fig. 11.7). In general relativity we will use notions such as the invariant interval, metrics, proper time, and proper length more than we did in our discussions of special relativity. So, let us review these ideas briefly and see that we can read effects such as time dilation, Lorentz contraction, and the relativity of simultaneity from expressions for the invariant interval. A problem familiar from elementary mechanics, such as the rotating reference frame, helps us move gradually into new, uncharted subjects.

Recall that the proper time $d\tau$ that passes on a clock attached to the moving particle can be calculated from the invariant interval ds,

$$ds^2 = c^2 dt^2 - dx^2 - dy^2 - dz^2. \tag{11.6}$$

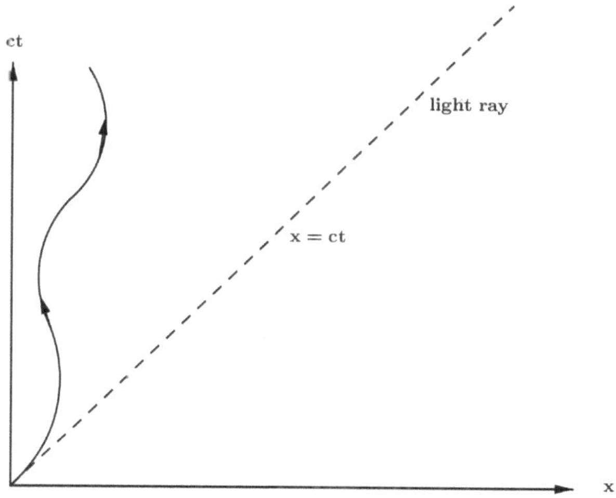

Figure 11.7 World line of a particle's motion on a Minkowski diagram.

Consider the meaning of the symbols in this formula. In a frame S, we imagine two events, one at $(ct,\ x,\ y,\ z)$ and another at $(ct + cdt,\ x + dx,\ y + dy,\ z + dz)$. ds is then the invariant interval between them. The events might be ticks on a clock, measurements of the ends of a rod, or whatever. The crucial point is that ds is the *same* in all reference frames, as we discussed in Section 4.3, so if we can calculate and understand it in one frame, we have its value in *all* frames. So, the proper time $d\tau$ that passes on a clock attached to the moving particle can be computed from ds by boosting to a frame S′ where the particle is at rest. In S', $dt' = d\tau$ (proper time), $dx' = dy' = dz' = 0$, so $ds^2 = c^2 d\tau^2$. This means that if the particle is moving along the x axis at velocity v, so $dx = v \cdot dt$, $dy = 0$, and $dz = 0$, then Eq. (11.6) reduces to

$$ds^2 = c^2 d\tau^2 = c^2 dt^2 - v^2 dt^2 = (c^2 - v^2) dt^2$$

or

$$d\tau = \sqrt{1 - v^2/c^2}\, dt \equiv dt/\gamma,$$

which is just the expression of time dilation—the moving clock runs slowly.

Minkowski diagrams, space–time pictures, are the natural arena for discussing dynamics because they show time and position information together. Because space and time mix under boosts, we must work in four-dimensional space–time. If a free particle's world line passes through $P_1 = (ct_1,\ x_1)$ and $P_2 = (ct_2,\ x_1)$, we know its velocity $v = (x_2 - x_1)/(t_2 - t_1)$ and its path.

As a first step toward developing relativistic particle motion in a gravitational field, consider relativistic force-free motion in a rotating reference frame. In rotating frames, there are centripetal and Coriolis "apparent" forces. Choose a cylindrical spatial coordinate system,

$$x = r\cos\varphi, \quad y = r\sin\varphi, \quad z = z$$

as shown in Fig. 11.8. Then the spatial distance becomes

$$dx^2 + dy^2 + dz^2 = dr^2 + r^2 d\varphi^2 + dz^2$$

and the space–time invariant interval reads

$$ds^2 = c^2 dt^2 - (dr^2 + r^2 d\varphi^2 + dz^2).$$

To describe a turntable that is rotating about the z axis at angular velocity ω, as in Fig. 11.9, we introduce a new azimuthal angle φ',

$$\varphi' = \varphi - \omega t, \tag{11.7}$$

so a point with fixed φ' has its φ increasing as ωt. This simple equation mixes the time coordinate with a spatial coordinate, so

$$d\varphi' = d\varphi - \omega dt$$

and the invariant interval written in terms of the rotating coordinates becomes

$$ds^2 = (c^2 - \omega^2 r^2)dt^2 - (dr^2 + r^2 d\varphi'^2 + 2\omega r^2 d\varphi' dt + dz^2) \tag{11.8}$$

and does not neatly separate into a spatial and temporal part.

Before dealing with ds^2 in its full glory, consider a clock at rest in the rotating reference frame and a distance r from the z axis. Two ticks of the

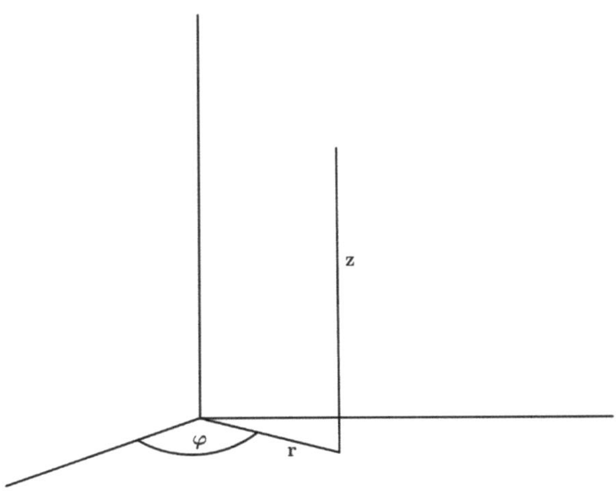

Figure 11.8 Cylindrical coordinate system used to parametrize motion on a relativistic turntable.

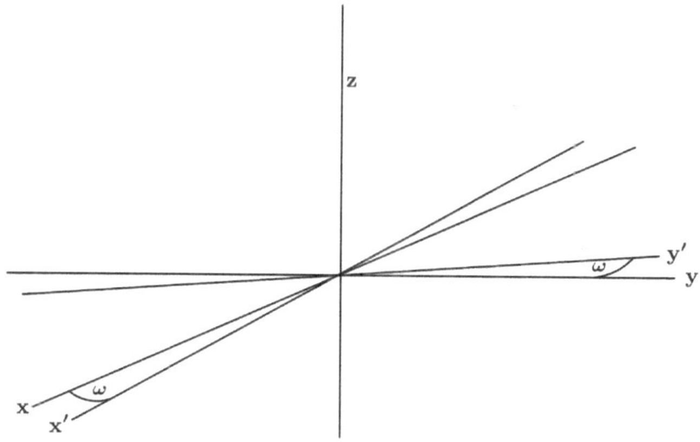

Figure 11.9 The primed coordinate system rotates at angular velocity ω relative to the unprimed system.

clock occur at a given r, φ', and z, so $dr = d\varphi' = dz = 0$, and Eq. (11.8) reduces to

$$ds^2 = c^2 d\tau^2 = (c^2 - \omega^2 r^2) dt^2.$$

So, the proper time kept by the clock is

$$d\tau = \sqrt{1 - \omega^2 r^2 / c^2}\, dt. \tag{11.9}$$

This result is the usual time dilation formula because the velocity of the clock relative to the inertial x, y, z frame is $v = \omega r$: as r increases, the clock's velocity (transverse) increases proportionally, and the time dilation effect is enhanced. We must restrict the possible values of ω and r so that $\omega r < c$, so that expressions such as Eq. (11.9) remain physical.

Now consider the cross term, $-2\omega r^2 d\varphi' dt$, which mixes time and space intervals. To appreciate what is going on, imagine setting up a cylindrically symmetrical gridwork of rods and synchronized clocks to make position and time measurements in the rotating reference frame. First, consider two clocks having the same r and φ' but different z values, as shown in Fig. 11.10. The clocks are rotating about the z axis at an angular velocity ω. We place a signal generator halfway between them, at rest in the rotating frame, to synchronize them. Because clocks 1 and 2 have velocities transverse to z, they receive the light pulses simultaneously in both the fixed (x, y, z) inertial frame and the rotating frame. The synchronization procedure produces clocks that are synchronized in both frames.

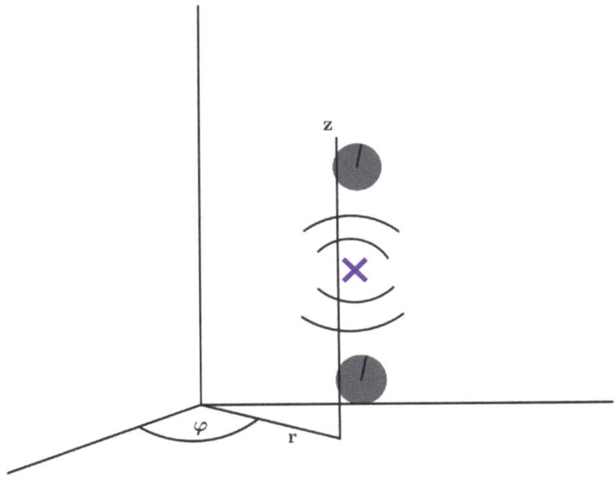

Figure 11.10 Synchronizing clocks at different heights on a rotating turntable.

However, clocks at the same r and z but different φ' suffer a different fate, as illustrated in Fig. 11.11, where the view is from above. We have placed our signal generator halfway between the two clocks at their common r and z values. But wait! The signal generator and the two clocks are each moving in separate directions, and each is experiencing a different centrifugal acceleration. We do not know how to handle this sort of

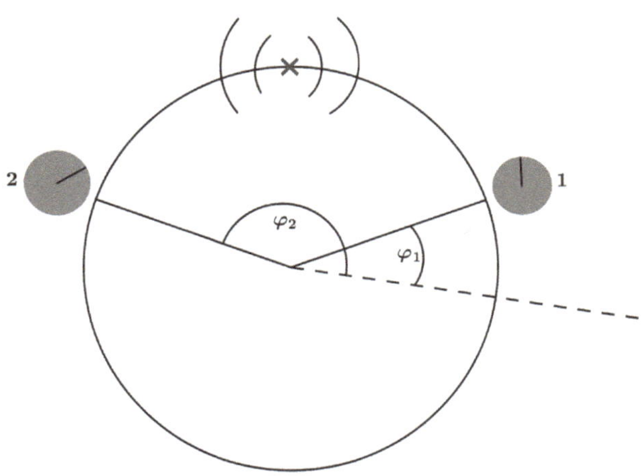

Figure 11.11 Synchronizing clocks at different angular positions on a rotating turntable.

situation using the principles of special relativity and the equivalence principle. However, if we move clocks 1 and 2 nearby, so they are separated by just an infinitesimal distance, we have a more secure, familiar situation. In this case, all three objects have essentially the same velocity, and they can be viewed from a locally inertial frame moving at their common velocity. Because clocks 1 and 2 are separated in their direction of motion, we know from our discussion of the relativity of simultaneity that they are *not* synchronized in the inertial frame defined by the coordinates (ct, x, y, z). The reason for this is clear—since light travels at the speed limit c with respect to any inertial observer, an observer at rest in the (ct, x, y, z) frame notes that clock 2 receives the light signal from the signal generator after clock 1 does because clock 2 is racing away from the source and clock 1 is racing toward the source. So, events that are synchronous in the rotating frame (clocks 1 and 2, for example) are not synchronous in the fixed inertial frame (ct, x, y, z) if they occur at different φ values. In fact, we know how large this effect is from our discussion of the relativity of simultaneity—it is the product of the velocity of the moving clocks times the distance between them in their rest frame divided by c^2. The relevant velocity is ωr. The relevant distance in the inertial frame (ct, x, y, z) is $rd\varphi$, which corresponds to a larger distance $\gamma(r)rd\varphi'$ in the clock's rest frame. Here $\gamma(r) = (1 - \omega^2 r^2/c^2)^{-1/2}$, so we have explicitly written γ as a function of r. (The notion of a local inertial frame is certainly important here.) So, $(\omega r)[\gamma(r)rd\varphi']/c^2$ is the time discrepancy that the frame (ct, x, y, z) notes on the clocks. But this is a time interval in the rotating frame, and we want the time difference in the fixed frame. This is given by multiplying by another factor of $\gamma(r)$ to account for time dilation, so the time difference is

$$\frac{\gamma^2(r)\omega r^2 d\varphi'}{c^2} = \frac{\omega r^2 d\varphi'}{c^2 - \omega^2 r^2}.$$

But the two clocks are synchronized in the rotating frame, so the time that passes there, dt', must be related to dt, the time that passes in the inertial frame, by

$$dt' = dt - \frac{\omega r^2}{c^2 - \omega^2 r^2} d\varphi'.$$

It must be that if we use this t' axis, the invariant interval will split into a spatial part and a temporal part. Substituting $dt = dt' + \omega r^2 d\varphi'/(c^2 - \omega^2 r^2)$ into Eq. (11.8), we find

$$ds^2 = (c^2 - \omega^2 r^2)dt'^2 - \left(dr^2 + \frac{c^2 r^2 d\varphi'^2}{c^2 - \omega^2 r^2} + dz^2\right) \tag{11.10}$$

and everything has worked out fine.

To interpret Eq. (11.10), we can compare it to the invariant interval of an inertial frame of reference chosen to approximate the transverse velocity ωr locally. For example, taking a clock at rest in the rotating frame, $ds^2 = (c^2 - \omega^2 r^2)dt'^2$, and comparing it to a clock at rest in a locally inertial frame. $ds^2 = c^2 d\tau^2$, we have time dilation again, $d\tau = \sqrt{1 - \omega^2 r^2/c^2}\,dt'$. Similarly, taking a meter stick pointing in the transverse direction at radius r, $ds^2 = -c^2 r^2 d\varphi'^2/(c^2 - \omega^2 r^2)$ and comparing that to a meter stick pointing in the same direction in a locally inertial frame, $ds^2 = -r^2 d\varphi^2$, we have $d\varphi = d\varphi'/\sqrt{1 - \omega^2 r^2/c^2}$, which is Lorentz contraction again because $rd\varphi$ is the proper length of the stick. These observations explain why the dt'^2 term in Eq. (11.10) has a prefactor $(c^2 - \omega^2 r^2)$ and the $d\varphi'^2$ term has that factor in the denominator. We see similar systematics in other invariant intervals in other applications for the same physical reason—the time dilation factor in special relativity is the same as the Lorentz contraction factor.

What kind of spatial geometry would an observer living on the turntable experience? The interval Eq. (11.10) does not resemble anything familiar. It is instructive to begin these considerations with a familiar example: the surface of a sphere. For a two-dimensional surface, such as a sphere, embedded in ordinary three-dimensional flat space, the metric is non-Euclidean, and the surface has intrinsic curvature as we will discuss in Section 12.1. There we construct a circle on the surface of the sphere of radius R and compare its circumference C to its geodesic radius a on the sphere itself. The curvature of the sphere causes C to be less than $2\pi a$. The Gaussian curvature of the spherical surface is,

$$K = \frac{3}{\pi} \lim_{a\to 0}\left(\frac{2\pi a - C}{a^3}\right) \tag{11.11}$$

and we will compute $K = 1/R^2$ in this case. So, smaller spheres have greater curvature that larger ones, an intuitively appealing answer. Eq. (11.11) will be derived in Chapter 12 where differential geometry and curvature will be introduced quantitatively.

In the four-dimensional space–time of general relativity, there is a generalization of curvature. Instead of one number K characterizing the curvature at a point, we need a whole collection of numbers, called the

Riemann curvature tensor, because the curvature in space—time depends on the direction in which we make measurements.

Consider a rotating coordinate system again and its spatial curvature. Imagine a circle of radius r whose center coincides with that of the turntable. In the nonrotating inertial frame, the circumference of the circle, C, is $2\pi r$. But what does an observer at rest at radius r in the *rotating* reference frame measure? When the observer at radius r' at rest in the rotating frame measures her circumference C', she places her meter stick down on the rotating turntable, making sure that it is perpendicular to the radius r'. An observer at rest in the nonrotating, inertial frame observes Lorentz contraction of this meter stick by a factor of $\gamma^{-1} = \sqrt{1 - v^2/c^2} = \sqrt{1 - \omega^2 r^2/c^2}$ and concludes that the number of meter sticks the rotating observer must place to go all the way around the circumference is $\gamma 2\pi r$. Both observers agree that $r' = r$ because the velocity of rotation is everywhere perpendicular to the spokes of the turntable. So, the space on the turntable is not flat, because the circumference C' is not 2π times the radius r'. We can calculate the curvature K using Eq. (11.11) and find for $v^2/c^2 \ll 1$ that $K = -3\, v^2/c^2 r^2$. Because $v = \omega r$ for the rotating turntable, we have the curvature at $r = 0$, $K = -3\, \omega^2/c^2$. Space on the rotating frame has a *negative* curvature because $C' > 2\pi r'$.

In summary, this effect comes from the metric, Eq. (11.10), for the turntable:

$$ds^2 = (c^2 - \omega^2 r^2)dt'^2 - \left(dr^2 + \frac{c^2 r^2 d\phi'^2}{c^2 - \omega^2 r^2} + dz^2\right). \tag{11.10}$$

We took a meter stick pointing in the transverse direction at radius r, $ds^2 = -c^2 r^2 d\varphi'^2/(c^2 - \omega^2 r^2)$ and compared that to a meter stick pointing in the same direction in a locally inertial frame, $ds^2 = -r^2 d\varphi^2$. So we found $d\varphi = d\varphi' / \sqrt{1 - \omega^2 r^2/c^2}$, and letting φ range from 0 to 2π, we saw that φ' changes by less, $\Delta\varphi' = 2\pi\sqrt{1 - \omega^2 r^2/c^2}$, and we have another indication of curvature.

A well-known consequence of the Lorentz contraction of the circumference of the relativistic turntable is "Thomas precession" [2] of atomic spectroscopy fame. Imagine that a tiny gyroscope is on the turntable at radius r. It rotates at an angular velocity $\omega = 2\pi/T$ (T is the period of rotation). When the turntable rotates through an angle φ, shown in Fig. 11.12A, the vector tangent to the rotating circumference rotates

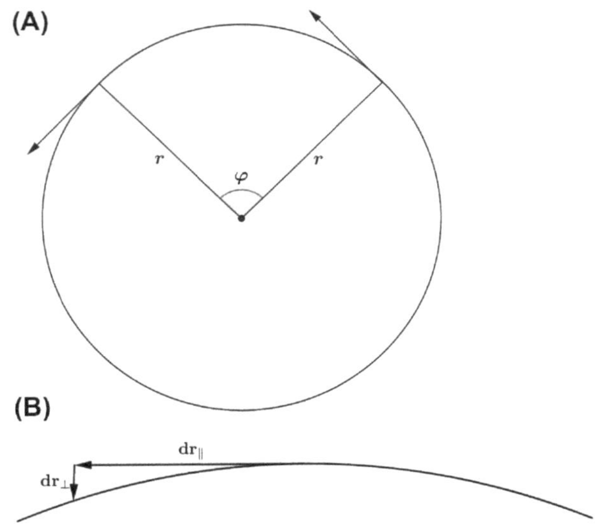

Figure 11.12 (A) Rotation of a vector tangent to the turntable. (B) Instantaneous transverse and longitudinal displacements of the tangent vector shown in (A).

through a greater angle α' because the circumference is contracted by a factor of $1/\gamma$ as measured by an observer who rotates with the turntable,

$$\alpha' = \frac{dr'_\perp}{dr'_\parallel} = \frac{dr_\perp}{dr_\parallel/\gamma} = \gamma\frac{dr_\perp}{dr_\parallel} = \gamma\varphi$$

where $dr_\perp$ and $dr_\parallel$ are shown in Fig. 11.12B. The difference of the angles in the inertial frame is $\Delta\alpha \equiv \alpha' - \varphi = \varphi(\gamma - 1)$. For a full rotation $\Delta\alpha = 2\pi(\gamma - 1)$ and the rate of precession of the gyroscope in the inertial frame is,

$$\frac{\omega_p}{\omega} = \frac{\Delta\alpha/T}{2\pi/T} = \gamma - 1$$

The angular precession in one revolution is,

$$2\pi(\gamma - 1) \approx 2\pi\left(\frac{1}{2}\frac{v^2}{c^2}\right) = \pi\frac{\omega^2 r^2}{c^2}$$

if $v^2/c^2 \ll 1$.

In atomic physics, this effect reduces the strength of the relativistic spin orbit coupling of spin-1/2 electrons orbiting a proton by a factor of two relative to a calculation where the electron travels on a linear trajectory past the proton. This subtle effect is discussed in textbooks on atomic spectroscopy and quantum mechanics, and it played an important role in the development of that subject [2]. The "best" derivation of the effect comes from the Dirac equation, which describes the quantum spin-1/2 electron relativistically. The relativistic energy of the stationary states of the electron bound in the electromagnetic field of the proton automatically incorporates Thomas precession since the Dirac equation has relativity built in.

There is much more to say about precession of gyroscopes in general relativity. The example of Thomas precession has just touched the surface of a broad and interesting field with many applications to the space science of orbiting vehicles.

The relativistic turntable is certainly instructive, but, since it rests on a simple coordinate transformation, it cannot illustrate all the basic principles of differential geometry and general relativity we are ultimately interested in. We will discuss curved surfaces and non-Euclidean geometry extensively in Chapter 12. But first let us consider the equivalence principle more critically.

11.3 TIDAL FORCES, NON-EUCLIDEAN GEOMETRY, AND THE MEANING OF "LOCAL INERTIAL REFERENCE FRAME"

There are aspects of the equivalence principle, which are quite subtle. The claim is that freely falling reference frames are *locally* inertial frames where the rules of special relativity hold. The key word is "local." What does that mean quantitatively? Let us answer the question with a thought experiment.

Suppose I am in the freely falling reference frame with four small balls. I am outside a planet of huge mass M with my four toys, and I orient myself so that my feet are nearest the planet, and I say that the planet is in the $-z$ direction. Newtonian mechanics will be adequate for the measurements and experiments we are about to describe. One of the balls is placed at my feet, one at the top of my head, another on my right side, and the last ball at my left side. They are all freely floating. If the reference frame were perfectly inertial, the four balls would retain their initial positions with respect to me at all times. What do I observe? The ball on my head accelerates away to larger z, the ball at my feet accelerates away to smaller z, and the balls at my right and left side accelerate toward me at half the rate of the first two balls! This surprises me. In fact, I feel rather poorly... I feel stretched in the z direction and squeezed in the transverse directions.

What is happening? Apparently the size of my body exceeds that of a local inertial frame centered at my midsection, and I feel the effects of gravity at my extremities. To describe the motions of the four balls, set up a Euclidean Cartesian coordinate system, x^i, $i = 1, 2, 3$, with its origin at the center of the planet. From Newton's perspective, the freely falling frame is accelerating toward the mass M. Call the acceleration of gravity at my midsection $\boldsymbol{g}$. It points in the $-z$ direction. Newton would describe the motions of my body at position $\boldsymbol{r}(t)$ with the second law,

$$\frac{d^2\boldsymbol{r}}{dt^2} = -\nabla\Phi(r) \tag{11.12}$$

where $\Phi(r) = -\frac{GM}{r}$ is Newton's gravitational potential and $r = \sqrt{x^2 + y^2 + z^2}$. Label the location of each ball with the vector $\boldsymbol{r} + \epsilon$, where $|\epsilon| \ll r$. The equation of motion of each ball then reads,

$$\frac{d^2}{dt^2}(\boldsymbol{r} + \epsilon) = -\nabla\Phi(\boldsymbol{r} + \epsilon) \tag{11.13}$$

Since ϵ is small we can expand the gravitational potential around the point $\boldsymbol{r}$,

$$\Phi(\boldsymbol{r} + \epsilon) = \Phi(r) + \epsilon\cdot\nabla\Phi(r) + \cdots$$

Finally we can take the difference of the two equations of motion Eqs. (11.12) and (11.13) to isolate the equation of motion for each ball to first order in ϵ,

$$\frac{d^2\epsilon}{dt^2} = -\nabla(\epsilon\cdot\nabla\Phi(r)) \tag{11.14}$$

We learn that the balls accelerate with respect to me by an amount that varies as the second derivatives of the potential. We can obtain an explicit formula by working out the derivatives,

$$\nabla\Phi(r) = \frac{GM}{r^3}\boldsymbol{r}$$

$$\nabla\left(\frac{GM}{r^3}\epsilon\cdot\boldsymbol{r}\right) = -3\frac{GM}{r^4}(\epsilon\cdot\boldsymbol{r})\widehat{\boldsymbol{r}} + \frac{GM}{r^3}\epsilon\cdot\widehat{\boldsymbol{r}}$$

So, finally, in Cartesian components,

$$\frac{d^2\epsilon^i}{dt^2} = \frac{GM}{r^5}\sum_k \epsilon^k\left(3x^k x^i - r^2\delta^{ki}\right) \tag{11.15}$$

We see that the right-hand side of this equation has the spatial distribution of a "quadrupole moment,"

$$Q^{(ij)} = \left(3x^i x^j - r^2\delta^{ij}\right)$$

We shall see more quadrupole moments in other gravitational phenomena below.

Now we can get our answers! Choose $x = 0$, $y = 0$, and $z = r$. Then, on substitution, into the equation of motion Eq. (11.15), we have,

$$\frac{d^2\epsilon^1}{dt^2} = -\frac{GM}{r^3}\epsilon^1, \quad \frac{d^2\epsilon^2}{dt^2} = -\frac{GM}{r^3}\epsilon^2, \quad \frac{d^2\epsilon^3}{dt^2} = +\frac{2GM}{r^3}\epsilon^3$$

And we read off the stretching and squeezing alluded to above: there is elongation in the z direction and squeezing in the transverse directions.

This exercise illustrates several points.

First, we have rediscovered gravitational tidal forces. Recall from your mechanics course that the Moon creates tidal forces on the Earth, and these forces are responsible for the daily high and low tides in a harbor such as London or New York. As the Earth rotates on its axis under the Moon, there are two high tides and two low tides each solar day, approximately. This is because of the elongation we discovered in $\epsilon^{(3)}$ and squeezing we discovered in the transverse directions $\epsilon^{(1)}$ and $\epsilon^{(2)}$. We visualize this effect in Fig. 11.13. In part A. of the figure, we show the forces exerted on the Earth by the Moon, and in part B. we *subtract* the force exerted on the *center* of the Earth so we concentrate on the "differential" force across the Earth. Here we see the pattern of forces derived in Eq. (11.15) that are described

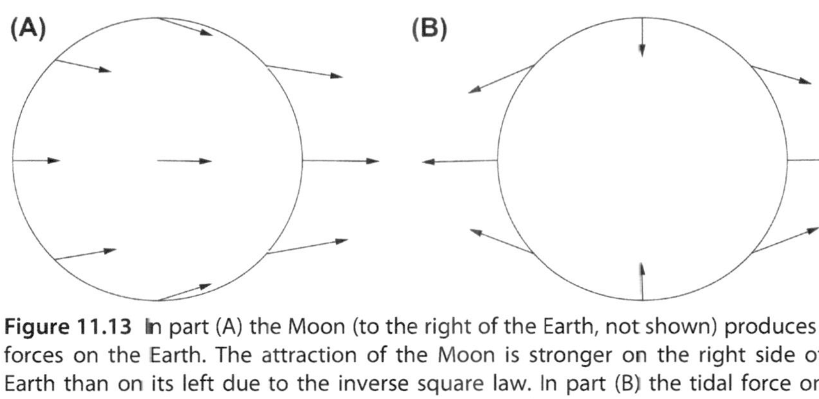

Figure 11.13 In part (A) the Moon (to the right of the Earth, not shown) produces tidal forces on the Earth. The attraction of the Moon is stronger on the right side of the Earth than on its left due to the inverse square law. In part (B) the tidal force on the center of the Earth has been subtracted from all the forces shown in part A. The results are the *differential* forces across the Earth.

by the quadrupole moment: there is elongation in the direction toward the Moon and squeezing in the transverse directions. We also see that the rotational frequency of the quadrupole moment is *twice* that of the Earth: there are two high tides and two low tides in a 24 h period, approximately.

We will have more to say about this later when we develop the equations of general relativity in Chapter 12.

11.4 GRAVITATIONAL REDSHIFT

11.4.1 A Freely Falling Inertial Frame

The gravitational redshift represents one of the simplest and decisive tests of the equivalence principle. We are interested in how light propagates in a gravitational field. Suppose that light of frequency ν_e is emitted upward from the surface of a planet to be observed some distance h away. What frequency ν_o does the observer measure? We know that light travels at the speed limit c in any inertial frame, and we know that a freely falling frame is inertial. So, consider first the propagation of the light signal from the surface of the planet from the perspective of an inertial frame falling freely in the approximately uniform gravitational field g as shown in Fig. 11.14. The light pulse travels a distance h in the time interval from $t = 0$ to $t = h/c$, and the falling frame has a velocity $v = gh/c$ downward relative to the observers at rest around the planet when the light pulse reaches its destination. The local falling frame is inertial, and the light pulse has frequency ν_e when it is emitted and when it is absorbed, as measured in the falling frame. But at $t = h/c$, the inertial frame is moving downward, opposite to the direction of propagation of the light pulse, so an observer at rest on the planet detects a Doppler shift toward the red when he observes the light pulse at $t = h/c$. The frequency shift is

$$\nu_o = \sqrt{\frac{1 - v/c}{1 + v/c}}\ \nu_e \approx \sqrt{1 - 2v/c}\ \nu_e \approx (1 - gh/c^2)\nu_e,$$

where we used the Doppler shift formula from Chapter 3 and $v = gh/c$, and linearized the formula for our present application, where $v/c \ll 1$.

The fractional change of the frequency is

$$\frac{\nu_o - \nu_e}{\nu_e} = \frac{\Delta\nu}{\nu} = -\frac{hg}{c^2}. \tag{11.16}$$

So, the observer at height h above the surface of the planet observes a slightly red-shifted wave.

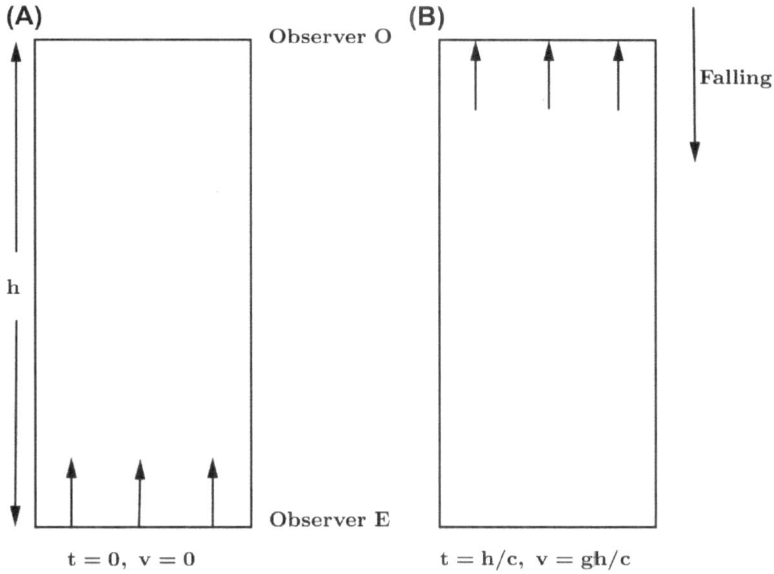

Figure 11.14 Propagation of a light signal from the surface of a planet from the perspective of an inertial frame falling freely in the approximately uniform gravitational field *g*. In part (A) the light signal leaves the planet's surface and in part (B) it is detected a height *h* above the surface.

Another way of presenting this result, which is more fundamental, is to say: *Identically constructed clocks run slower in lower gravitational potentials.* Because frequency varies as the reciprocal of time, $\nu = 1/t$, we can express Eq. (11.16) as a fractional time difference,

$$\frac{\Delta t}{t} = \frac{hg}{c^2}. \tag{11.17}$$

In other words, an observer E at height 0 could send light signals to observer O at height h once a second. Eq. (11.16) then states that observer O detects these signals more widely spaced in time, at a diminished or red-shifted frequency. So, observer O concludes that the clock at height 0 is running more slowly than hers, according to Eq. (11.17).

Just as the observer at height 0 sent light signals to an observer at height h to compare clocks, their roles could be interchanged. Then the same argument, modified to account for the fact that now the freely falling frame is accelerating in the same direction as the light ray, so the Doppler shift formula applies with v replaced by $-v$, predicts that the lower observer detects a blue-shifted light ray. The lower observer concludes that the clock in the higher gravitational potential runs faster than his by a fractional change of $-hg/c^2$.

11.4.2 An Accelerating Spaceship

Another way to use the equivalence principle to analyze the gravitational redshift is to replace the problem in a gravitational field with one in an accelerating reference frame (e.g., a spaceship), as shown in Fig. 11.15. The spaceship of height h is in a background inertial frame, so light travels at velocity c there regardless of its origins. Light leaves the base of the ship (E), travels up during a time $t = h/c$, and is received at the tip of the ship (O). But O is receding at velocity $v = at = gh/c$, so it detects the frequency of light

$$\nu_o = \sqrt{\frac{1 - v/c}{1 + v/c}}\ \nu_e \approx \sqrt{1 - 2v/c}\nu_e \approx (1 - gh/c^2)\nu_e$$

in agreement with our earlier argument.

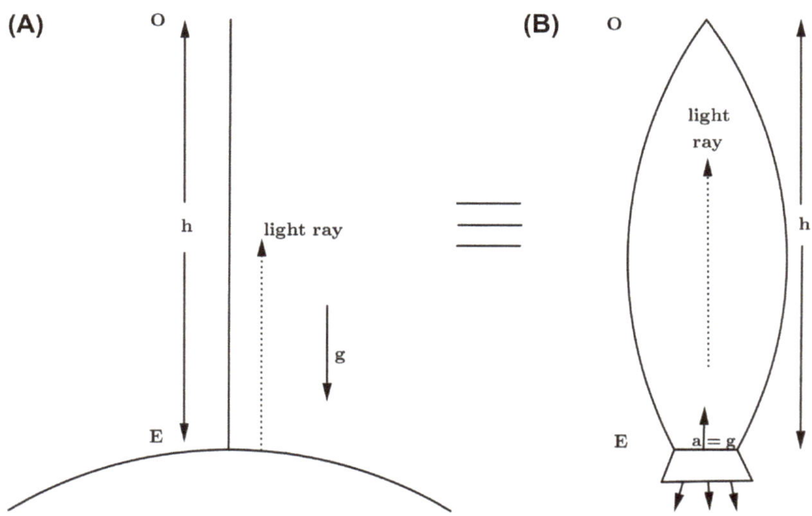

Figure 11.15 Gravitational redshift from the perspective of an accelerating frame of reference. Part (A) shows the physical situation on the surface of a planet and part (B) shows the equivalent environment in an accelerating spaceship.

11.4.3 Gravitational Redshift and the Relativity of Simultaneity

The gravitational redshift is certainly more fundamental than the Doppler shift formulas, so let us derive the result directly from more basic features of relativity. Observer O wants to compare the amount of time that passes at her height h in the gravitational field to the amount that passes at height 0 where observer E resides. Suppose that O has a clock; call it clock 2. Observer E also has a clock with him, call it clock 1, but suppose that light beams cannot be sent between the two observers. How can the observers compare the operation of clocks 1 and 2? Well, observer O could drop clock 2 to observer E, and the clocks could be compared at height 0. This is a good idea because when the observer O drops her clock 2, clock 2 is freely falling in gravity, and it is in an inertial frame of reference where space and time follow the rules of special relativity that we know so well. (In fact, suppose as usual that there is a synchronized gridwork of clocks in this inertial, freely falling frame. We will need to consult them later in this argument.) It is important that all the clocks involved be constructed identically, as usual. Observer E at height 0 can then note the hands on clock 2 when it reaches him at time t. The time t is determined by the fact that the falling clock accelerates at a rate g through a distance h, so $gt^2/2 = h$. So, $t = \sqrt{2h/g}$ and the velocity of clock 2 when it passes observer E at height 0 is $v = gt = \sqrt{2hg}$. But observer E is not really interested in the time on clock 2; he wants to know how much time has passed at height h. Observer O at height h can look at a freely falling clock at height h that was synchronized with clock 2 in that freely falling inertial frame and read off the amount of time that has passed. Now for the crux of the matter! Observer E, being a good student of relativity, knows that a freely falling clock at height h is *not* synchronized with clock 2 when measured from *his* perspective because of the relative motion between the frame where he is at rest, the surface of the planet, and the freely falling frame—the relativity of simultaneity states that E measures such a clock to be ahead of clock 2 by a time interval xv/c^2, where x is just the height h. But $x = vt/2$ because the acceleration is approximately constant, so the time difference is $xv/c^2 = v^2t/2c^2$. Therefore, observer E states that a time $t + v^2t/2c^2$ has passed at height h while a time interval t has passed at $h = 0$. The total time interval T detected by observer E is, then,

$$T = t + \frac{v^2 t}{2c^2} = \left(1 + \frac{v^2}{2c^2}\right)t = \left(1 + \frac{1}{2}\cdot\frac{2hg}{c^2}\right)t = \left(1 + \frac{hg}{c^2}\right)t.$$

So, observer E concludes that a clock at height h runs more quickly—more time has passed there—with a fractional difference of $\Delta t/t = hg/c^2$. In other words, observer E concludes that clocks at higher gravitational potential run faster. This result is in agreement with our previous conclusion. The relativity of simultaneity, the fact that clocks that are synchronized in one inertial frame are measured to be out of synchronization when measured by an observer in relative motion, is the fundamental idea behind the gravitational redshift, correctly concludes observer E.

Now that our two observers E and O can compare time measurements at different gravitational potentials, they can turn to length measurements. Observer E can measure observer O's meter sticks similarly to how she measured O's clock readings. We suppose again that observer O is at height $z = h$ above observer E who is at height $z = 0$. Observer O aligns her meter stick vertically and drops it. After a time $t = \sqrt{2h/g}$ it is at height $z = 0$ and observer E measures its length. Since it is traveling at velocity $v = \sqrt{2hg}$ relative to observer E, observer E measures its length contracted by a factor of $\sqrt{1 - v^2/c^2} = \sqrt{1 - 2hg/c^2} \approx 1 - hg/c^2$. Similarly, if the dropped meter stick were aligned in the transverse direction, observer E would measure its length unaffected by the gravitational field.

These considerations suggest that in a radially symmetric, weak gravitational potential $V(r)$, the invariant interval would read,

$$ds^2 = \left(1 + \frac{2V(r)}{c^2}\right)c^2 dt^2 - \left(1 - \frac{2V(r)}{c^2}\right)dr^2 - r^2\left(d\theta^2 + \sin^2\theta d\varphi^2\right) + O\left(\frac{1}{c^4}\right)$$

This metric will be derived from the Einstein field equations of general relativity in the next chapter where strong gravity solutions will also be presented. It is useful in calculating the bending of light as it passes by stars, the precession of planets and satellites in elliptical orbits around the stars, etc.

11.4.4 A Rotating Reference Frame

Our exercise in rotating reference frames affords yet another example of the gravitational redshift. The proper time interval $d\tau$ was related to the time interval that passes in the fixed inertial frame as $d\tau = \sqrt{1 - \omega^2 r^2/c^2}dt$. A mass at rest in the rotating frame experiences a centripetal acceleration $\omega^2 r$, which is generated from a potential $V(r) = -\omega^2 r^2/2$, because $\omega^2 r = -(d/dr)(-\omega^2 r^2/2)$. So, we can write $d\tau = \sqrt{1 + 2V(r)/c^2}dt$ and the proper time intervals at different radii r_1 and r_2 are related,

$$\frac{d\tau_1}{d\tau_2} = \sqrt{\frac{1 + 2V(r_1)/c^2}{1 + 2V(r_2)/c^2}}.$$

Therefore, the frequency ν_1 of an oscillator at r_1 is related to that at r_2 by

$$\frac{\nu_1}{\nu_2} = \sqrt{\frac{1 + 2V(r_2)/c^2}{1 + 2V(r_1)/c^2}}.$$

If both $V(r_1)/c^2$ and $V(r_2)/c^2$ are much less than unity, we can linearize this expression and find

$$\frac{\nu_1}{\nu_2} \approx \left[1 + \frac{V(r_2) - V(r_1)}{c^2}\right]$$

or

$$\frac{\nu_1 - \nu_2}{\nu_2} \approx \frac{V(r_2) - V(r_1)}{c^2}. \tag{11.18}$$

In the case of the rotating reference frame, $V(r) = -\omega^2 r^2/2$. So, if a pulse of light is emitted radially from r_1 and is observed in the rotating frame at $r_2 \neq r_1$, the observer at r_2 detects a frequency-shifted pulse. Clearly, this is just an example of the transverse Doppler shift of special relativity because $v_t = \omega r$ is the transverse velocity of the rotating reference frame at radius r.

Collecting our various examples, a general discussion of the gravitational redshift in a nonuniform gravitational field V predicts

$$\frac{\Delta\nu}{\nu} \approx -\frac{\Delta V}{c^2}, \tag{11.19}$$

where $\Delta\nu$ is the frequency difference and ΔV is the potential difference between the point of detection and emission of the light wave. We can write this result in differential form,

$$\frac{d\nu}{\nu} = -\frac{dV}{c^2}$$

and integrate it between point A where the frequency is ν_A and the potential is V_A and point B where the frequency is ν_B and the potential is V_B,

$$\frac{\nu_B}{\nu_A} = \frac{e^{-V_B/c^2}}{e^{-V_A/c^2}} = e^{-(V_B - V_A)/c^2}$$

This result will prove useful in applications.

We can further summarize these exercises by incorporating the gravitational potential into the space–time metric. In a weak gravitational field $V(r)/c^2 \ll 1$, the invariant interval becomes,

$$ds^2 = (1 + 2V(r)/c^2)c^2dt^2 + \cdots$$

to first order in $V(r)/c^2$. The Minkowski metric of special relativity had $g_{00} = 1$ but weak gravitational effects alter it to read,

$$g_{00} = 1 + 2V(r)/c^2 + \cdots$$

And in strong gravitational fields,

$$ds^2 = e^{2V/c^2}c^2dt^2 + \cdots, \qquad g_{00} = e^{2V/c^2}$$

Chapter 12 will expand on these ideas.

11.4.5 A Famous Experimental Test of Gravitational Redshift

It is interesting that the best experimental determination of gravitational redshift comes from terrestrial, controlled experiments that rely on the Mössbauer effect to measure tiny frequency shifts rather than from observations of distant stars. Recall the Mössbauer effect discussed in Section 6.6. The experiment involved an emitting atom within a regular crystal array. The emitting atom shares its recoil energy and momentum with its entire crystal environment, so its recoil velocity is reduced to essentially zero, and all the energy between the quantum energy levels in the atom shows up as the energy of the emitted light wave. If the light emitted from such a crystal is incident on another identical crystal, absorption is possible because the energy of the light exactly matches an energy difference in the spectrum of the receiving atom, which can absorb the light without recoiling. If either the absorbing crystal or emitting crystal has a small relative velocity, the attendant Doppler shift shifts the energy of the light enough to make resonant absorption impossible. (Recall that in quantum physics, the frequency of light is related to its energy by the Planck relation $E = h\nu$, where h is Planck's constant, $h = 6.627 \cdot 10^{-34}$ J-s, so we can speak equally well in terms of frequency ν or energy when discussing emission and absorption of photons by atoms.) In fact, the energy levels of atoms are not infinitely precise—each energy level has a natural width that is a consequence of the Heisenberg uncertainty relation. Mössbauer was then able to study the profiles of spectral lines, obtaining their widths and detailed shapes by using his crystals and the physics of the Doppler shift.

R. V. Pound, G. A Rebka, and J. L. Snider used the Mössbauer effect to measure the redshift of light near the surface of Earth [3]. They placed a source of ^{57}Co 22.6 m higher than a receiver. The equivalence principle predicts a fractional frequency shift of $\Delta\nu/\nu_e \approx 2.46 \cdot 10^{-15}$, which is very tiny: three orders of magnitude smaller than the intrinsic natural uncertainty in the frequency of the emitted light $\Delta\nu_{QM}$, $\Delta\nu_{QM}/\nu \approx 1.13 \cdot 10^{-12}$. To obtain an observable effect, they gave the source a sinusoidal velocity $u = u_o \cos \omega t$ to induce a controlled, mechanical Doppler shift, $\Delta\nu_D/\nu = -(u_o/c) \cos \omega t$. Then the resonance absorption cross section depends on $\Delta\nu + \Delta\nu_D$, and. if $\Delta\nu_D \gg \Delta\nu$, an exercise in quantum mechanics shows that the cross section has a term linear in $\cos \omega t$ with a strength proportional to $\Delta\nu$, the gravitational redshift. By isolating this characteristic external frequency ω in the cross-sectional measurements, the experimenters confirmed the prediction of general relativity to good precision, about 10%. It is interesting that this terrestrial experiment is much more decisive than astronomical data of light emitted from distant stars because of all the inherent uncertainties in such observations (the Doppler effects of the movement of the star and the movement of the emitting atom in the turbulent hot surface of the star must be accounted for somehow).

11.4.6 Gravitational Redshift and Energy Conservation

We end this discussion with an illustration of the gravitational redshift that derives it from energy conservation. We show that the gravitational redshift is necessary for the internal consistency of relativistic dynamics—without it we could construct a perpetual motion machine!

Consider a pulley on the surface of Earth supporting two observers at the ends of a string as shown in Fig. 11.16 [4].

Let the difference of heights of the observers be h, as usual. Let the lower observer shine a flashlight at the higher observer, who sees the light, so it is absorbed on his retina. (Notice the similarity of this argument to our original derivation of $E = mc^2$, except the apparatus is now vertical and in an ordinary gravitational field.) Let the light have energy E, so from special relativity we know that it has a mass equivalent of E/c^2. So, if the light travels from the lower to the higher observer, then mass E/c^2 has been transferred between the observers; the upper one is now slightly heavier than the lower one, so it sinks and does work equal to the change in the potential energy, $hg \cdot E/c^2 = E \cdot \Delta V/c^2$. Our pulley system is now back to its original configuration, and we can repeat the process as many times as we wish and have an inexhaustible source of work.

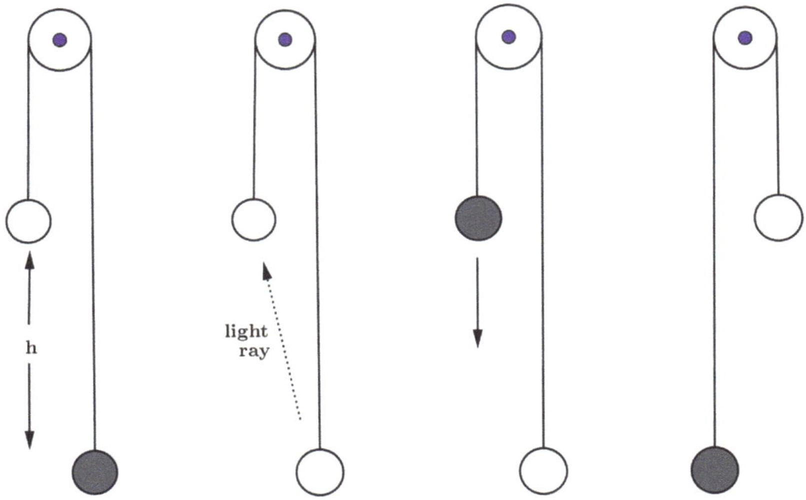

Figure 11.16 The conjectured operation of a perpetual motion machine which is, in fact, undone by the gravitational redshift.

We have done it! We have made a perpetual motion machine! Well, except for one thing—we forgot about the gravitational redshift! The light detected by the higher observer in Fig. 11.16 has a smaller frequency, given by our gravitational redshift formula, $\Delta\nu/\nu = -\Delta V/c^2$. How can this get us out of our conundrum? If we return to Eq. (6.20), it is easy to see (as we verify later) that the energy that light carries transforms between frames exactly as its frequency. If that is the case, then the energy the light deposits on the retina of the higher observer is diminished by $E\cdot\Delta V/c^2$, which is exactly the work we were hoping to get out of our machine! So, in reality, when the light from the lower observer reaches the higher one, its energy is diminished by the gravitational redshift by just the amount we wanted to generate. So, yet another perpetual motion machine design bites the dust!

Let us check that light's frequency and energy transform identically in special relativity. Recall from Eq. (6.20) that if we know the energy E and x component of the relativistic momentum p_1 in frame S, then the energy in frame S' is $E' = \gamma(E - vp_1)$. But for light propagating in the x direction, $p_1 = E/c$, so the transformation law becomes

$$E' = \gamma\left(E - \frac{vE}{c}\right) = \gamma\left(1 - \frac{v}{c}\right)E = \sqrt{\frac{1 - v/c}{1 + v/c}}E,$$

which we recognize as the Doppler shift formula for the light's frequency. And so the proposed perpetual motion machine fails miserably.

The fact that light's frequency ν and energy E transform identically under boosts is important in the quantum theory of light we have touched on in our discussion of relativistic collisions and quantum energy levels. Planck's equation, $E = h\nu$ (where h is Planck's constant), is consistent with special relativity because E and ν share the same transformation law.

This observation suggests yet another way of viewing the gravitational redshift—it is just a consequence of energy conservation! When a photon, a quantum of light energy, travels from the observer on the surface of a planet where its frequency is ν_e and its height is 0, to an observer where its frequency is ν_o and its height is h, the total energy, accounting for the gravitational potential, must be conserved:

$$E_e = E_o + hg\frac{E_o}{c^2}.$$

The change in potential energy has been written as $hg(E_o/c^2)$, which is the change in the potential hg in the uniform gravitational field times the mass equivalent of the energy E_o there. Solving for E_o,

$$\frac{E_o}{E_e} = \frac{1}{1 + hg/c^2} \approx 1 - \frac{hg}{c^2},$$

which we can write as a fractional change in energy, which is also the fractional change in frequency,

$$\frac{E_o - E_e}{E_e} = \frac{\nu_o - \nu_e}{\nu_e} \approx -\frac{hg}{c^2},$$

and we have derived Eq. (11.16) again from a different perspective!

This result gives us a nice alternative view of gravitational redshift. Why does the frequency of light change as it propagates away from the surface of a celestial body? Because it propagates to a new location where its gravitational potential is larger, so its relativistic energy, and hence its frequency, must be diminished accordingly!

11.5 THE TWINS AGAIN

When we last left our twins, Mary and Maria, they were coping with the fact that Mary is 4 years older than Maria after her space trip. In that discussion, we found that from Maria's perspective Mary ages an unexpectedly large amount when Maria jumps from the outgoing to the incoming rocket at the midpoint of her trip. Recall from that discussion that the lines of constant time in the frame of the outgoing rocket are at widely different angles from the lines of constant time in the frame of the incoming rocket, as shown in the Minkowski diagram, Fig. 3.19. In fact, we saw that Maria measures that 6.4 years passes on Mary's clock during Maria's turnaround!

By using the device of two rockets, we have been able to analyze the twin paradox without explicitly considering space and time measurements in an accelerated reference frame. Now that we know that accelerated reference frames are equivalent to environments in a gravitational field, we can face the problem head on. Our only limitation is that our discussion of general relativity so far is good only for weak gravitational fields, or, equivalently, small values of v/c. We will do better when we present the full apparatus of general relativity, Einstein's field equations in Chapter 12. That work will justify this application where large gravitational effects occur.

View the trip from Maria's frame. From Maria's perspective, Mary goes out and back. By the equivalence principle, then, Maria's acceleration can be replaced by a gravitational field. We already know that the important portion of the trip is the period during which Maria detects Mary's reversal. Call the distance between the sisters vT at this point, where T is the time they have been traveling apart at velocity v, according to Mary. At this point, the turnaround, Maria actually turns on her rocket motors and experiences an acceleration a for a time t. However, invoking the equivalence principle, we can replace the rocket environment by one in which there is a gravitational field in Maria's frame that produces an acceleration a, as shown in Fig. 11.17. The gravitational potential difference between the sisters is distance times acceleration, avT, which causes Mary's clock (her heartbeat) to gain the total amount $(avT/c^2)t$, according to our formulas developed for the gravitational redshift in Section 11.4.3. Finally, t must be long enough to accommodate the reversal from the outgoing speed v to the incoming speed $-v$, so $at = 2v$. Therefore, the total time gained by Mary's clock is $2Tv^2/c^2$.

Does this result agree with our previous analysis of the twin paradox? Return to Fig. 3.19 and compute the time that Maria states passes on Mary's clock during the turnaround. We pull out our trusty relativity of simultaneity formula, which reminds us that Maria measures clocks at rest in Mary's frame that are separated by a distance x to be out of synchronization by an amount xv/c^2. So, if we change from the outgoing to the incoming rocket, causing v, to change by Δv, the time that passes on Mary's clock is

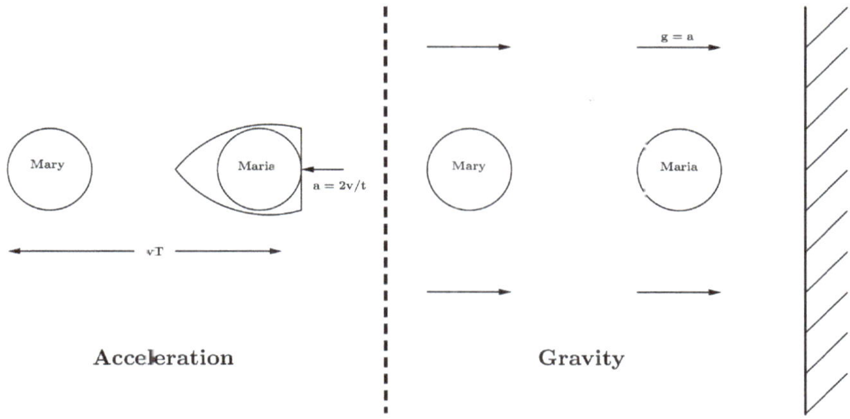

Figure 11.17 General relativistic analysis of the twin paradox. From Maria's perspective Mary ages 6.4 years during Maria's turnaround.

$x\Delta v/c^2$. But x is the distance between the sisters in Mary's frame, which is vT in our discussion here. And finally, because the velocity v reverses, $\Delta v = 2v$. So, our earlier relativity of simultaneity discussion predicted that Mary's clock, as measured by Maria, gains a time of $2Tv^2/c^2$, in complete agreement with the general relativity result obtained here! Using the numbers of our previous discussion in Section 3.4, Fig. 3.19, $T = 5$ and $v/c = 0.8$, so Mary's time gain is $2 \cdot 5 \cdot 0.8^2$, which is 6.4 years, as we found earlier.

11.6 AN AGING ASTRONAUT

When an astronaut circles the globe on the space shuttle, does she age faster or slower than we who stay at home on Earth?

There are two effects to consider. First, she is moving in an Earth orbit, circling Earth every 90 min or so. Therefore, special relativity, time dilation in particular, predicts that we measure her clocks as running slowly. Second, she resides at a higher altitude where the gravitational potential, $V(r) = -GM/r$, is higher, so general relativity predicts that her clocks run faster than ours. These two effects compete. Which wins out?

First, her velocity in a low Earth orbit is rather modest on the scale of the speed of light c. The orbit is almost circular with a radius r, and

Newtonian kinematics and dynamics apply quite accurately. The centrifugal acceleration outward balances the gravitational acceleration inward, so

$$\frac{v^2}{r} = \frac{GM}{r^2},$$

where v is the astronaut's velocity and M is the mass of Earth. It is more convenient to write this in terms of the acceleration of gravity g, $9.8\ \mathrm{m/s^2}$, observed on the surface of Earth,

$$g \equiv \frac{GM}{r_o^2},$$

where r_o is the radius of Earth. So,

$$v^2 = \frac{gr_o^2}{r}.$$

Substituting in numbers for the low Earth orbit, we find that $v \approx 7700\ \mathrm{m/s}$, so $v^2/c^2 \approx 6.6 \cdot 10^{-10}$ and all relativistic effects will be very tiny—but measurable by modern techniques. Comparing the astronaut's proper time τ to the time that passes on Earth t, taking just the relative velocity into account, gives

$$t = \gamma\tau = \frac{\tau}{\sqrt{1 - v^2/c^2}} \approx \left(1 + \frac{v^2}{2c^2}\right)\tau.$$

So,

$$\frac{\Delta t}{\tau} \approx \frac{v^2}{2c^2} = \frac{gr_o^2}{c^2 r} \tag{11.20a}$$

is the extra time that passes on clocks on the surface of Earth when a time τ passes on the astronaut's wristwatch.

Now we need the contribution to $\Delta t/\tau$ due to the gravitational potential difference between the astronaut and we, who stay at home. This is given by our red-shift formula, Eq. (11.18),

$$\frac{\Delta t}{\tau} \approx \frac{V(r_o) - V(r)}{c^2} = \frac{(-GM/r_o + GM/r)}{c^2}.$$

So,

$$\frac{\Delta t}{\tau} \approx \frac{gr_o^2}{c^2}(1/r - 1/r_o). \tag{11.20b}$$

Because r is always greater than r_o, Δt is negative, as expected—general relativity causes clocks at lower potentials to run slower, so the clock on Earth falls behind.

Finally, adding Eq. (11.20a) and Eq. (11.20b), we get our full answer,

$$\frac{\Delta t}{\tau} \approx \frac{gr_o^2}{c^2}\left(\frac{3r_o}{2r} - 1\right) \approx 7 \cdot 10^{-10}\left(\frac{3r_o}{2r} - 1\right). \tag{11.21}$$

What a funny result! We can enhance or reverse the relative aging processes by adjusting the height of the astronaut's orbit—if the orbit is high ($r > 3r_o/2$), the gravitational effects win and a person on the surface ages more slowly than the astronaut, but if the orbit is low ($r < 3r_o/2$), the velocity effects win and a person on the surface ages faster than the astronaut. Of course, these effects are truly tiny for earthly conditions. But they are observable using atomic clocks, which have accuracies greater than one part in 10^{10}.

11.7 BENDING OF LIGHT IN A GRAVITATIONAL FIELD

As we discussed earlier, the equivalence principle implies that *all* physical phenomena experience the acceleration of gravity. This point applies to wave phenomena, as well as to particles with rest masses. The original quantitative "calculation" of the gravitational bending of light was done in a nonrelativistic setting in 1801 by Johann Georg von Soldner, a German mathematician, following Newton's "corpuscular" theory of light. It is interesting that the relativistic calculation gives *twice* as large a deflection and is in agreement with modern high-tech experiments.

Before we review the relativistic calculation that uses the weak field metric discussed in Section 11.4.3, let us consider the nonrelativistic calculation. Imagine a particle of rest mass m that glances by a star of mass M and radius R as shown in Fig. 11.18. For a star similar to the Sun, the deflection angle χ proves to be very small. The particle feels the radial force of gravity and, as is clear from the symmetry in the picture, the star imparts a net transverse momentum to the particle, which bends its trajectory as shown. The calculation is successful only if m, the mass of the particle, cancels out of the calculation: the effect must be universal and independent of m. This occurs because, if we use Newton's theory of gravity, the particle experiences a force $\mathbf{f}$ proportional to its mass m, which produces an acceleration given by $\mathbf{f} = m\mathbf{a}$, predicting an acceleration $\mathbf{a}$ independent of m.

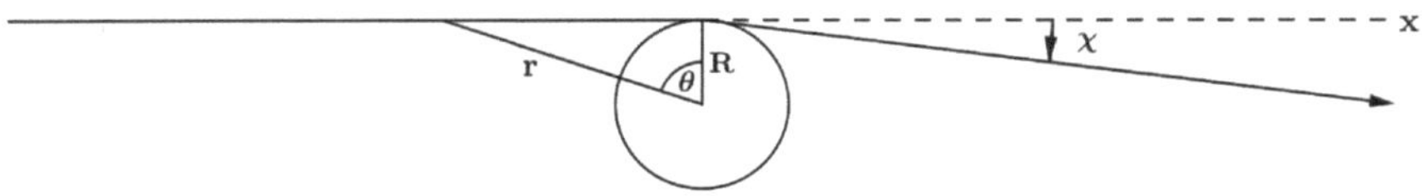

Figure 11.18 Kinematic set up for a Newtonian calculation of the bending of light ("corpuscular" light for Newton). The closest approach of the light beam to the center of the planet is R and it bends through an angle χ.

Following the original Newtonian calculation, the gravitational force F produces a change in the particle's transverse momentum P_T,

$$dP_T = F\cos\theta \ dt. \tag{11.22}$$

We read from Fig. 11.18 that $x = R\tan\theta$, so we can trade the linear position of the particle for the angle θ. The angle θ varies from $-\pi/2$ to $\pi/2$ during the process. In addition, because the deflection is small we have to good approximation $dx = cdt$, where we are also supposing that the particle's speed is essentially c. Now, $dx = R\sec^2\theta \ d\theta$, so Eq. (11.22) becomes using $R = r\cos\theta$,

$$P_T = \int_{-\pi/2}^{\pi/2} \frac{GMm}{cR}\cos\theta \ d\theta = \frac{2GMm}{cR}. \tag{11.23}$$

The deflection angle is approximately

$$\chi = \frac{P_T}{P} = \frac{2GMm/cR}{mc}. \tag{11.24}$$

We see that m cancels out and the final answer is

$$\chi = \frac{2GM}{c^2R} \text{ radians}. \tag{11.25}$$

Substituting in the parameters for the Sun, we predict $\chi = 0.87$ s of arc, a tiny but measurable effect. Note that the effect is suppressed by two powers of the speed of light.

This curious calculation leaves several points unanswered. Treating light as a particle obeying Newtonian kinematics, on the one hand, and traveling at the relativistic speed limit, on the other, is peculiar and ad hoc. In addition, in this classical physics setting, light is a wave phenomenon so the applicability of this calculation is not really clear. The calculation is historically significant, however, because it illustrates Newton's corpuscular theory of light. It misses the real answer of general relativity by a factor of 2.

Let us calculate the deflection of light by the gravitational attraction of the Sun by modeling light as a wave and treating space–time through general relativity [5]. Consider a light front propagating through a gravitational field described to $O\left(\frac{1}{c^4}\right)$ by the weak field metric motivated in Section 11.4.3,

$$ds^2 = \left(1 - \frac{2GM}{c^2 r}\right)c^2 dt - \left(1 + \frac{2GM}{c^2 r}\right) dr^2 - r^2\left(d\theta^2 + \sin^2\theta \ d\varphi^2\right). \tag{11.26}$$

Our calculation will be accurate just to first order in $1/c^2$. In an inertial reference frame, light travels on straight lines at the speed limit c. Along the world line, the invariant interval ds^2 vanishes. But because ds^2 is an invariant, $ds^2 = 0$ holds in the rest frame of the Sun as well as in a freely falling inertial frame. So, inspecting Eq. (11.26) with $ds^2 = 0$, we see that the velocity of light, as measured on the axes ct, x, y, and z, will be r dependent. So, when a wave front passes the Sun, different parts of it propagate at different speeds, and the wave front changes direction. The situation is shown in Fig. 11.19. The deflection $\Delta\chi$ is, reading from the figure,

$$\Delta\chi \approx \frac{v(y + \Delta y)\Delta t - v(y)\Delta t}{\Delta y} = \frac{\partial v}{\partial y}\Delta t.$$

So,

$$\frac{d\chi}{dx} \approx \frac{\partial v}{\partial y}\frac{dt}{dx} = \frac{1}{v}\frac{\partial v}{\partial y}.$$

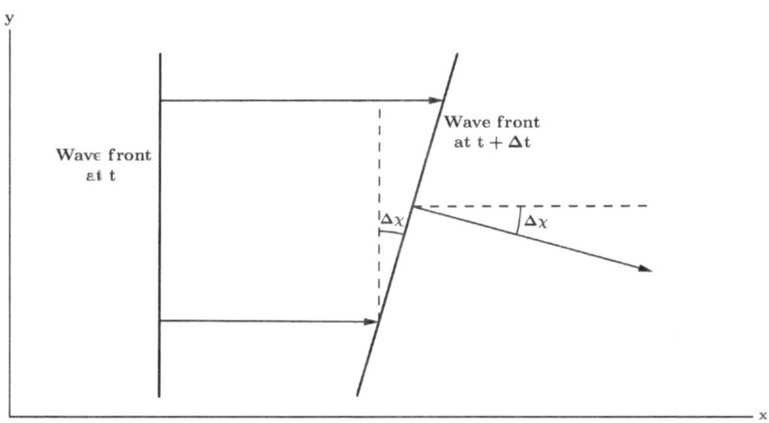

Figure 11.19 The bending of a light wave as it passes near a mass M in general relativity.

The rate of change of $\partial v/\partial y$ is very small, of order $1/c^2$, so the deflection will also be very small. To calculate the deflection, we need the speed v to first order in $1/c^2$. To obtain this, write ds^2 in terms of dt and dx because the light rays are propagating in the x direction, and deflection occurs only because $\partial v/\partial y$ is nonzero:

$$ds^2 = \left(1 - \frac{2GM}{c^2 r}\right) c^2 dt^2 - \left(1 + \frac{2GM}{c^2 r}\right)\left(\frac{dr}{dx}\right)^2 dx^2 - r^2\left(\frac{d\theta}{dx}\right)^2 dx^2.$$

Call the closest approach of the light ray to the Sun R as shown in Fig. 11.20.

$$r = \sqrt{x^2 + y^2} \quad \cos\theta = \frac{y}{r}.$$

The derivatives we need in the metric, to effectively rewrite it in Cartesian coordinates, starting from polar coordinates, are

$$\frac{dr}{dx} = \frac{x}{r} \quad \frac{d\theta}{dx} = -\frac{y}{r^2}.$$

Substituting into the expression for the invariant interval,

$$ds^2 = \left(1 - \frac{2GM}{c^2 r}\right) c^2 dt^2 - \left(1 + \frac{2GM}{c^2 r}\right) \frac{x^2}{r^2} dx^2 - r^2 \frac{y^2}{r^4} dx^2.$$

So,

$$ds^2 = \left(1 - \frac{2GM}{c^2 r}\right) c^2 dt^2 - \left(1 + \frac{2GMx^2}{c^2 r^3}\right) dx^2.$$

Setting $ds^2 = 0$ we obtain $v = dx/dt$, to first order in $1/c^2$,

$$v = \left[1 - \frac{GM}{c^2 r}\left(1 + \frac{x^2}{r^2}\right)\right] c.$$

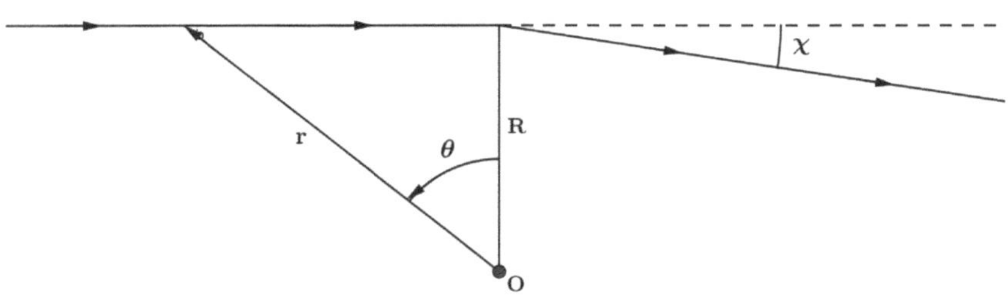

Figure 11.20 Kinematic set up for the calculation of the bending of light as it passes near a mass M centered at the origin O.

To calculate the deflection, we need $(1/v)$ $(\partial v/\partial y)$ to first order in $1/c^2$,

$$\frac{1}{v}\frac{\partial v}{\partial y} \approx \frac{GM}{c^2}\left(\frac{3x^2 y}{r^5} + \frac{y}{r^3}\right).$$

Integrating from $-\infty$ to $+\infty$ gives the full deflection,

$$\chi \approx \frac{GM}{c^2}\int_{-\infty}^{\infty}\left(\frac{3x^2 y}{r^5} + \frac{y}{r^3}\right) dx.$$

Integrate by using the variable θ, which varies from $-\pi/2$ to $\pi/2$. From Fig. 11.20, this gives

$$x = R\tan\theta, \quad R = r\cos\theta, \quad y \approx R.$$

So,

$$\chi \approx \frac{GM}{c^2 R}\int (3\sin^2\theta + 1)\cos\theta \, d\theta$$

$$\chi = \frac{4GM}{c^2 R}$$

which is twice the Newtonian result, as promised!

This result played a very important role in the historical development and acceptance of general relativity. Better and better measurements of χ are actively being pursued.

11.8 SIMILARITIES AND DIFFERENCES OF ELECTROMAGNETISM AND GRAVITY

It is interesting to consider the similarities between the two field theories considered in this book, electromagnetism and general relativity. We will also return to this subject after we have developed general relativity for strong gravity in the next chapter.

Our investigations in electromagnetism started with Coulomb's law, the force between two charges q_1 and q_2 a distance r apart,

$$\boldsymbol{F}(\boldsymbol{r}) = k\frac{q_1 q_2}{r^2}\hat{\boldsymbol{r}} \tag{11.27}$$

and our investigations in gravity started with Newton's law of the gravitational force between two masses m_1 and m_2 a distance r apart,

$$\boldsymbol{F}_g(\boldsymbol{r}) = -G\frac{m_1 m_2}{r^2}\hat{\boldsymbol{r}} \tag{11.28}$$

We rewrote Coulomb's law as a local differential equation on our way to developing the field theory of electromagnetism,

$$\nabla \cdot \boldsymbol{E}(\boldsymbol{r}) = -\nabla^2 V(\boldsymbol{r}) = 4\pi k\rho(\boldsymbol{r}) \tag{11.29}$$

where ρ is the charge per unit volume, charge density, at the position $\boldsymbol{r}$ and $V(\boldsymbol{r})$ is the electrostatic potential, $\boldsymbol{E} = -\nabla V(\boldsymbol{r})$.

In the case of gravity we introduce the potential $V_g(\boldsymbol{r})$ due to the presence of particle m_2 at the origin and write $\boldsymbol{F}_g(\boldsymbol{r}) = -m_1 \nabla V_g(\boldsymbol{r})$. Then the same steps that led from Eqs. (11.27)–(11.29) produce the differential equation,

$$\nabla^2 V_g(\boldsymbol{r}) = 4\pi G\rho_g(\boldsymbol{r}) \tag{11.30}$$

where the source of the gravitational field is the mass density $\rho_g(\boldsymbol{r})$.

Both forces $\boldsymbol{F}(\boldsymbol{r})$ and $\boldsymbol{F}_g(\boldsymbol{r})$ are long range, falling off as r^{-2}. In the case of electromagnetism, we found that the long-range character of the electrostatic force led to the fact that the *dynamical* electromagnetic field satisfies the wave equation,

$$\left(\nabla^2 - \frac{\partial^2}{c^2\partial t^2}\right)\boldsymbol{E}(\boldsymbol{r}, t) = \left(\nabla^2 - \frac{\partial^2}{c^2\partial t^2}\right)\boldsymbol{B}(\boldsymbol{r}, t) = 0$$

in free space. This predicted that electromagnetic radiation exists, that electromagnetic waves travel at the speed limit and that the photon has a vanishing rest mass. This development suggests that the gravitational potential will be replaced by the gravitational field when we pass from Newton's to general relativistic space–time. We have already seen that Newton's gravitational potential enters general relativity through deviations in the space–time metric $g_{\mu\nu}$ from the flat Minkowski metric of special relativity. So, we expect that fluctuations in the gravitational field will manifest themselves in general relativity as fluctuations in the metric, and they will propagate at the speed limit and will satisfy the wave equation in Minkowski space–time if they are small. The associated quantum, the "graviton," will be massless just like the photon. These educated "guesses" are, in fact, true, but their verification requires investigations into general relativity, the subject of Chapter 12.

The electrodynamic force law states that like charges repel and unlike charges attract. For example, electrons attract their antiparticle relative, the positron, but repel each other. This is a characteristic of the spin-1 nature of the photon, as you will learn in field theory. However, Newton's law of gravity states that masses attract one another and only positive masses are known to exist. The universal attraction of masses, actually energies in the relativistic version of Newton's law, the Einstein field equations, is an essential element of the equivalence principle. The equality of inertial and gravitational masses is the basis of interpreting gravity as an aspect of geometry. These ideas generalize to field theories because field theory predicts that the masses of particles and their associated antiparticles are identical: particle and antiparticles must have opposite charges, but they must have identical *positive* inertias. We have seen in our studies of magnetism that the screening characteristic of electrodynamics is essential to its phenomenology. For example, the discovery and relevance of magnetism in practical situations relies on the fact that in most situations matter is effectively neutral and static electric forces are absent, leaving us able to find magnetic effects even though they are typically much smaller effects, suppressed by several powers of v/c. In these cases, the positive charges of nuclei are screened by the negative charges of mobile electrons. There is no such screening in gravity, so its various velocity-dependent forces are overwhelmed by static gravitational attractions between slowly moving objects.

At the quantum level, the spin-1 nature of the photon explains why like charges repel and unlike charges attract. What does the universal nature of gravity's attraction tell us at the quantum level? In electrodynamics the carrier of the force is the spin-1 photon and in gravitation the carrier of the force is the graviton, the quantum description of gravitational radiation that we will study in Chapter 12. The graviton is described by a traveling wave in the metric $g_{\mu\nu}$ analogous to the photon, which is a traveling wave in the electric and magnetic fields. The graviton carries spin 2, and its spin-2 nature predicts the universal attraction of the theory. If the graviton is traveling through empty Minkowski space, it travels at the speed limit, as will be discussed further in Chapter 12. Since both the photon and the graviton travel at the speed limit, they have some additional features in common: they both have but *two* degrees of freedom! In the case of electromagnetic radiation, we have already seen that there are two polarization states, which describe it: right-handed or left-handed polarized waves. Contrast this to the spin-1 state of a *massive* particle described by quantum mechanics: it has three spin states, which are described by its spin projection in some direction, say the z axis, of -1, 0, or $+1$. This gives three degrees of freedom. Why does radiation only have two such states, -1 and $+1$? The special feature of radiation is, of course, that it travels at the speed limit, which is a Lorentz invariant notion. So, its spin state must also be a Lorentz invariant. This will be true if the spin points either in the same direction as the propagation of the wave $\boldsymbol{k}$ or in the opposite direction, $-\boldsymbol{k}$. We saw this in our discussion of electromagnetic waves in Chapter 9, Section 2 where we found that right- and left-handed circularly polarized electromagnetic waves were the only possibilities because of the transversality conditions. The same argument applies to the graviton: it will propagate with its spin 2 aligned with its direction of motion or antialigned. In both cases, the radiated field has two degrees of freedom, even though they have different spin values! We will check this point when we discuss gravitational radiation and the LIGO experiment in Chapter 12.

Now let us discuss the differences between Eqs. (11.27) and (11.28) and between Eqs. (11.29) and (11.30). First, the charges q_1 and q_2 in Eq. (11.27) are Lorentz invariants; they are the same in all frames. The masses, m_1 and m_2 in Eq. (11.28), are the rest masses of the two particles at fixed positions in the static force law. We need to ask how they transform if we boost Eq. (11.28) or Eq. (11.30) to a frame with velocity $\boldsymbol{v}$. In the case of electromagnetism, the charge density $\rho(\boldsymbol{r})$ was identified as the zeroth component of a four-vector $J_\mu = \rho_0 v_\mu = (\gamma\rho_0 c,\ \gamma\rho_0 \boldsymbol{v})$ where ρ_0 is the rest frame charge density and $\gamma = \gamma(\boldsymbol{v})$. The fact that the source of the electric field $\boldsymbol{E}$ was the zeroth component of a four-vector, the charge current, was critical in the derivation of Maxwell's equations from Gauss' law and special relativity.

In the case of gravity, Eqs. (11.28) and (11.30), we have the energy density acting as the source of the gravitational field. What are the properties of energy-per-volume under boosts? It is here that gravity and electromagnetism diverge. Energy is the zeroth component of the energy—momentum four-vector $p_\mu = (E/c,\boldsymbol{p})$. Under a boost, volume^{-1},$1/V$, transforms as γ, which matches the behavior of the zeroth component of a four-vector such as the four-velocity $v_\mu = (\gamma c,\ \gamma \boldsymbol{v})$. So, the right-hand side of the equation for the gravitational field, Eq. (11.30), is proportional to the *product* of the zeroth components of *two* four-vectors. In a general inertial frame the right-hand side will be proportional to the product $p_\mu v_\nu$. If we have an extended source of gravity, perhaps many particles i, then the right-hand side of Eq. (11.30) would be proportional to $\sum \mu_0^i c^2 v_\mu^i v_\upsilon^i$ where μ_0^i is the rest mass of particle i. Mathematical objects that are labeled by several four-vector indices and transform under boosts accordingly are called *tensors* as we discussed in Chapter 8—10. They will play central roles in differential geometry and general relativity. We are familiar with the electromagnetic field tensor, $F_{\sigma\rho}$, for example, from Chapter 8. The tensor uncovered here is $T_{\mu\nu}$, the energy—momentum tensor. The energy—mass relation of special relativity implies that *all* forms of energy—momentum should contribute to $T_{\mu\nu}$ in addition to the masses that contribute to Newton's law of gravity. For example, the energy—momentum carried by the electromagnetic field contributes to $T_{\mu\nu}$. We will pursue these ideas for gravity in Chapter 12.

The fact that gravity couples to energy—momentum while electromagnetism couples to charge, a conserved quantity that is only carried by massive matter fields produces more differences between these theories. In the case of electromagnetism, this property gave rise to the linear superposition principle, which helped us solve electrodynamics problems. Gravity does not enjoy the linear superposition principle, and this fact makes it much more challenging. For example, the gravity field itself carries energy—momentum, so it couples to itself! The theory is intrinsically nonlinear! The Newtonian limit of the theory is, in fact, linear, but this is an exception to the rule. We will see this explicitly when we solve for the gravitational field outside a spherically symmetric, static mass and discover a black hole. A dramatic implication of the nonlinearity of gravity is the following: let us say you are inside a black hole and try to escape by flailing about or turning on a rocket motor. Do you succeed in evading being sucked in? On the contrary, the extra energy you expend *increases* your attraction to the black hole and as a consequence you are sucked in more forcefully and your proper lifetime is diminished! In the language of the fundamentals of general relativity, the gravitons that generate the attraction between bodies attract any object that carries energy—momentum and thus attract one another and generate an apparently very singular theory. This avalanche of interactions is one reason why physicists have yet to understand the quantized version of the theory!

Let us return to our task of comparing and contrasting classical electromagnetism and classical gravity. We begin with static problems and consider the electrostatic potential far from a collection of charges and the analogous problem in gravity, the gravitational potential far from a collection of masses. Suppose there are charges e_a at points $\boldsymbol{r}_a$ near the origin of a convenient coordinate system. Then the electrostatic potential is given by a linear superposition of Coulomb potentials,

$$V(R) = k\sum_a \frac{e_a}{|\boldsymbol{R} - \boldsymbol{r}_a|} \tag{11.31}$$

We use the uppercase notation $\boldsymbol{R}$ to indicate the point where the potential is measured and to emphasize that $R = |\boldsymbol{R}|$ is much larger than the largest $|\boldsymbol{r}_a|$. In this case it is useful to Taylor expand $\frac{1}{|\boldsymbol{R}-\boldsymbol{r}_a|}$ around the origin,

$$\frac{1}{|\boldsymbol{R}-\boldsymbol{r}|} = \frac{1}{R} - \sum_j x_j \frac{\partial}{\partial x_j}\frac{1}{R} + \frac{1}{2}\sum_{jk} x_j x_k \frac{\partial^2}{\partial x_j \partial x_k}\frac{1}{R} + \cdots \tag{11.32}$$

where $\boldsymbol{R} = (x_1, x_2, x_3)$. Taking the derivatives and doing some straightforward algebra,

$$\frac{1}{|\boldsymbol{R}-\boldsymbol{r}|} = \frac{1}{R} + \left(\sum_j x_j n_j\right)\frac{1}{R^2} + \left(\frac{1}{2}\sum_{jk}(3x_j x_k - r^2\delta_{jk})n_j n_k\right)\frac{1}{R^3} + \cdots \tag{11.33}$$

where $\boldsymbol{n}$ is the unit vector pointing in the $\boldsymbol{R}$ direction, $\boldsymbol{n_i} = (x, y, z)/R$. If we apply this expansion to Eq. (11.31) we find,

$$V(\boldsymbol{R}) = k\left(Q\frac{1}{R} + D\frac{1}{R^2} + Q^{(2)}\frac{1}{R^3} + \cdots\right) \tag{11.34}$$

where $Q = \sum_a e_a$ is the total charge of the source, $D = \sum_i d_i \boldsymbol{n_i}$; $d_i = \sum_a e_a x_i^{(a)}$ is the dipole moment of the source, and $Q^{(2)} = \frac{1}{2}\sum_{jk} D_{jk}\boldsymbol{n_j}\boldsymbol{n_k}$ and $D_{jk} = \sum_a e_a\left(3x_j^{(a)}x_k^{(a)} - r^{(a)2}\delta_{jk}\right)$ is the quadrupole moment of the charge distribution.

Eq. (11.34) is particularly useful because it organizes the potential as increasing moments of the charge distribution and correlates these moments to inverse powers of R, indicating how significant they are far from the source. For example, if the source is net neutral, $Q = \sum_a e_a = 0$, then the "monopole" term in Eq. (11.34) is absent and the next term, the dipole moment of the distribution dominates the field's behavior at large R. Since D is a relatively simple aspect of the charge distribution, it is easy to estimate $V(\boldsymbol{R})$ for large R without knowing every detail of the source. The dipole and the quadrupole moments have particular properties, which lead to particular sensitivities in $V(\boldsymbol{R})$ to the orientation of the source. For example, $D = \boldsymbol{d}\cdot\boldsymbol{n}$, so $V(\boldsymbol{R})$ varies as the cosine of the angle between the source's dipole moment $\boldsymbol{d}$ and the orientation $\boldsymbol{n}$ of the observation position $\boldsymbol{R}$. Textbooks in electromagnetism develop these points in detail. If the source is neutral, then it is easy to show that $\boldsymbol{d}$ is an intrinsic property of the charge distribution, i.e., it is independent of the origin of the coordinate system in which it is defined. If the charge distribution consists of two equal and opposite charges, then $\boldsymbol{d}$ is the product of the positive charge times the vector from the negative to the positive charge.

How does this compare with the same exercise in Newtonian gravity $V_g(\boldsymbol{R})$? Clearly the arithmetic details are the same with the replacement $e_a \rightarrow m_a$. This replacement leads to two important, and perhaps surprising, differences. First, the strength of the R^{-1} term in the expansion of $V_g(\boldsymbol{R})$ is the total mass of the distribution, $M = \sum_a m_a$. Unlike electrostatics where bulk matter is often neutral $Q = \sum_a e_a = 0$ and the R^{-1} term is absent, in gravity the R^{-1} term is *always* present. Since each m_a is positive, there is no "screening" in gravity. Furthermore, we can always choose the origin of the coordinate system at the center of mass of the distribution, so the dipole term vanishes identically, $\boldsymbol{r}_{cm} = \sum_a m_a \boldsymbol{r}^{(a)} = 0$. If there were no external forces on the collection of masses, then if the center of mass were initially at rest at $\boldsymbol{r}_{cm} = \sum_a m_a \boldsymbol{r}^{(a)} = 0$, it stays at rest at $\boldsymbol{r}_{cm} = 0$ forever. This means that the dipole term vanishes, and the first correction to the $\frac{M}{R}$ term in the potential is the *quadrupole* term $\frac{Q^{(2)}}{R^3}$. It is the quadrupole moments of mass distributions that determine tidal forces in planetary systems and the gravitational radiation emitted from binary neutron star systems, as we shall soon see.

The differences between electromagnetism and gravity do not end here. Now let us think about how one might observe a traveling electromagnetic wave and contrast that to how one could detect a traveling gravity wave. In the electromagnetic case, we have the Lorentz force law $\boldsymbol{F} = d\boldsymbol{p}/dt = e(\boldsymbol{E} + \boldsymbol{u} \times \boldsymbol{B})$. If the charged particle is at rest in frame S, one could observe a traveling $\boldsymbol{E}$ and $\boldsymbol{B}$ wave by observing the resulting force, a change in the particle's momentum. No subtlety here. Traveling electromagnetic waves induce electric currents in circuits, etc.

Now consider the same situation in the case of gravity in a curved space. How could we use the kinematic state of an *isolated* point particle to detect a passing gravity wave? A gravity wave cannot be detected locally! Such a possibility would violate the equivalence principle! To understand this, recall that in a local region *all* the effects of gravity can be transformed away—we can jump into a local inertial reference frame, and the effects of gravity disappear. Therefore, the gravity wave *cannot* be detected by observing what is happening in a strictly local region, in particular, by observing one isolated particle. This means that a single isolated particle does not change its position relative to the free-falling coordinate labels (t, x, y, z). Subtle! We shall see this principle expressed in the equations of motion of point masses in general relativity in Chapter 12, and especially in Appendix G, Section 3.

So, we can only detect gravity waves by recording their influence on *two* or more separated test particles! We have already seen in Section 3 above that tidal forces are caused by the spatial variations in gravitational potentials. In Chapter 12, we will see that oscillating quadrupole moments of mass distributions produce gravity waves. The gravity waves influence matter through tidal forces—the characteristic stretching and squeezing illustrated earlier. A gravity wave detector must consist of freely falling masses over extended regions of space–time.

In Chapter 12, we will learn that there are two characteristic quadrupole moments of a mass distribution whose oscillations can produce propagating gravitational waves. One is the + polarization, which is a "breathing" mode in which "stretching" in one direction is coordinated with "compression" in the perpendicular direction as shown in Fig. 11.21. The second polarization, the × polarization, has the same characteristics but is rotated by 45 degrees. We have seen these patterns in our discussions of tidal forces and will see them again when we study gravitational waves in Chapter 12.

Compare this to the dominant radiation patterns in electromagnetism. Oscillating dipole moments of charge distributions produce the strongest radiation: linearly polarized traveling electric and magnetic waves. The two

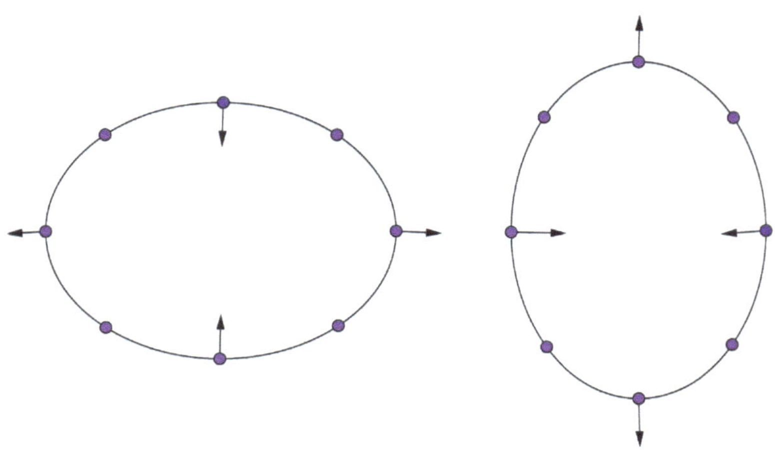

Figure 11.21 The + polarization of a quadrupole moment. This is a "breathing" mode as shown at two times, one on the left side and the second on the right side of the figure. There is "stretching" in one direction, which is coordinated with "compression" in the perpendicular direction.

independent linear polarization states are oriented at 90 degrees. Such waves carry considerable energy, which we have calculated in problem sets using the Larmor formula. Although a pointlike local energy density of gravity is inconsistent with the equivalence principle, gravity waves carry energy and momentum that can be expressed using the space–time derivatives of the traveling gravitational wave. In fact, when black holes collide and merge, a great deal of mass is converted to energy and that energy is radiated away. The first such astronomical event observed in 2015 by the gravity wave detector "Advanced LIGO" converted three solar masses into gravitational wave energy. In the vicinity of the Earth, the gravity wave, which originated more than 1 billion light years away, had a tiny amplitude, which could be treated as a slight perturbation on the local Minkowski metric.

These remarks remind us of another important difference between electromagnetism and gravity. How are the detectors of electromagnetic and gravitational radiation different? Typical detectors of light respond to light's *intensity*, which is proportional to the square of its amplitude. The intensity of light typically decreases as the square of the distance from the source of the radiation. However, the situation in gravity is different. Detectors of gravity waves measure the amplitude of the wave, which typically decreases as the inverse of the distance to the source. We will see this explicitly in the sections of Chapter 12, which focus on LIGO. This fact is critical in the observational science of gravitational waves since events that produce copious gravitational waves are rare and gravitational effects are so intrinsically weak.

All these ideas will be expanded on in Chapter 12 where these and other aspects of gravity and gravitational waves will be derived and discussed.

11.9 MAKING THE MOST OUT OF TIME

Let us take a brief interlude from our serious studies and illustrate some features of general relativity and the gravitational redshift in an unusual setting. The gravitational redshift means that clocks run quickly in high gravitational potentials, and they run slowly in low gravitational potentials.

Suppose there is a colony of tough but gentle space creatures, called Scruffs, who live in a region of the universe where the gravitational potential varies rapidly in space. The Scruffs, who live in nuclear family units, are able to arrange their environment to their needs. In fact, they use the gravitational potential to control time for their day-to-day convenience.

For example, each family of Scruffs has a room in its house where the gravitational potential is very negative. Scruffs call this room "time out," and whenever their baby boy is naughty, they give him a "time out" by putting him unceremoniously into this dreaded place. When the baby is put into "time out," all his actions slow down and are much more tolerable. Even the baby's whining is now heard at a lower, more tolerable frequency. This is a parent's dream come true!

The tricks do not end here. In the attic, each family has a "play room" where the gravitational potential is very high. Scruff teenage daughter, who is always behind on her homework, jumps into this "play room" when she needs extra time to finish her physics homework before its due date. Her father uses the room when he needs extra time to complete a project for his pushy boss.

Actually, life in a Scruff house can be quite stressful. Because there are regions where the gravitational potential varies rapidly in space, Scruffs experience strong forces as they move about. For example, as they put their noisy son into "time out," the gravitational potential varies rapidly across his body, pulling one side much more than the other. (These effects are called tidal forces and are similar to, but highly magnified compared with, the forces the Moon exerts on Earth. These tidal forces—the fact that the Moon pulls the side of Earth nearest to it more than the side further from it—flatten Earth slightly and cause high and low tides, a familiar effect that was first calculated by Isaac Newton.) Luckily, the Scruffs have evolved into tough little beings who can withstand these stretching and compressing forces. Even their pets take advantage of their environment. The family Jat, a long and lazy creature, naps with its head in the "play room," so it can have a relatively long snooze, while its belly is in "time out" so it can savor its lunch. The only problem with the Jat is that no one can tell how old it is, because from the Scruffs' perspective, its various parts are aging at alarmingly different rates!

PROBLEMS

11-1. A sodium lamp emits light in its rest frame with a wavelength of 5890 Å. If the lamp is placed on a turntable and is rotating at a speed of $0.2c$, what wavelength would an observer fixed at the center of the turntable measure?

11-2. Calculate the gravitational redshift for a spectral line emitted at rest on the surface of the Sun and subsequently detected on Earth.

11-3. Identical atomic clocks are placed on two Boeing 747s that circle the globe at the equator, one traveling west and the other traveling east. The planes fly at an altitude of 10 km and at a speed of 0.24 km/s. After their trips around the world, the clocks are compared with one another as well as with a clock that remained behind at the airport. (The rotational speed of the surface of Earth is approximately 0.5 km/s.)

a. What are the differences in the readings of the three clocks predicted by special relativity?

b. What are the differences in the readings of the three clocks predicted by the gravitational redshift?

c. Combine the results of parts (a) and (b) to find the actual differences in the clocks' readings.

11-4. Let us compare the strength of static gravitational forces to electrostatic forces.

Suppose we have an electron and a proton at rest in the lab. Calculate the ratio of the electrostatic and the gravitational forces between them. Evaluate the ratio numerically and comment on it.

REFERENCES

[1] W. Rindler, Essential Relativity, Springer-Verlag, Berlin, 1971.
[2] J.D. Jackson, Classical Electrodynamics, John Wiley & Sons, New York, 1962.
[3] R.V. Pound, G.A. Rebka Jr., Apparent weight of photons, Phys. Rev. Lett. 4 (1960) 337. R.V. Pound, J.L. Snider, Effect of gravity on gamma radiation, Physics Rev. B140, 788(1965).
[4] O.R. Frisch, Time and relativity, Contemp. Phys. 3 (1961) 16, 3, 194(1962).
[5] R. Adler, M. Bazin, M. Schiffer, Introduction to General Relativity, McGraw-Hill, New York, 1965.

CHAPTER 12

Curvature, Strong Gravity, and Gravitational Waves

Contents

12.1 A LOOK AT CURVED SURFACES: A FIRST STEP TO THE MATHEMATICS OF GENERAL RELATIVITY

The language of general relativity is differential geometry, Riemannian geometry, in fact. Geodesic paths in space—times with intrinsic curvature are the central concepts. Let us begin by considering these ideas for surfaces embedded in three-dimensional Euclidean space and later turn to their roles in general relativity and modern theoretical physics.

Everyone knows how to set up Cartesian coordinates in a flat three-dimensional Euclidean space, as shown in Fig. 12.1.

Now consider two points, P_1 and P_2, which are very close: one at (x, y, z) and the second at $(x + dx, y + dy, z + dz)$. The vector between the points is written as $d\mathbf{r} = \mathbf{i}\ dx + \mathbf{j}\ dy + \mathbf{k}\ dz$. The length squared of this

Special Relativity, Electrodynamics, and General Relativity
ISBN 978-0-12-813720-8
https://doi.org/10.1016/B978-0-12-813720-8.00012-X

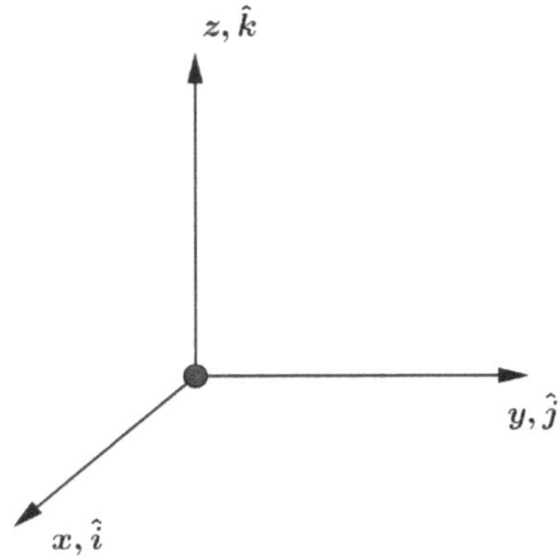

Figure 12.1 Cartesian coordinates and basis vectors for a three-dimensional Euclidean space.

infinitesimal vector is given by the dot product of the vector with itself, and using the fact that the unit vectors **i**, **j**, and **k** are mutually orthogonal we have

$$d\mathbf{r}^2 = dx^2 + dy^2 + dz^2,$$

which reproduces the Pythagorean theorem. This sort of space, called a metric space because it has a notion of distance, is Euclidean because the Pythagorean theorem holds. It is also called flat.

We can write the metric $d\mathbf{r}^2$ in any convenient coordinate system. For example, in the cylindrical coordinates discussed and illustrated in Section 11.2, the metric reads

$$d\mathbf{r}^2 = dr^2 + r^2 d\varphi^2 + dz^2.$$

This expression is more complicated than the expression for the metric in Cartesian coordinates, but the underlying geometry is the same old three-dimensional, flat Euclidean space. The point is that in differential geometry you must look beyond the appearances of the formulas to uncover intrinsic aspects of the space or surface you are studying. Do not be fooled by how expressions are parameterized. The emphasis in modern differential geometry is on intrinsic aspects of spaces and the relations between them. Modern notation, which we do not use here, avoids coordinates much more than classical approaches and just manipulates mathematical objects with intrinsic geometric significance.

One such quantity in three-dimensional space is the length $d\mathbf{r}^2$ between two points. In the context of relativity, the invariant interval that establishes the metric is such a quantity in our space–time world of four-dimensional Minkowski diagrams. For special relativity, the invariant interval reads

$$ds^2 = c^2 dt^2 - dx^2 - dy^2 - dz^2 \qquad (12.1)$$

It is important because of its invariance properties—it is the same in all inertial frames. As we have seen, we can use the metric to derive time dilation, Lorentz contraction, and the Lorentz transformation laws. So, the invariance of ds^2 characterizes relativity in a very crisp fashion. Because most of us are particularly familiar with the Euclidean metric, which is always positive, it takes some care to call ds^2 a metric as well. For example, two space–time events that are connected by a light ray have $ds^2 = 0$, as we have discussed and illustrated before in Minkowski diagrams. If events are simultaneous but spatially separated in one frame, then their interval is negative and can be interpreted as the negative of a proper length. If, however, the two events are separated in time but occur at the same spatial point, their interval is positive and can be interpreted as a proper time.

It is instructive to consider curved surfaces embedded in three-dimensional Euclidean space. Consider the surface shown in Fig. 12.2, the two-dimensional coordinate mesh on the surface (u^1, u^2), and the tangent vectors $(\mathbf{e}_1, \mathbf{e}_2)$. The vector between two nearby points on the surface can be written $d\mathbf{r} = du^1\mathbf{e}_1 + du^2\mathbf{e}_2$, where du^1 and du^2 represent the coordinate mesh differences along the coordinate directions $(\mathbf{e}_1, \mathbf{e}_2)$ on the surface (the superscripts on du^1 and du^2 denote directions, not algebraic

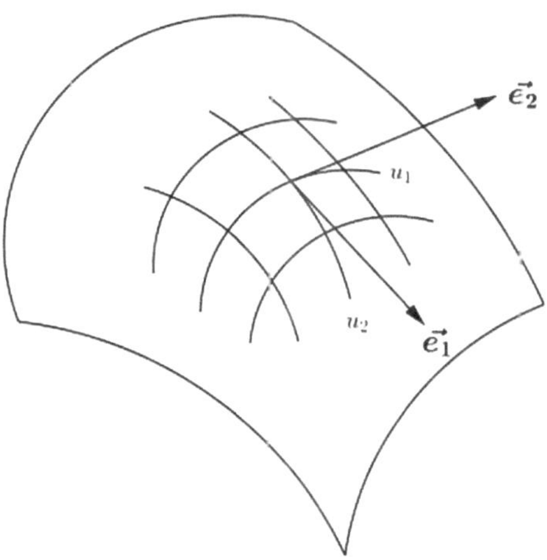

Figure 12.2 A coordinate mesh on a two-dimensional surface. $\mathbf{e}_1$ points along the lines of constant u^2 and $\mathbf{e}_2$ points along the lines of constant u^1.

powers) and $e_1 = \partial r/\partial u^1$ and $e_2 = \partial r/\partial u^2$. The distance between the points is

$$d\mathbf{r}^2 = \sum_{ij} \mathbf{e}_i \cdot \mathbf{e}_j \, du^i du^j,$$

where the sum over the indices runs over $i = 1, 2$ and $j = 1, 2$. It is conventional to call the collection of dot products of the tangent vectors, $\mathbf{e}_1$ and $\mathbf{e}_2$, the metric tensor, $g_{ij} = \mathbf{e}_i \cdot \mathbf{e}_j$. Note that because $d\mathbf{r}$ is confined to a curved surface where the tangent vectors $\mathbf{e}_1$ and $\mathbf{e}_2$ may not be orthogonal, the expression for $d\mathbf{r}^2$ has a cross term $2\mathbf{e}_1 \cdot \mathbf{e}_2 \, du^1 \, du^2$. The angle between the tangent vectors and their lengths will change as we move around the surface. $d\mathbf{r}^2$ has physical significance while (u^1, u^2) is just a convenient coordinate mesh.

It is clear that there is enormous freedom in choosing coordinate patches on a smooth surface. How do we distinguish between the intrinsic properties of the surface and the properties of the coordinate mesh? This was the question posed by Carl Fredrich Gauss, the great mathematician who was one of the pioneers of the field. He asked the question: Is it possible to tell if a space or surface is Euclidean or is curved just from measurements within it, intrinsic measurements? Gauss answered the question yes and gave a definition of curvature that produces the same result no matter what local coordinates we use on the surface. We shall focus on Gaussian curvature below.

Take a familiar example—a sphere. This surface has constant positive curvature. What do we mean by that? Instead of giving a formal answer here, let us contrast the features of the sphere to a flat two-dimensional space such as a plane (e.g., a piece of paper). If we have two points on a plane, the curve of shortest distance between them is a straight line. On the sphere, a curve of shortest distance is a great circle, called a geodesic, familiar from air travel around Earth. Using these geodesics, we can then form closed paths and compare their properties to familiar constructions on the plane. For example, the sum of the interior angles of a triangle drawn on a plane is 180 degrees. If we make the same construction on the sphere using its geodesics, great circles, we find that the "deficit angle," the difference of the sum of the interior angles and 180 degrees, is positive. The deficit angle can be used as a definition of intrinsic curvature, as we will learn below when the Gauss–Bonnet theorem is introduced. Gauss liked this idea so much that he applied it to the space in which we live. He assumed that light travels on geodesics in our world (good!), and he used three mountains near him as vertices of his geodesic triangle. He found that the sum of the interior angles was 180 degrees to good accuracy, so he concluded that space in the vicinity of Earth is well approximated by Euclidean space.

A quantitative way of defining curvature is to consider a circle on the surface of interest and make it small, with a geodesic radius of distance a. Now measure the circumference, C. If a and C are related by a factor of 2π, then the surface is flat. If the circumference is less than that, the surface has a positive curvature. And if the circumference is greater than that, the surface has a negative curvature. Let us do this exercise using the sphere shown in Fig. 12.3.

The circumference of the circle shown is $C = 2\pi R \sin(a/R)$, where R is the radius of the sphere. Now define the Gaussian curvature to be proportional to the difference between $2\pi a$ and C,

$$K = \frac{3}{\pi} \lim_{a \to 0} \left(\frac{2\pi a - C}{a^3} \right). \tag{12.2}$$

Using the formula for the circumference and letting a be small so one can expand the sine, $\sin x = x - x^3/6 \ldots$ when $x \ll 1$, one computes $K = 1/R^2$, which further motivates this definition. The result $K = 1/R^2$ shows that small spheres have more curvature than large ones (very reasonable). Clearly in the case of the sphere, the curvature does not change as we move around the surface, but other surfaces can have a nontrivial function $K(u^1, u^2)$. The student should check that a surface in the shape of a saddle has a negative

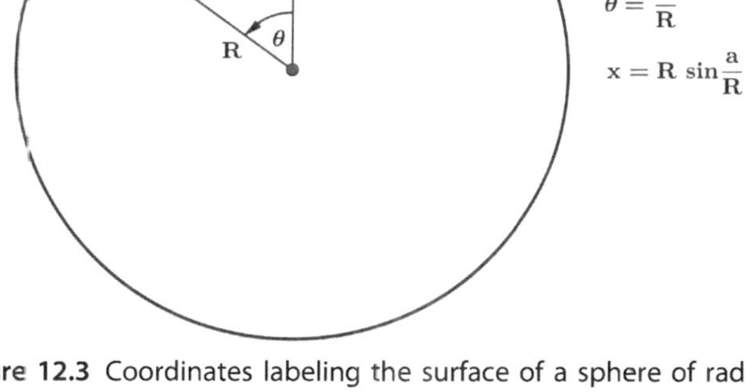

Figure 12.3 Coordinates labeling the surface of a sphere of radius *R*.

curvature and that a cylinder, which can be cut in the direction along its axis and laid flat on a plane, has a vanishing curvature.

Another procedure to measure the intrinsic curvature of a surface is to calculate the rate at which its geodesics converge or diverge. In flat Euclidean space, geodesics are straight lines which, if they are initially parallel, remain parallel forever, and straight lines that intersect diverge at a linear rate. On the surface of a sphere, geodesics are great circles. In Fig. 12.4 we show two great circles. Notice that they are parallel at the equator but intersect at the North and South Poles. The rate at which they converge can be computed using the geometry shown in the figure. The distance between the geodesics at polar angle θ is,

$$\varepsilon = \varphi R \sin\frac{s}{R}$$

where φ is the opening angle between the two great circles at the North Pole and the arc length s parameterizes the curves. Note that ε satisfies the "equation of motion,"

$$\frac{d^2\varepsilon}{ds^2} = -K\varepsilon \tag{12.3}$$

where K is the Gaussian curvature, $K = 1/R^2$. At the equator, $\theta = \pi/2$, the tangents to the great circles are parallel. Nonetheless, the "acceleration" $d^2\varepsilon/ds^2 < 0$ there and the geodesics head toward one another and intersect at the South Pole.

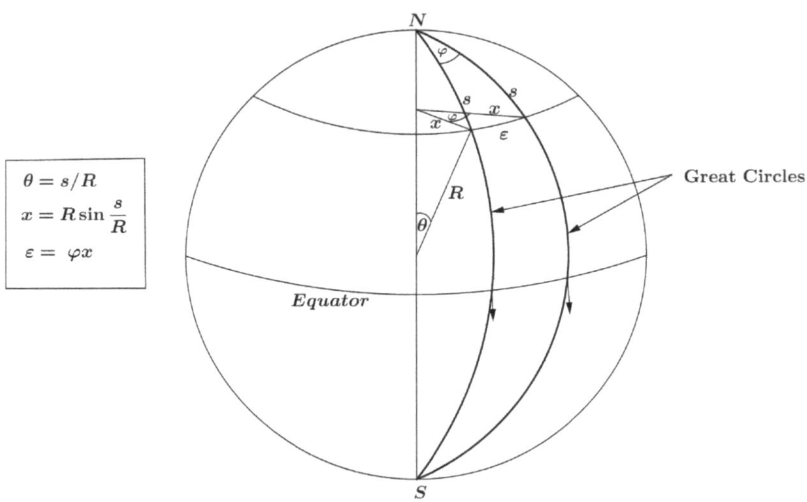

Figure 12.4 Two great circles on a sphere of radius *R*.

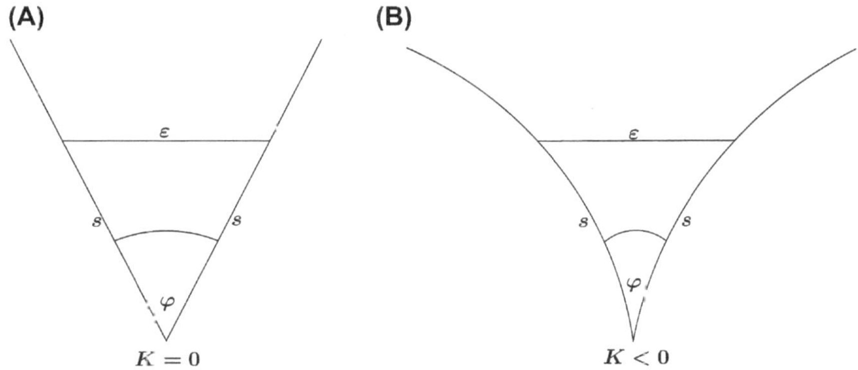

Figure 12.5 Distance between geodesics, ε, on a surface of vanishing curvature, part (A), and a surface of negative curvature, part (B).

If we made the same construction in flat space and considered two straight lines that intersected at point P with an opening angle φ, then $d^2\varepsilon/ds^2 = 0$, and we infer that $K = 0$, as shown in part a. of Fig. 12.5. Similarly, if we considered two intersecting geodesics on a saddle we would compute $d^2\varepsilon/ds^2 > 0$ corresponding to $K < 0$ and the curves would diverge as shown in part B. of Fig. 12.5.

We will discuss much more about all this below. Eq. (12.3) will be derived in more generality after ideas such as geodesic coordinate meshes have been introduced in a discussion of curves and surfaces in three dimensions. The general relativistic analogue of ε will appear when we discuss the equation of motion of the "geodesic deviation," tidal forces, and gravitational waves. In a higher dimensional curved space where the curvature varies from point to point, K will be replaced by the Riemann curvature tensor.

Now let us present some elements of classical differential geometry. We will consider curves and surfaces embedded in three-dimensional Euclidean space. Our objectives include the development of an intuition into the major concepts of classical differential geometry such as metrics, tangent spaces, and intrinsic properties of curves and surfaces such as tangential curvature and Gaussian curvature. We will see that tangent spaces (planes in our illustrations) play the role of freely falling, locally inertial reference frames of general relativity. Our later discussions of general relativity will not use embedding spaces but the discussion here will help us visualize surfaces and tangent spaces and help us gain a perspective on the more abstract concepts in four-dimensional curved space—time.

Consider a curve in three-dimensional Euclidean space [1]. We label points along the curve by their arc length, a scalar parameter s, $\boldsymbol{x} = \boldsymbol{x}(s)$. The curve has a tangent vector at any point along it, $\boldsymbol{t} = d\boldsymbol{x}/ds \equiv \boldsymbol{x}'$, where the "prime" indicates differentiation with respect to s. The tangent vectors have unit length, $\boldsymbol{t}\cdot\boldsymbol{t} = 1$. If we differentiate this constraint with respect to the arc length, we find $\boldsymbol{t}\cdot\boldsymbol{t}' = 0$ where $\boldsymbol{t}' = d\boldsymbol{t}/ds = d^2\boldsymbol{x}/ds^2$. So, $\boldsymbol{t}'$ is perpendicular to the curve, $\boldsymbol{t}$, point by point along the curve. If we introduce a unit vector $\widehat{\boldsymbol{n}}$ in the direction of $\boldsymbol{t}'$, we can define the curve's curvature κ_C,

$$d^2\boldsymbol{x}/ds^2 = \boldsymbol{t}' = \kappa_C\widehat{\boldsymbol{n}} \tag{12.4}$$

so $\kappa_C^2 = \boldsymbol{t}'\cdot\boldsymbol{t}'$. One defines $R = \kappa_C^{-1}$ and refers to R as the radius of curvature of the curve at P. Further analysis shows that the magnitude of R is the radius of the circle passing through three consecutive points along the curve. It is instructive to illustrate these concepts for a spiral winding around the z-axis, for example. See the problem set and later discussions for additional examples. The simplest illustration is a circle of radius R in the x–y plane. Let s be the arc length as shown in Fig. 12.6. Then the position of a point on the circle is $\boldsymbol{x} = (x, y) = \left(R\sin\frac{s}{R}, R\cos\frac{s}{R}\right)$ and $d^2\boldsymbol{x}/ds^2 = \frac{1}{R}\left(-\sin\frac{s}{R}, -\cos\frac{s}{R}\right) = \frac{1}{R}\widehat{\boldsymbol{n}}$ and we identify $\kappa_C = 1/R$.

Now consider a surface embedded in a three-dimensional Euclidean space. We put a coordinate mesh (u, v) on the surface and locate a point on

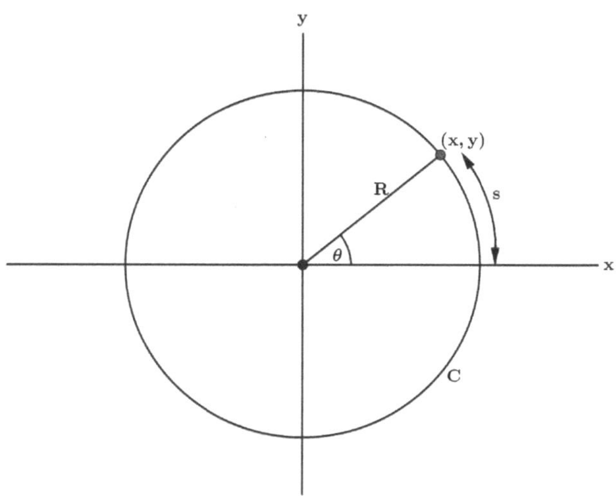

Figure 12.6 A circle in two dimensions illustrating the curvature $\kappa_C = 1/R$.

the surface with Cartesian coordinates, $x_i = x_i(u, v)$. The surface of a sphere of radius R parameterized by spherical coordinates (ϑ, φ) is a familiar example,

$$x = x_1 = R \sin\theta \cos\varphi, \quad y = x_2 = R \sin\theta \sin\varphi, \quad z = x_3 = R \cos\theta$$

as shown in Fig. 12.7.

Now return to our general discussion and consider a coordinate mesh as shown in Fig. 12.8. Note that at the point P the vector $\partial \boldsymbol{x}/\partial u \equiv \boldsymbol{x}_u$ is tangent to the curve $v =$ constant, and $\partial \boldsymbol{x}/\partial v \equiv \boldsymbol{x}_v$ is tangent to the curve

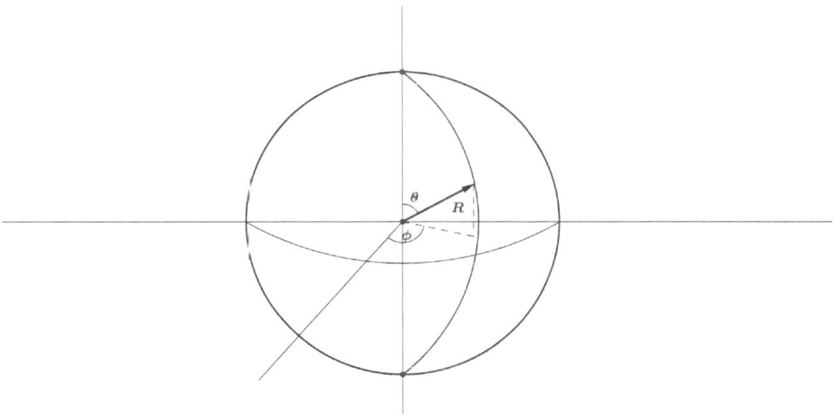

Figure 12.7 Spherical coordinates parameterizing the surface of a sphere of radius *R*.

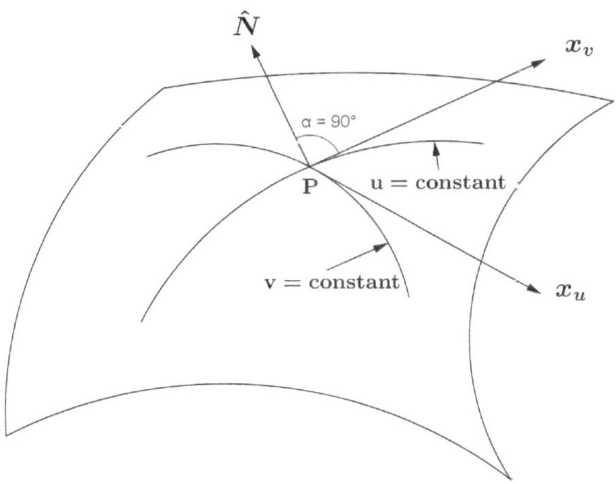

Figure 12.8 A coordinate mesh on a two-dimensional surface showing the vectors $\partial \boldsymbol{x}/\partial u \equiv \boldsymbol{x}_u$ and $\partial \boldsymbol{x}/\partial v \equiv \boldsymbol{x}_v$ that span the tangent space at point *P* and the normal $\boldsymbol{N}$ to the surface.

$u =$ constant. The mesh is an adequate parametrization of the surface if $\boldsymbol{x}_u$ and $\boldsymbol{x}_v$ do not vanish and have different directions. As shown in the figure, the cross product $\boldsymbol{x}_u \times \boldsymbol{x}_v$ is a natural normal vector to the surface at P.

We will want to transform this description of the surface to another curvilinear coordinate system,

$$u = u(\overline{u}, \overline{v}) \quad v = v(\overline{u}, \overline{v})$$

so $\boldsymbol{x} = \boldsymbol{x}(\overline{u}, \overline{v})$. The new tangent vectors, $\boldsymbol{x}_{\overline{u}}$ and $\boldsymbol{x}_{\overline{v}}$, can be calculated in terms of the original tangent vectors $\boldsymbol{x}_u$ and $\boldsymbol{x}_v$ using the chain rule,

$$\boldsymbol{x}_{\overline{u}} = \boldsymbol{x}_u \frac{\partial u}{\partial \overline{u}} + \boldsymbol{x}_v \frac{\partial v}{\partial \overline{u}}, \quad \boldsymbol{x}_{\overline{v}} = \boldsymbol{x}_u \frac{\partial u}{\partial \overline{v}} + \boldsymbol{x}_v \frac{\partial v}{\partial \overline{v}}$$

One can also calculate the normal vector $\boldsymbol{x}_{\overline{u}} \times \boldsymbol{x}_{\overline{v}}$, and find,

$$\boldsymbol{x}_{\overline{u}} \times \boldsymbol{x}_{\overline{v}} = \frac{\partial(u, v)}{\partial(\overline{u}, \overline{v})} \boldsymbol{x}_u \times \boldsymbol{x}_v$$

So, the new normal vector points in the same direction as the original, as must be the case, but its relative normalization is changed by a factor of,

$$\frac{\partial(u, v)}{\partial(\overline{u}, \overline{v})} \equiv \det \begin{vmatrix} \dfrac{\partial u}{\partial \overline{u}} & \dfrac{\partial v}{\partial \overline{u}} \\ \dfrac{\partial u}{\partial \overline{v}} & \dfrac{\partial v}{\partial \overline{v}} \end{vmatrix} \neq 0$$

which is the Jacobian of the transformation. This determinant must be nonzero for an adequate parametrization of the surface.

The metric on the surface will be the source of the intrinsic properties of the surface that we are most interested in. We start by considering a curve on the surface written in parametric form,

$$C: \quad u = u(t), \quad v = v(t)$$

Then $d\boldsymbol{x}/dt$ is tangent to the curve and, therefore, to the surface. It can be written in terms of the coordinate mesh,

$$\frac{d\boldsymbol{x}}{dt} = \boldsymbol{x}_u \frac{du}{dt} + \boldsymbol{x}_v \frac{dv}{dt}$$

or in a form independent of the parameter t,

$$d\boldsymbol{x} = \boldsymbol{x}_u du + \boldsymbol{x}_v dv$$

The distance element is then,

$$ds^2 = d\boldsymbol{x} \cdot d\boldsymbol{x} = \sum_i dx_i dx_i = g_{uu} du^2 + 2 g_{uv} du dv + g_{vv} dv^2 \tag{12.5a}$$

where

$$g_{uu} = \boldsymbol{x}_u \cdot \boldsymbol{x}_u, \quad g_{uv} = \boldsymbol{x}_u \cdot \boldsymbol{x}_v, \quad g_{vv} = \boldsymbol{x}_v \cdot \boldsymbol{x}_v \tag{12.5b}$$

provides the metric in the notation of classic differential geometry. ds^2 is called the "first fundamental form." The metric satisfies positivity in the form,

$$\det(g) = \det\begin{vmatrix} g_{uu} & g_{uv} \\ g_{uv} & g_{vv} \end{vmatrix} > 0$$

which follows from the positivity of,

$$(\boldsymbol{x}_u \times \boldsymbol{x}_v)\cdot(\boldsymbol{x}_u \times \boldsymbol{x}_v) = (\boldsymbol{x}_u \cdot \boldsymbol{x}_u)(\boldsymbol{x}_v \cdot \boldsymbol{x}_v) - (\boldsymbol{x}_u \cdot \boldsymbol{x}_v)^2 = \det(g) > 0$$

where we applied the vector identity, $(\boldsymbol{a} \times \boldsymbol{b})\cdot(\boldsymbol{c} \times \boldsymbol{d}) = (\boldsymbol{a}\cdot\boldsymbol{c})\,(\boldsymbol{b}\cdot\boldsymbol{d}) - (\boldsymbol{a}\cdot\boldsymbol{d})\,(\boldsymbol{b}\cdot\boldsymbol{c})$. Vector identities and properties of cross products are reviewed in Appendix D.

The tangent plane at the point $\boldsymbol{x}$ is given by,

$$\boldsymbol{X} = \boldsymbol{x} + \lambda\boldsymbol{x}_u + \eta\boldsymbol{x}_v$$

where λ and η are scalar parameters. The unit normal to the surface at $\boldsymbol{x}$ is,

$$\widehat{\boldsymbol{N}} = \frac{\boldsymbol{x}_u \times \boldsymbol{x}_v}{|\boldsymbol{x}_u \times \boldsymbol{x}_v|} = \frac{\boldsymbol{x}_u \times \boldsymbol{x}_v}{\sqrt{\det(g)}} \tag{12.6}$$

as shown in Fig. 12.8.

The determinant of the metric also figures in the area of a surface patch,

$$A = \iint \sqrt{\det(g)}\, du dv \tag{12.7}$$

This formula follows from the observation that the area patch dA on the surface is given by the area of the parallelogram bounded by the vectors $\boldsymbol{x}_u du$ and $\boldsymbol{x}_v dv$, so $dA = |\boldsymbol{x}_u du \times \boldsymbol{x}_v dv| = |\boldsymbol{x}_u \times \boldsymbol{x}_v| du dv = \sqrt{\det(g)}\, du dv$.

The second fundamental form of differential geometry focuses on the surface's curvature. It is obtained by considering the curvature of a curve C constrained to the surface and passing through the point P on the surface. Following our earlier discussion of curves is three dimensions, we decompose the curve's curvature vector $d\boldsymbol{t}/ds$ into a component normal to the surface and a component lying in the tangent plane,

$$d\boldsymbol{t}/ds = \boldsymbol{k}_n + \boldsymbol{k}_g = \kappa\widehat{\boldsymbol{N}} + \boldsymbol{k}_g \tag{12.8}$$

where we introduced a curvature parameter κ. The curve inherits this component of curvature from its embedding on the surface. The vector

$\boldsymbol{k}_g$ is called the tangential curvature vector. It can be written as a coefficient, the tangential curvature κ_g, times a unit vector $\widehat{\boldsymbol{u}}$ in the tangent plane. $\boldsymbol{k}_g = \kappa_g \widehat{\boldsymbol{u}}$ is also perpendicular to the curve's tangent $\boldsymbol{t}$, in light of the discussion preceding Eq. (12.4). The direction of $\widehat{\boldsymbol{u}}$ is chosen by convention so that the sense $\boldsymbol{t} \rightarrow \widehat{\boldsymbol{u}}$ is the same as $\boldsymbol{x_u} \rightarrow \boldsymbol{x}_v$.

Since $\widehat{\mathbf{N}} \cdot \boldsymbol{t} = 0$, we have by differentiation along the curve parameterized by its arc length s that $\frac{d}{ds}\left(\widehat{\mathbf{N}} \cdot \boldsymbol{t}\right) = 0$, so,

$$\frac{d\boldsymbol{t}}{ds} \cdot \widehat{\mathbf{N}} = \kappa = -\boldsymbol{t} \cdot \frac{d\widehat{\mathbf{N}}}{ds} = -\frac{d\boldsymbol{x}}{ds} \cdot \frac{d\widehat{\mathbf{N}}}{ds} = -\frac{d\boldsymbol{x} \cdot d\widehat{\mathbf{N}}}{d\boldsymbol{x} \cdot d\boldsymbol{x}} \tag{12.9}$$

The numerator,

$$\begin{aligned} -d\boldsymbol{x} \cdot d\widehat{\mathbf{N}} &= -(\boldsymbol{x}_u du + \boldsymbol{x}_v dv) \cdot \left(\widehat{\mathbf{N}}_u du + \widehat{\mathbf{N}}_v dv\right) \\ &= b_{uu} du^2 + 2b_{uv} du dv + b_{vv} dv^2 \end{aligned} \tag{12.10a}$$

where,

$$b_{uu} = -\boldsymbol{x}_u \cdot \widehat{\mathbf{N}}_u, \quad 2b_{uv} = -\left(\boldsymbol{x}_u \cdot \widehat{\mathbf{N}}_v + \boldsymbol{x}_v \cdot \widehat{\mathbf{N}}_u\right), \quad b_{vv} = -\boldsymbol{x}_v \cdot \widehat{\mathbf{N}}_v \tag{12.10b}$$

is the "second fundamental form" of the surface. So, the curvature κ can be written as the ratio of the two fundamental forms,

$$\kappa = -\frac{d\boldsymbol{x} \cdot d\widehat{\mathbf{N}}}{d\boldsymbol{x} \cdot d\boldsymbol{x}} = \frac{b_{uu} du^2 + 2b_{uv} du dv + b_{vv} dv^2}{g_{uu} du^2 + 2g_{uv} du dv + g_{vv} dv^2} \tag{12.11}$$

The curvature κ depends on the direction the curve $C(t)$ takes through the point P. To calculate κ as a function of this direction, it is advantageous to cast κ as the solution of a generalized eigenvalue problem: In Eq. (12.11) multiply through by the first fundamental form and use index notation of u^i, $i = 1, 2$ instead of u and v,

$$\sum_{ij} (b_{ij} - \kappa g_{ij}) du^i du^j = 0$$

where,

$$b_{ij} = \begin{pmatrix} b_{uu} & b_{uv} \\ b_{uv} & b_{vv} \end{pmatrix}, \quad g_{ij} = \begin{pmatrix} g_{uu} & g_{uv} \\ g_{uv} & g_{vv} \end{pmatrix}$$

We can recast this problem as a conventional eigenvalue problem by multiplying through by the inverse of g_{ij}, $(g^{-1})_{ij} \equiv g^{ij}$,

$$\sum_j g^{ij} g_{jk} = \delta^i_k$$

Recall from Appendix C that it is easy to find the inverse of a 2×2 symmetric matrix,

$$g^{ij} = \frac{1}{\det(g)} \begin{pmatrix} g_{vv} & -g_{uv} \\ -g_{uv} & g_{uu} \end{pmatrix}$$

Note also that $\det(g^{ij}) = 1/\det(g_{ij})$ as the reader can check from this formula for g^{ij}.

Next, call κ_α and $v^i_{(\alpha)}$ the eigenvalue and associated eigenvector (direction) for $\alpha = 1, 2$,

$$\sum_j (b_{ij} - \kappa_\alpha g_{ij}) v^j_{(\alpha)} = 0$$

Multiply through by g^{ki},

$$\sum_{ij} g^{ki} (b_{ij} - \kappa_\alpha g_{ij}) v^j_{(\alpha)} = 0$$

$$\sum_j \left(b^k_j - \kappa_\alpha \delta^k_j \right) v^j_{(\alpha)} = 0$$

where we have been careful with raising indices using g^{ki} and have distinguished between covariant and contravariant indices, as introduced in Section 6.5. We learn from this last equation that κ_α for $\alpha = 1, 2$ are the eigenvalues of the matrix b^k_j. In the basis $v^j_{(\alpha)}$, the matrix b^k_j is diagonalized,

$$b^k_j = \begin{pmatrix} \kappa_1 & 0 \\ 0 & \kappa_2 \end{pmatrix}$$

So, $\det\left(b^k_j\right) = \kappa_1 \kappa_2$. The eigenvectors $v^j_{(\alpha)}$ are called the "directions of principal curvature" and κ_α are called the "principal curvatures," and the two eigenvectors $v^j_{(\alpha)}$ comprise an orthogonal basis. The product $\kappa_1 \kappa_2$ defines the Gaussian curvature of the surface,

$$K = \kappa_1 \kappa_2 = \det\left(b^k_j\right) = \det\left(g^{ki} b_{ij}\right) = \det\left(g^{ki}\right) \det(b_{ij}) = \frac{\det(b)}{\det(g)}$$

$$= \frac{b_{uu} b_{vv} - b^2_{uv}}{g_{uu} g_{vv} - g^2_{uv}} \tag{12.12}$$

which will prove to be a useful formula.

Additional analysis shows that K can actually be written just in terms of the first fundamental form and its derivatives with respect to u and v. The proof uses the fact that the second fundamental form is $-d\boldsymbol{x}\cdot d\widehat{\mathbf{N}}$ and $\widehat{\mathbf{N}}$ can be written in terms of vectors lying in the tangent plane, $\widehat{\boldsymbol{N}} = (\boldsymbol{x}_u \times \boldsymbol{x}_v)/\det(g)$. Problem (12.3) leads one through the algebra and culminates with a very useful and famous formula for the curvature K in terms of the components of the metric and its derivatives. We will see that this is a very important conceptual point in the field!

Gauss was led to the curvature K because he was interested in finding intrinsic properties of surfaces. Gauss' pioneering research was guided by a deep and intuitive understanding of geometry. Gauss showed that K was a "bending invariant": it is unchanged by deformations of the surface that do not involve stretching, shrinking, or tearing. Bending leaves the distance between points on the surface unchanged and also leaves the direction between two tangent directions at a point unchanged. One can show that another bending invariant is the tangential curvature κ_g [1]. Gauss' proof was, in fact, very similar in spirit to the modern approaches we usually associate with later mathematicians such as Riemann, one of Gauss' doctoral students. All the concepts of geodesics, Christoffel symbols, and curvature tensors appear in Gauss' work, hidden somewhat by outdated notation. All these topics will reappear when we develop Einstein's field equation for general relativity because intrinsic curvature of space–time is the central concept of the theory: mass–energy creates intrinsic curvature in space–time!

Let us continue our discussion of classical differential geometry by introducing the ideas of geodesic curves, derive Eq. (12.2), which provides the geometric interpretation of curvature and finally present the Gauss–Bonnet theorem.

By definition geodesic curves on the surface are those that have $\kappa_g = 0$. This means that the geodesic's curvature vector is everywhere normal to the surface. In this sense, a geodesic on the surface is "as straight" as possible.

There is a geodesic pointing in every tangent direction at any point of a surface, a fact which will follow from the differential equation for geodesic curves that we will derive in the next section of this chapter. Consider the example of "great circles" on a sphere. Starting from the North Pole, it is clear that there is a great circle heading in every direction. These are the lines of constant longitude on a spherical globe of the world. Using this property we can set up geodesic coordinates. They prove to be very convenient in theory and applications. Consider a curve C_0 and the family of geodesics that are perpendicular to C_0 as shown in Fig. 12.9. Denote

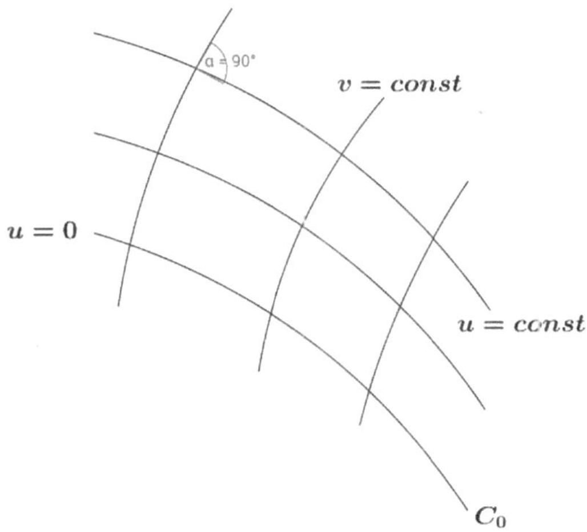

Figure 12.9 Geodesic coordinate mesh on a two-dimensional surface. The curves $u =$ constant and $v =$ constant are chosen mutually orthogonal.

geodesics as the curves $v =$ constant. Then the curves $u =$ constant and $v =$ constant are chosen mutually orthogonal so the metric is diagonal,

$$ds^2 = g_{uu}du^2 + g_{vv}dv^2$$

This expression can be simplified further by defining $w = \int_0^w \sqrt{g_{uu}}du$, so

$$ds^2 = dw^2 + g_{vv}dv^2$$

So, if a point moves along a geodesic the distance, arc length, along it is just w, $s = \int_{w_1}^{w_2} dw = w_2 - w_1$, which means that the segments on all the geodesics $v =$ constant included between two orthogonal trajectories are equal. The converse is also true. This leads to the term "geodesic parallels."

This metric is a generalization of plane polar coordinates in two-dimensional Euclidean space, $ds^2 = dr^2 + r^2 d\varphi^2$. In our case $r^2 \to g_{\varphi\varphi}$ and $v \to \varphi$, so we write on the curved surface $ds^2 = dr^2 + g_{\varphi\varphi}d\varphi^2$.

Let us get a better understanding of the curvature of the surface by concentrating on a small patch around a point P which we label with $r = 0$. The Gaussian curvature for this simple metric $ds^2 = dr^2 + g_{\varphi\varphi}d\varphi^2$ is just (see the Problem 12.3),

$$K = -\frac{1}{\sqrt{g_{\varphi\varphi}}}\frac{\partial^2 \sqrt{g_{\varphi\varphi}}}{\partial r^2} \tag{12.13}$$

Let us expand $\sqrt{g_{\varphi\varphi}}$ around point P labeled $r = 0$,

$$\sqrt{g_{\varphi\varphi}} = r + a_2 r^2 + a_3 r^3 + \ldots \tag{12.14}$$

where we are fitting our curved surface to flat space $ds^2 = dr^2 + r^2 d\varphi^2$ with curvature corrections parameterized by the coefficients a_2, a_3, etc. Substituting Eq. (12.14) into Eq. (12.13),

$$\frac{\partial^2 \sqrt{g_{\varphi\varphi}}}{\partial r^2} = -K\sqrt{g_{\varphi\varphi}}$$

$$2a_2 + 6a_3 r + \ldots = -K(r + \ldots)$$

Matching coefficients of r, we learn that $a_2 = 0$ and $a_3 = -\frac{1}{6}K$,

$$ds^2 = dr^2 + r^2\left(1 - \frac{1}{3}Kr^2\right)d\varphi^2 \tag{12.15}$$

where we squared $\sqrt{g_{\varphi\varphi}}$ and found $g_{\varphi\varphi} = r^2 - \frac{1}{3}Kr^4 + \ldots$. Eq. (12.15) shows the influence of the curvature on length measurements near $r = 0$. Consider a circumference curve C having constant r by letting φ range from 0 to 2π. Then C = length of the circumference, shown in Fig. 12.10, is given by,

$$C = \int_0^{2\pi} \sqrt{g_{\varphi\varphi}}d\varphi = \int_0^{2\pi}\left(r - \frac{1}{6}Kr^3\right)d\varphi = 2\pi r - \frac{\pi}{3}Kr^3$$

and we see that the presence of K in this formula corrects the circumference–radius relation, $C = 2\pi r$, of Euclidean geometry,

$$K = \frac{3}{\pi}\lim_{r\to 0}\left(\frac{2\pi r - C}{r^3}\right)$$

which reproduces Eq. (12.2).

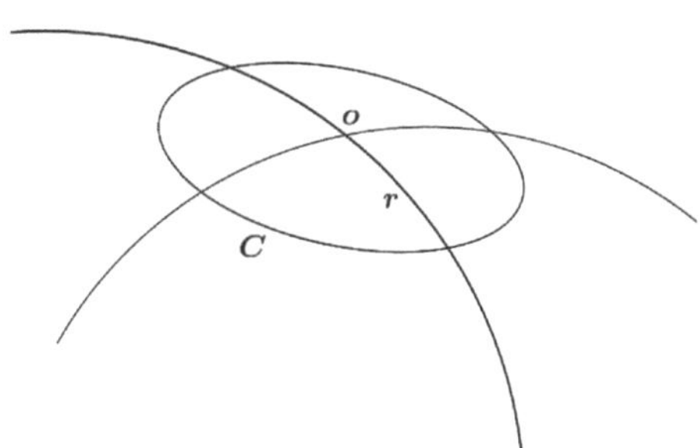

Figure 12.10 The circumference C of a curve at geodesic distance r from a point O on a two-dimensional surface.

We can also do this exercise for small patches of areas near $r = 0$, using Eq. (12.7),

$$A = \int_0^r \int_0^{2\pi} \sqrt{g_{\varphi\varphi}}\, dr d\varphi$$

and find,

$$K = \frac{12}{\pi} \lim_{r \to 0} \left(\frac{\pi r^2 - A}{r^4} \right)$$

Historians of mathematics believe that these expressions were the guiding lights behind Gauss' great inventions that established the field of differential geometry.

We end this discussion with an integral relation between the curvature κ_g of curves and the curvature K of surfaces. It is the Gauss—Bonnet theorem [1]. In its simplest form we consider a smooth, simply connected region R on the surface, which is bounded by a smooth curve C as shown in Fig. 12.11. Then the Gauss—Bonnet theorem reads,

$$\oint \kappa_g ds + \oiint K dA = 2\pi \tag{12.16}$$

This wonderful result indicates that we can measure the surface curvature inside a region by integrating the curvature of the curve C that bounds it. We will see that it also relates local properties of a surface to its global topological character.

Consider a simple illustration of the Gauss—Bonnet theorem. Imagine a spherical cap $\boldsymbol{S}$ as show in Fig. 12.12. The circular curve $\mathbf{C}$ on the sphere is at

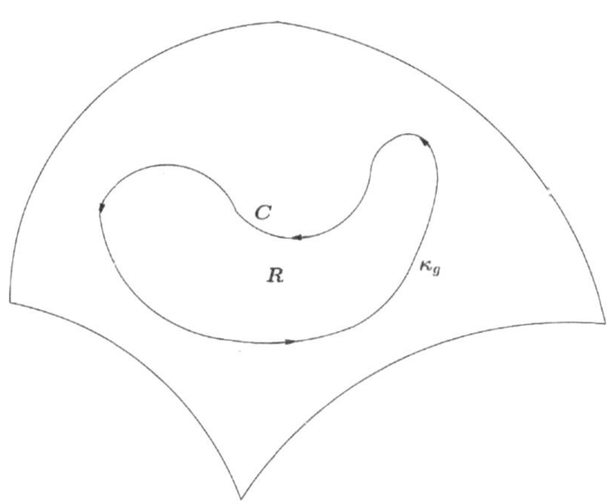

Figure 12.11 A smooth simply connected region R on a surface that is bounded by a smooth curve C.

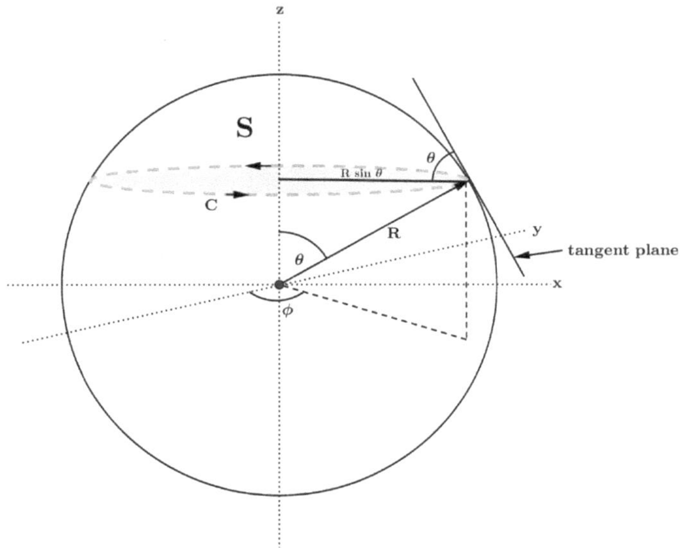

Figure 12.12 The "cap" of the sphere is labeled "***S***"—it is the surface of the sphere north of the curve ***C***, which has radius $R \sin \theta$.

an azimuthal angle θ. The "cap" of the sphere is labeled "$\boldsymbol{S}$" in the figure—it is the surface of the sphere north of the curve $\mathbf{C}$. Note that the area of the cap $\boldsymbol{S}$ is $A = \int_0^\theta \int_0^{2\pi} R\, d\theta\, R \sin \theta d\phi = 2\pi R^2 \int_0^\theta \sin \theta\, d\theta = 2\pi R^2(1 - \cos \theta)$. So, the surface curvature term in the Gauss–Bonnet theorem is $\oiint K dA = \frac{1}{R^2} \oiint dA = 2\pi(1 - \cos \theta)$. The line integral of the tangential curvature can be computed by first noting that the curvature κ_C of $\mathbf{C}$ is the reciprocal of its radius, $1/(R \sin \theta)$. To calculate κ_g we must project the curvature vector of magnitude κ_C onto the tangent plane. Noting the orientation of the tangent plane in the figure and that the curvature vector of C lies along the radius $R \sin \theta$, we find $\kappa_g = \cos \theta/(R \sin \theta)$. So, the line integral is $\oint \kappa_g\, ds = \frac{\cos \theta}{R \sin \theta}(2\pi R \sin \theta) = 2\pi \cos \theta$. Collecting terms, $\oiint K\, dA + \oint \kappa_g\, ds = 2\pi(1 - \cos \theta) + 2\pi \cos \theta = 2\pi$, which reproduces Eq. (12.16) perfectly! If we had chosen $\mathbf{C}$ to be the equator, which is a great circle with $\kappa_g = 0$, then $\oiint K dA = \frac{1}{R^2} \oiint dA = \frac{1}{R^2}\left(\frac{1}{2} \cdot 4\pi R^2\right) = 2\pi$, which is a particularly simple case.

It is useful to consider generalizations of this theorem to curves that are only piecewise differentiable: let C consist of M arcs of smooth curves making exterior angles $\varphi_1, \varphi_2, \ldots, \varphi_M$ at the vertices $A_1, A_2, \ldots, A_M$

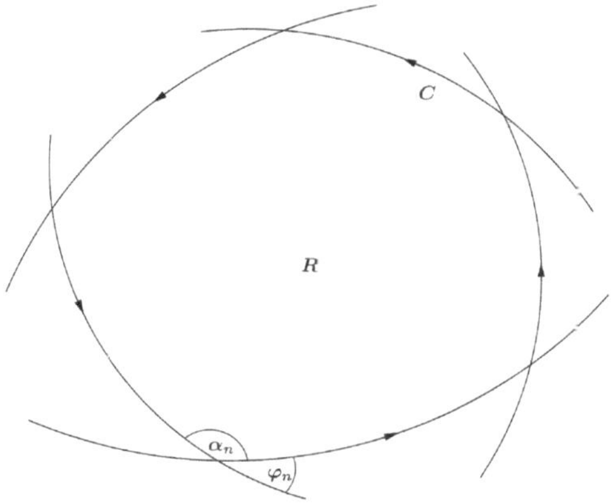

Figure 12.13 Generalization of the Gauss–Bonnet theorem to a curve C, which is only piecewise differentiable: C consist of M arcs of smooth curves making exterior angles $\varphi_1, \varphi_2, \ldots, \varphi_M$.

where the arcs meet as shown in Fig. 12.13. In this case the Gauss–Bonnet theorem reads,

$$\oint \kappa_g \, ds + \oiint K \, dA = 2\pi - \sum_1^M \varphi_i$$

Some special cases of this general result are quite informative. Suppose that the boundary curve C consists of M geodesic lines so $\kappa_g = 0$. We can express the answer in terms of interior angles α_i instead of the exterior angles φ_i, $\alpha_i = \pi - \varphi_i$, as shown in Fig. 12.13,

$$\oiint KdA = 2\pi - \sum_1^M (\pi - \alpha_i) = 2\pi - M\pi + \sum_1^M \alpha_i$$
$$= \sum_1^M \alpha_i - (M-2)\pi$$

For the case $M = 3$ we have a geodesic triangle, and the right-hand side of the Gauss–Bonnet theorem is the excess of the sum of its interior angles over π radians. For a surface with positive (negative) curvature, we learn that the sum of the triangle's interior angles is greater (less) than the sum of the interior angles in a triangle drawn in flat space. If we apply this theorem to a sphere of radius R, we find that the area of a spherical triangle equals the product of its interior angular excess times R^2.

Finally, if we apply the Gauss–Bonnet theorem Eq. (12.16) to a surface, which is topologically equivalent to a sphere, we find $\oiint KdA = 4\pi$. The proof is the result of adding the Gauss–Bonnet theorem applied to the

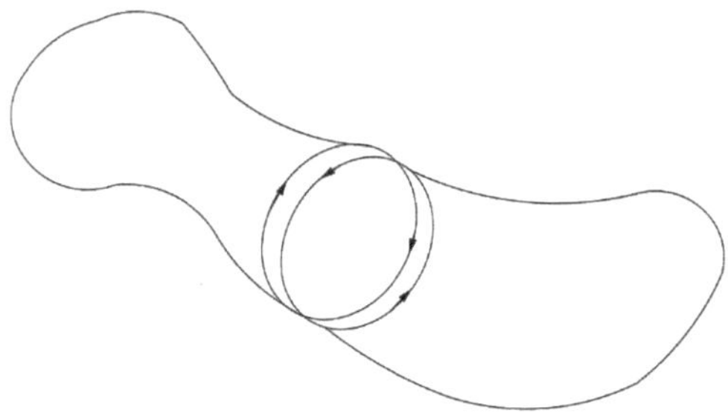

Figure 12.14 Application of the Gauss–Bonnet Theorem to a surface that is topologically equivalent to a sphere.

closed surface cut into two parts as shown in Fig. 12.14. The result shows that the integral of the Gaussian curvature over the entire surface is both a bending invariant and a topological invariant as well. The generalization of this result to a smooth orientable, compact surface having N handles, is that the right hand side, 4π, is replaced by $4\pi(1 - N)$. This connection between local geometry and global topology was very influential in the development of the subject.

The integral curvature $\oint K dA$ has an appealing intuitive interpretation, introduced by Gauss. Suppose that the integration region is R and it is bounded by a curve C. Imagine the surface normals emanating from the surface within the region R. Translate them to a common origin O preserving their angles, and suppose that they pierce a unit sphere in a region R_1, which is bounded by a curve C_1. Then the claim is that the curvature is,

$$K = \lim_{\Delta A \to 0} \frac{\Delta A_s}{\Delta A} \tag{12.17}$$

where ΔA_s is the area on the unit sphere, which is the image of the area ΔA on the original surface. The proof of this relation is interesting. Introduce coordinates (u, v) on the unit sphere such that a point on the original surface and its spherical image have the same (u, v). The equation of C_1 is $\widehat{\mathbf{N}} = \widehat{\mathbf{N}}(s)$, where s is the arc length of C. The element of the arc length of C_1 is $ds_1^2 = d\widehat{\mathbf{N}} \cdot d\widehat{\mathbf{N}}$, and the element of area dA_S of R_1 is, as discussed after Eq. (12.7),

$$dA_s = \sqrt{(\mathbf{N}_u \times \mathbf{N}_v)\cdot(\mathbf{N}_u \times \mathbf{N}_v)}\, dudv$$

But $\mathbf{N}_u \times \mathbf{N}_v = K\boldsymbol{x}_u \times \boldsymbol{x}_v$ (see Problem 12.4), so

$$dA_s = |K|\sqrt{(\boldsymbol{x}_u \times \boldsymbol{x}_v)\cdot(\boldsymbol{x}_u \times \boldsymbol{x}_v)}\, dudv = |K| dA$$

so the integral (absolute) curvature of a region on the surface equals the area of is spherical image (Gauss) and Eq. (12.17) follows.

It will be interesting to see these concepts again in the context of general relativity, curved space–time in four dimensions. We will see that general relativity is founded on Riemann's approach to differential geometry, which will be introduced along our journey. This is because Riemann's formulation only deals with intrinsic properties of the spaces it studies. In this sense its foundation begins with Gauss' finest result in classical differential geometry, that the Gaussian curvature is an intrinsic feature of two-dimensional surfaces. The discussion of differential geometry above is meant to provide an intuitive, visual background to the more abstract but powerful and economical version pioneered by Riemann, under Gauss' guidance, and to provide some motivation for some of the constructions underpinning the more modern approach.

In Riemannian geometry in four-dimensional space–time, the mathematical constructions will produce four-vectors and tensors. Following our discussion of special relativity in tensor notation, this will allow us to write down equations that are truly covariant, i.e., true in every coordinate system. In special relativity, we constructed laws that were true in all *inertial* reference frames, which were related by Lorentz transformations (boosts). In general relativity, we encompass all frames, both inertial and noninertial. The Lorentz transformation of special relativity, $x'^{\mu} = \sum_{\sigma} L^{\mu}{}_{\sigma} x^{\sigma}$, will be generalized to local coordinate transformations.

There are several reasons for making this ambitious step. First, physical laws should be independent of the coordinates used to parameterize space–time. This parametrization can vary as we move through space–time, so the global transformations of Lorentz transformations ("global" indicates that the coefficients $L^{\mu}{}_{\sigma}$ are constants, they have no dependence on x^{σ}) must be replaced by local transformations. Another reason is that general relativity is primarily a theory of gravity, and gravity, as we have already seen in Chapter 11, is incorporated into the structure of space–time. Lying at the center of general relativity is Einstein's field equation, which states that energy–momentum density creates local curvature in four-dimensional space–time.

The language of general relativity will be that of Riemannian geometry. Geodesic curves and space–time curvature will be central concepts, but we need to express them in equations that are free of specific choices of coordinate systems. This requirement brings four-vectors and tensors to the foreground. We shall find generalizations of geodesic curves, surfaces with Gaussian curvature and tangent spaces, and the equations will be tensor generalizations of the illustrations discussed above. Hopefully, those illustrations will motivate the intricate tensor equations to come!

This book sketches the conceptual growth of Newton's ideas into Einstein's ideas of physical laws. These developments were paralleled by earlier developments in mathematics taking that field from classical differential geometry, whose pinnacle was achieved by Gauss, to the formulation of modern differential geometry developed by his student, Riemann. Once Gauss proved his Theorema Egregium, that K is an intrinsic property of a surface, the challenge was to formulate differential geometry using intrinsic concepts from the start. This was the theme of Riemann's thesis. This mathematical development provided the language needed to formulate Einstein's ideas in general relativity and establish general covariance in the theory step-by-step.

12.2 THE EQUATION OF MOTION OF PARTICLES IN CURVED SPACE–TIME

Consider a curved space, such as the surface of a sphere or a saddle, etc. The physics in such a space will involve local differential equations and will be based on quantities and relationships, which are intrinsic, i.e., independent of the particular coordinate system used. The coordinate system is, of course, chosen to be convenient for writing down physical quantities and relationships. But to find and express intrinsic relationships, it helps enormously to use quantities that have simple transformation properties. We have seen this philosophy at work in special relativity, and it becomes even more important in general relativity where we want the theory to be invariant under general coordinate transformations, not just Lorentz transformations that are linear transformations with specific constant coefficients. For example, four-vectors were particularly useful in special relativity because relationships between four-vectors retained their form under transformations to different frames of reference and invariant quantities could be constructed by general procedures. For example, we expressed conservation laws using four-vectors when we considered collisions and were guaranteed that these relationships were true in all frames of reference if they were true in one. We could also develop invariants by taking the "inner" products of four-vectors and derive frame-independent results.

We need to generalize these ideas to curved spaces and will face several challenges. Some of the procedures done in "flat" Minkowski space where the invariant interval reads $ds^2 = dt^2 - dx^2 - dy^2 - dz^2$ in Cartesian coordinates will not work in curved space. Others can be generalized to curved space, but that goal will require new constructions.

When we transform from one coordinate system designated by (t, x, y, z) to another (t', x', y', z'), the transformation will depend on the location involved. We will use more four-vector notation for clarity and generality. Say that x^μ transforms into x'^μ. Suppose we have a scalar function $\phi(x)$, where x is short hand for the four-vector x^μ. Since x and x' represent the same point in space−time, $\phi(x) = \phi'(x')$. We can differentiate this relation using the chain rule, which is reviewed in Appendix C,

$$\frac{\partial \phi(x)}{\partial x^i} = \frac{\partial \phi'(x'(x))}{\partial x^i} = \sum_k \frac{\partial \phi'}{\partial x'^k} \frac{\partial x'^k}{\partial x^i} \tag{12.18a}$$

If we denote $A_i = \frac{\partial \phi}{\partial x^i}$, then Eq. (12.18a) reads,

$$A_i = \sum_k \frac{\partial x'^k}{\partial x^i} A'_k \tag{12.18b}$$

and a quantity with this transformation law is called a "covariant" vector. In special relativity the matrix of coefficients $\frac{\partial x'^k}{\partial x^i}$ were constants, as pointed out in Chapter 6, but in curved spaces this is not true. This means that if we take differentials of Eq. (12.18b) we find a complicated expression,

$$dA_i = \sum_k \frac{\partial x'^k}{\partial x^i} dA'_k + \sum_k A'_k d\left(\frac{\partial x'^k}{\partial x^i}\right)$$

$$dA_i = \sum_k \frac{\partial x'^k}{\partial x^i} dA'_k + \sum_{k,l} A'_k \frac{\partial^2 x'^k}{\partial x^i \partial x^l} dx^l \tag{12.19}$$

So, even though $A_i(x)$ is a covariant vector, $dA_i(x)$ is not. This is true because the matrix of coefficients $\frac{\partial x'^k}{\partial x^i}$ are in general not constants. We learn that $dA_i(x)$ is not a covariant vector because it compares A_i at x with A_i at $x + dx$, and these quantities have different transformation laws.

How do we deal with this problem? We know the answer from our earlier discussion of locality in field theory—it is necessary that the two vectors to be compared be located at the *same* point in space−time. In other words, we must "transport" one of the vectors to the point where the second one is located before we compare them. Let us consider contravariant, upper index, vectors first. If we used Cartesian coordinates in flat space, the "transport" will not contribute to dA^i. In other words, it will move the vector parallel to itself. For this reason, the "transport" process is called "parallel translation." So, to compare two infinitesimally separated vectors, one of them must be subjected to parallel transport to the point where the second one is located. Consider a vector A^i at x. Then at $x + dx$ it is equal to $A^i + dA^i$. Imagine parallel transporting A^i to $x + dx$. Call δA^i the change in the components of A^i. Then the difference DA^i between the two vectors now located at the same point $x + dx$ is [2],

$$DA^i = dA^i - \delta A^i \tag{12.20}$$

The form of δA^i is easy to write down in general. This infinitesimal difference will be a sum of terms each one containing a factor of dx^k and A^i. We will call the coefficients Γ^i_{kl} and write,

$$\delta A^i = -\sum_{k,l} \Gamma^i_{kl} A^k dx^l \tag{12.21}$$

The Γ factors are called Christoffel symbols, famous in classical differential geometry. We will learn how to calculate them later in this and following sections. They help us construct a "covariant" derivative D which, when it acts on a contravariant vector, produces a contravariant vector,

$$DA^i = \sum_l \left(\frac{\partial A^i}{\partial x^l} + \sum_k \Gamma^i_{kl} A^k \right) dx^l \tag{12.22a}$$

$$D_l A^i = \partial_l A^i + \sum_k \Gamma^i_{kl} A^k \tag{12.22b}$$

Eq. (12.22b) will play a critical role as we write expressions in curvilinear coordinates.

This exercise in differential geometry guides us to the equation of motion of massive point particles in curvilinear coordinates. Begin with the observation that we can obtain the equation from that of special relativity because at each point in our curved space–time we can choose a coordinate system which is freely falling and is described by special relativity with a flat, constant Minkowski metric. In special relativity in force-free space–time, a particle moves along a straight line at constant velocity, so $du^\mu/d\tau = 0$ or $du^\mu = 0$ where $u^\mu = dx^\mu/d\tau$, as discussed in Chapter 8. The differential of the velocity vanishes. Now express this equation in a general curvilinear coordinate system. The geometric idea is that the velocity does not change as the particle moves: so the ordinary differential in the flat coordinate system becomes the covariant differential in the curvilinear system,

$$Du^\mu = 0 \tag{12.23}$$

which reads in explicit terms,

$$du^\mu + \sum_{\nu,\sigma} \Gamma^\mu_{\nu\sigma} u^\nu dx^\sigma = 0 \tag{12.24}$$

Dividing through by $d\tau$,

$$\frac{d^2x^\mu}{d\tau^2} + \sum_{\nu,\sigma} \Gamma^\mu_{\nu\sigma} \frac{dx^\nu}{d\tau} \frac{dx^\sigma}{d\tau} = 0 \tag{12.25}$$

Since this equation corresponds to a straight line in the locally inertial reference frame, it is called a "geodesic" line in the curvilinear reference frame. If we look back at the previous chapter, we see that it resembles Newton's second law written in a noninertial frame where the velocity-dependent centripetal and Coriolis forces appear as "apparent" forces: the term $\sum_{\nu,\sigma} \Gamma^\mu_{\nu\sigma} \frac{dx^\nu}{d\tau} \frac{dx^\sigma}{d\tau}$ expresses "apparent" forces in four-dimensional curved space–time as we shall see in considerable detail below in general arguments, illustrations and problems.

Now it is an easy and interesting exercise to write Newton's second law and the Lorentz force law of electrodynamics in curvilinear coordinates, suitable for applications in general relativity. Newton's second law was written in four-vector form in Chapter 8. In Minkowski space–time,

$$f^\mu = m\frac{d^2x^\mu}{d\tau^2} = \frac{d}{d\tau}p^\mu \tag{12.26}$$

where τ is the proper time, p^μ is the energy–momentum four-vector, $p^\mu = mu^\mu = mdx^\mu/d\tau$ and f^μ is the four-vector force. To write Eq. (12.26) in curvilinear coordinates, the ordinary differentials become covariant differentials, which we write out in terms of Christoffel symbols,

$$\frac{1}{m}f^\mu = \frac{Du^\mu}{D\tau} = \frac{du^\mu}{d\tau} + \sum_{\nu,\sigma} \Gamma^\mu_{\nu\sigma} u^\nu u^\sigma = \frac{d^2x^\mu}{d\tau^2} + \sum_{\nu,\sigma} \Gamma^\mu_{\nu\sigma} u^\nu u^\sigma \tag{12.27a}$$

The Lorentz force law is a special case of this expression where $f^\mu = q\sum_\lambda u_\lambda F^{\lambda\mu}$, as we learned in Chapter 10, where we wrote $\boldsymbol{f} = q(\boldsymbol{E} + \boldsymbol{u} \times \boldsymbol{B}) = d\boldsymbol{p}/dt$ in Minkowski covariant form,

$$\frac{q}{m}\sum_\lambda u_\lambda F^{\lambda\mu} = \frac{d^2x^\mu}{d\tau^2} + \sum_{\nu,\sigma} \Gamma^\mu_{\nu\sigma} u^\nu u^\sigma \tag{12.27b}$$

Once we learn to calculate the Christoffel symbols, we shall be able to write the Lorentz force law in curved space–time. The four-vector character of the covariant derivative, D_l, Eq. (12.22b), makes all of this easy.

Let us return to our general discussion of curvilinear coordinates.

Parallel transport and the meaning of covariant differentiation are particularly easy to understand and visualize if we consider a curved space embedded in a flat Euclidean space. Imagine a sphere or a torus sitting in the lab. In the ordinary, flat Euclidean space we set up Cartesian coordinates and unit vectors along each coordinate direction in the usual fashion. A vector can be written in this coordinate system with mutually orthogonal constant unit vectors ($\boldsymbol{a}_\alpha$, $\alpha = 1, \ldots, N$) as,

$$\boldsymbol{V} = \sum_\alpha V^\alpha \boldsymbol{a}_\alpha \tag{12.28}$$

For example, in three dimensions, $\boldsymbol{r} = x\widehat{\boldsymbol{i}} + y\widehat{\boldsymbol{j}} + z\widehat{\boldsymbol{k}}$, in standard notation. If $\boldsymbol{V}$ is the four-velocity of a particle, which experiences no forces, then its equation of motion reads,

$$\boldsymbol{A} = \frac{d\boldsymbol{V}}{d\tau} = 0 \tag{12.29}$$

As the particle moves it sweeps out a straight line, which we parameterize with τ.

The curved space embedded in the Euclidean space has basis vectors ($\boldsymbol{b}_\beta$, $\beta = 1, \ldots, M$), which generally depend on their position and have variable orientations. Think of the surface of a sphere of unit radius. Any point on it is located with the angles (θ, ϕ) of ordinary spherical coordinates. There is a tangent plane at the point (θ, ϕ), which swivels about as (θ, ϕ) changes. The tangent plane is spanned by the vectors $\boldsymbol{b}_1 = \boldsymbol{b}_\theta = \partial\boldsymbol{r}/\partial\theta$ and $\boldsymbol{b}_2 = \boldsymbol{b}_\phi = \partial\boldsymbol{r}/\partial\phi$. The normal to the sphere also depends on (θ, ϕ) and is given by a unit vector $\widehat{\boldsymbol{n}} = \partial\boldsymbol{r}/\partial r$.

Now return to the general case. There is a vector $\boldsymbol{V}$ within the curved space, and we can write it in terms of the basis ($\boldsymbol{b}_\beta$, $\beta = 1, \ldots, M$),

$$\boldsymbol{V} = \sum_\beta V^\beta \boldsymbol{b}_\beta \tag{12.30}$$

Now suppose that $\boldsymbol{V}$ sweeps out a curve parameterized by τ. The rate of change of $\boldsymbol{V}$ *along the curve* is,

$$\frac{d\boldsymbol{V}}{d\tau} = \sum_\beta \frac{dV^\beta}{d\tau}\boldsymbol{b}_\beta + \sum_\beta V^\beta \frac{d\boldsymbol{b}_\beta}{d\tau} \tag{12.31}$$

where the second term shows that the curvilinear coordinates have τ dependence, i.e., the basis vectors vary as the point of interest moves along the curve. Recall that when one solves geometry or physics problems in three dimensions using curvilinear coordinates such as spherical or cylindrical coordinates, terms of this sort contribute. They are in fact the extra ingredients in expressions for the gradient, divergence, or Laplacian in such coordinate systems.

Next we can use the chain rule to write $\frac{d\boldsymbol{b}_\beta}{d\tau}$ in terms of the orientation of the basis in curved space and its movement along the curve,

$$\frac{d\boldsymbol{b}_\beta}{d\tau} = \sum_\alpha \frac{\partial \boldsymbol{b}_\beta}{\partial x^\alpha} \frac{dx^\alpha}{d\tau} \tag{12.32}$$

The changing orientation of the basis can be expanded in the basis ($\boldsymbol{b}_\beta$, $\beta = 1, \ldots, M$),

$$\frac{\partial \boldsymbol{b}_\beta}{\partial x^\alpha} = \sum_\gamma \Gamma^\gamma_{\beta\alpha} \boldsymbol{b}_\gamma \tag{12.33}$$

where we have introduced the Christoffel symbols again. We will show below that they are the same quantities introduced earlier. Eq. (12.33) gives a more concrete definition of the Christoffel symbols than the previous more abstract but general Eq. (12.21). We see that the Christoffel symbols record the fact that the basis vectors of a curvilinear coordinate system change their orientation as a point moves along a curve, and this affects the rate of change of the components of the vector of interest.

Now we can substitute into Eq. (12.31) and relabel the summation indices to our advantage,

$$\frac{d\boldsymbol{V}}{d\tau} = \sum_\beta \frac{dV^\beta}{d\tau} \boldsymbol{b}_\beta + \sum_{\alpha\beta\gamma} V^\beta \Gamma^\gamma_{\beta\alpha} \frac{dx^\alpha}{d\tau} \boldsymbol{b}_\gamma$$

$$\frac{d\boldsymbol{V}}{d\tau} = \sum_\beta \left(\frac{dV^\beta}{d\tau} + \sum_{\alpha\gamma} \Gamma^\beta_{\gamma\alpha} \frac{dx^\alpha}{d\tau} V^\gamma \right) \boldsymbol{b}_\beta \tag{12.34}$$

The quantity in parenthesis is the covariant derivative introduced earlier,

$$\frac{DV^\beta}{D\tau} = \frac{dV^\beta}{d\tau} + \sum_{\alpha\gamma} \Gamma^\beta_{\gamma\alpha} \frac{dx^\alpha}{d\tau} V^\gamma \tag{12.35}$$

Since $\frac{dV^\beta}{d\tau} = \sum_\alpha \frac{\partial V^\beta}{\partial x^\alpha} \frac{\partial x^\alpha}{\partial \tau} = \sum_\alpha \frac{\partial V^\beta}{\partial x^\alpha} u^\alpha$, we can write Eq. (12.35),

$$\frac{DV^\beta}{D\tau} = \sum_\alpha u^\alpha D_\alpha V^\beta \tag{12.36}$$

and identify the covariant derivative introduced in Eq. (12.22),

$$D_\alpha V^\beta = \partial_\alpha V^\beta + \sum_\gamma \Gamma^\beta_{\gamma\alpha} V^\gamma \tag{12.37}$$

If we apply Eq. (12.35) to the four-velocity of the particle $V = u = dx/d\tau$ then the vanishing of its acceleration $d^2x/d\tau^2 = 0$ becomes,

$$\frac{Du^\beta}{D\tau} = 0 = \frac{d^2x^\beta}{d^2\tau} + \sum_{\alpha\gamma} \Gamma^\beta_{\gamma\alpha} \frac{dx^\alpha}{d\tau} \frac{dx^\gamma}{d\tau} \tag{12.38}$$

which we recognize as the equation for a geodesic derived earlier.

In our applications to relativity, we begin with the absence of acceleration in a local inertial frame, $\frac{dV}{d\tau} = 0$. This maps onto the equation of motion, Eq. (12.38), in the curvilinear coordinate system that describes the same physics in a noninertial frame of reference where the gravitational effects are experienced as the curvature of space−time. Eq. (12.38) describes the motion of massive particles in a gravitational field. That field is represented by a nontrivial metric and connections, the Christoffel symbols that enter here. We will see how all this works out as we develop the physics of general relativity.

Note that only the symmetric piece of $\Gamma^\beta_{\gamma\alpha}$ contributes to Eq. (12.38), so we treat the Christoffel symbols in our applications to relativity as explicitly symmetric in the lower indices, $\Gamma^\beta_{\gamma\alpha} = \Gamma^\beta_{\alpha\gamma}$.

Let us illustrate some of the ideas here by considering a familiar example: Newtonian mechanics in two-dimensional polar coordinates. We want to calculate the Christoffel symbols and use them to write Newton's second law in plane polar coordinates. The problem set also explores these examples in greater detail and from several different perspectives. In this problem the force of gravity is inserted explicitly into the equation of motion. We will contrast this later with similar problems in general relativity where the "force" emerges from the curvature of space−time through the Christoffel symbols.

Consider a nonrelativistic particle in a gravitational potential generated by a mass M at the origin,

$$\frac{d^2\boldsymbol{r}}{dt^2} = -\frac{GM}{r^2}\widehat{\boldsymbol{r}}$$

This is a problem with a central potential so the motion of the particle in the potential will be planar. The particle's velocity in plane polar coordinates is $u^i = \left(\frac{dr}{dt}, \frac{d\theta}{dt}\right)$, and the equation of motion reads,

$$\frac{du^i}{dt} + \sum_{kl} \Gamma^i_{kl} u^k u^l = -\frac{GM}{r^2}\delta^i_1 \tag{12.39}$$

We need the Christoffel symbols. Let us calculate them for plane polar coordinates using the geometric approach,

$$\partial_\beta \boldsymbol{e}_\alpha = \sum_\gamma \Gamma^\gamma_{\alpha\beta} \boldsymbol{e}_\gamma \tag{12.33}$$

In Cartesian coordinates $\boldsymbol{r} = r \cos\theta\, \boldsymbol{i} + r \sin\theta\, \boldsymbol{j}$ and the coordinate vectors of plane polar coordinates are,

$$\frac{\partial \boldsymbol{r}}{\partial r} = \boldsymbol{e}_r = \cos\theta\, \boldsymbol{i} + \sin\theta\, \boldsymbol{j} \quad \frac{\partial \boldsymbol{r}}{\partial \theta} = \boldsymbol{e}_\theta = -r \sin\theta\, \boldsymbol{i} + r\cos\theta\, \boldsymbol{j}$$

Then one can calculate, following Problem 12.9,

$$\frac{\partial \boldsymbol{e}_r}{\partial r} = 0, \quad \frac{\partial \boldsymbol{e}_r}{\partial \theta} = \frac{1}{r}\boldsymbol{e}_\theta, \quad \frac{\partial \boldsymbol{e}_\theta}{\partial r} = \frac{1}{r}\boldsymbol{e}_\theta, \quad \frac{\partial \boldsymbol{e}_\theta}{\partial \theta} = -r\boldsymbol{e}_r$$

We can read off the Christoffel symbols from these results using Eq. (12.33),

$$\Gamma^\theta_{r\theta} = \Gamma^\theta_{\theta r} = \frac{1}{r}, \quad \Gamma^r_{\theta\theta} = -r$$

with the other Christoffel symbols vanishing. The equations of motion Eq. (12.39) become,

$$\frac{d^2 r}{dt^2} - r\left(\frac{d\theta}{dt}\right)^2 = -\frac{GM}{r^2}$$

$$\frac{d^2\theta}{dt^2} + \frac{2}{r}\frac{dr}{dt}\frac{d\theta}{dt} = 0$$

These are the results we wanted. The radial equation shows how the Christoffel symbols for the curvilinear coordinate system, plane polar coordinates, brings in the expected centripetal "apparent" force, discussed in Chapter 11. Similarly, the Christoffel symbols bring in the Coriolis "apparent" force into the equation of motion for the angular variable. The reader should recognize this equation as the conservation law for the orbital angular momentum per mass, $r^2\frac{d\theta}{dt} = L$.

Let us end this introduction to parallel transport and geodesics with some observations, which will be useful in applications. Note that if a vector is parallel transported along a geodesic, the angle between the transported vector and the tangent to the geodesic remains constant. Similarly, if two vectors V^α and U^β are parallel transported along a curve C, then $\sum_{\alpha\beta} g_{\alpha\beta} U^\alpha V^\beta$, their inner product, remains constant along C.

12.3 COVARIANT DERIVATIVES AND COVARIANT VECTOR FIELDS

In the text above, we considered covariant derivatives acting on contravariant vectors. We also need to differentiate covariant vectors. The general form of the relation must be,

$$D_\mu W_\sigma = \partial_\mu W_\sigma + \sum_\gamma \widetilde{\Gamma}^{\gamma}_{\mu\sigma} W_\gamma$$

where we have borrowed the general structure of the equation from the discussion above for contravariant vectors, but the coefficients in the parallel transport of covariant vectors, $\widetilde{\Gamma}$, must be determined.

To relate $\widetilde{\Gamma}$ to Γ, let us exploit the fact that $\sum_\gamma V_\gamma W^\gamma$ is a scalar function if V_γ and W^γ are four-vectors. For scalar functions D_μ reduces to ordinary ∂_μ,

$$D_\mu\left(\sum_\gamma V_\gamma W^\gamma\right) = \partial_\mu\left(\sum_\gamma V_\gamma W^\gamma\right) = \sum_\gamma (\partial_\mu V_\gamma) W^\gamma + \sum_\gamma V_\gamma(\partial_\mu W^\gamma) \tag{12.40}$$

But the covariant derivative satisfies,

$$D_\mu\left(\sum_\gamma V_\gamma W^\gamma\right) = \sum_\gamma (D_\mu V_\gamma) W^\gamma + \sum_\gamma V_\gamma(D_\mu W^\gamma)$$

just like ordinary differentiation, so,

$$\begin{aligned} D_\mu\left(\sum_\gamma V_\gamma W^\gamma\right) &= \sum_\gamma \left(\partial_\mu V_\gamma + \sum_\alpha \widetilde{\Gamma}^{\alpha}_{\mu\gamma} V_\alpha\right) W^\gamma \\ &\quad + \sum_\gamma V_\gamma \left(\partial_\mu W^\gamma + \sum_\alpha \Gamma^{\gamma}_{\mu\alpha} W^\alpha\right) \end{aligned} \tag{12.41}$$

Comparing Eqs. (12.40) and (12.41), we learn that,

$$0 = \sum_{\alpha\gamma}\left(\widetilde{\Gamma}^{\alpha}_{\mu\gamma} V_\alpha W^\gamma + \Gamma^{\gamma}_{\mu\alpha} V_\gamma W^\alpha\right)$$

Since this expression must be true for all V_α and W^β, we learn that the coefficients must satisfy,

$$\widetilde{\Gamma}^{\alpha}_{\mu\gamma} = -\Gamma^{\alpha}_{\mu\gamma}$$

In summary, for each upper index,

$$D_\mu V^\sigma = \partial_\mu V^\sigma + \sum_\gamma \Gamma^\sigma_{\mu\gamma} V^\gamma$$

and for each lower index we have now,

$$D_\mu W_\sigma = \partial_\mu W_\sigma - \sum_\gamma \Gamma^\gamma_{\mu\sigma} W_\gamma$$

Further manipulations show that these rules generalize to tensors with upper and lower indices.

For example, one can show that,

$$D_\mu T^\rho_\sigma = \partial_\mu T^\rho_\sigma - \sum_\gamma \Gamma^\gamma_{\mu\sigma} T^\rho_\gamma + \sum_\gamma \Gamma^\rho_{\mu\gamma} T^\gamma_\sigma$$

There are other properties of the covariant derivative we need throughout the text and some we have already anticipated.

1. When the covariant derivative acts on a scalar function, it reduces to the ordinary gradient,

$$D_\rho f(x) = \partial_\rho f(x)$$

2. The covariant derivative satisfies the "Leibniz rule" of ordinary differentiation. For example, the ordinary gradient satisfies,

$$\partial_\mu (T_{\alpha\beta} V_{\rho\gamma}) = (\partial_\mu T_{\alpha\beta}) V_{\rho\gamma} + T_{\alpha\beta}(\partial_\mu V_{\rho\gamma})$$

and the covariant derivative does also,

$$D_\mu (T_{\alpha\beta} V_{\rho\gamma}) = (D_\mu T_{\alpha\beta}) V_{\rho\gamma} + T_{\alpha\beta}(D_\mu V_{\rho\gamma}) \tag{12.42}$$

The proof of Eq. (12.42) notes that it is true in the tangent space where the Christoffel symbols vanish in Cartesian coordinates and the covariant derivatives reduce to ordinary derivatives. But Eq. (12.42) is an expression involving four-vectors when it is written in terms of D_μ, so if it is true in one coordinate system it is true in all. Note that the four-vector character of the covariant derivative is essential in this slick line of argument.

12.4 THE EQUIVALENCE PRINCIPLE, METRIC COMPATIBILITY, AND CHRISTOFFEL SYMBOLS

We introduced tangent spaces when we reviewed classical differential geometry. The tangent space is the best linear approximation to the curved space at and near a point P. For a surface embedded in Euclidean space a local coordinate mesh can be chosen so that its metric satisfies,

$$g_{jk} = \delta_{jk} + O\left(\frac{|\Delta x|^2}{R^2}\right)$$

where $|\Delta x|^2$ is the distance from P on the surface and R is a measure of the curvature of the surface at P, such as a principle radius of curvature there. One says that the surface is "locally flat."

In the case of relativity we have,

$$g_{\alpha\beta}(P) = g^{(0)}_{\alpha\beta}(P), \quad \partial_\gamma g_{\alpha\beta}(P) = 0, \quad \partial_\gamma \partial_\mu g_{\alpha\beta}(P) \neq 0 \tag{12.43}$$

where $g^{(0)}_{\alpha\beta}$ is the Minkowski metric. The equation $g_{\alpha\beta}(P) = g^{(0)}_{\alpha\beta}(P)$ is the first part of an expression of the equivalence principle: the freely falling frame at P is described locally by an inertial reference frame of special relativity. The second part of the statement is $\partial_\gamma g_{\alpha\beta}(P) = 0$ in the freely falling frame. This means that there are no net forces in this frame at P: freely falling frames are locally inertial. This is the essence of the equivalence principle, and it allows gravity to be interpreted as an aspect of geometry. The last expression, $\partial_\gamma \partial_\mu g_{\alpha\beta}(P) \neq 0$, means that the curvature of space–time is a frame-independent physical effect, which cannot be transformed away by a choice of coordinates. More about this will be discussed later when we develop the Riemann tensor.

Eq. (12.43) leads to a formula for the Christoffel symbols in terms of the metric and its derivatives. We learned above that statements involving ordinary derivatives of tensors in the freely falling frame become statements about the covariant derivative of those tensors in a general coordinate frame. So the second equality in Eq. (12.43) becomes,

$$D_\gamma g_{\alpha\beta}(P) = 0$$

This relation holds for any arrangement of the tensor indices. Using the results of the previous section on the covariant derivative, we can write this expression out explicitly,

$$\partial_\rho g_{\alpha\beta} - \sum_\mu \Gamma^\mu_{\rho\alpha} g_{\mu\beta} - \sum_\mu \Gamma^\mu_{\beta\rho} g_{\alpha\mu} = 0$$

$$\partial_\alpha g_{\beta\rho} - \sum_\mu \Gamma^\mu_{\alpha\beta} g_{\mu\rho} - \sum_\mu \Gamma^\mu_{\beta\alpha} g_{\rho\mu} = 0$$

$$\partial_\beta g_{\rho\alpha} - \sum_\mu \Gamma^\mu_{\beta\rho} g_{\mu\alpha} - \sum_\mu \Gamma^\mu_{\alpha\rho} g_{\beta\mu} = 0$$

We can isolate a Christoffel symbol by subtracting the second and third equations from the first one and use the fact that the Christoffel symbol and the metric tensor are symmetric in their two lower indices to find,

$$2\sum_{\mu}\Gamma^{\mu}_{\rho\beta}g_{\alpha\mu} + \partial_{\alpha}g_{\rho\beta} - \partial_{\rho}g_{\beta\alpha} - \partial_{\beta}g_{\alpha\rho} = 0$$

Finally we multiply through by the inverse of $g_{\alpha\mu}$, i.e., $g^{\gamma\mu}$, and obtain the result,

$$\Gamma^{\mu}_{\rho\beta} = \frac{1}{2}\sum_{\alpha}g^{\mu\alpha}\left(\partial_{\rho}g_{\beta\alpha} + \partial_{\beta}g_{\alpha\rho} - \partial_{\alpha}g_{\rho\beta}\right) \tag{12.44}$$

So, the Christoffel symbols can be computed in terms of the metric and its derivatives. The presence of $g^{\mu\alpha}$, the inverse of the metric, in Eq. (12.44) indicates the formula is also nonlinear in the metric itself.

The equation $D_{\gamma}g_{\alpha\beta}(P) = 0$, which is called "metric compatibility," is particularly useful when manipulating equations. It means that raising and lowering indices pass through the covariant derivative. For example,

$$\sum_{\beta}g_{\alpha\beta}D_{\rho}V^{\beta} = \sum_{\beta}D_{\rho}\left(g_{\alpha\beta}V^{\beta}\right) = D_{\rho}V_{\alpha}$$

Later in this chapter we will introduce the Einstein field equations, which express the idea that energy–momentum distributions warp space–time and produce its curvature. This curvature means that there is a metric tensor $g_{\mu\nu}$, which depends on x. From $g_{\mu\nu}$ one can calculate the Christoffel symbols and compute the motion of massive particles using Eq. (12.25). As the massive particles move, the energy–momentum tensor varies and the gravitational field changes, which alter the motion of the massive particles, etc. Einstein's equation for the gravitational field and the equation of motion Eq. (12.25) are a coupled system of differential equations which in principle solve the problem of the cosmos.

12.5 THE CURVATURE OF SPACE–TIME

Now let us return to parallel transport and relate it to the curvature of space–time. In a flat Minkowski space–time $du^{\mu} = 0$ in a force-free environment. We learned that this implies that $Du^{\mu} = 0$ in curved space–time. This means that $u^{\mu} = dx^{\mu}/d\tau$ parallel transports along its trajectory from x to $x + dx$. We already observed that under parallel displacement of two vectors, the angle between them remains constant. Therefore, the angle between a general vector which undergoes parallel displacement along a geodesic and the tangent to the geodesic remains unchanged, i.e., the components of the vector along the geodesic remain unchanged all along the path.

It is also clear that parallel displacement of a vector along a curve between two points will depend on the path chosen [2]. In particular, if a vector is parallel displaced along a closed path, on returning to its initial point it will *not* coincide with its initial orientation. It is easy to demonstrate this by visualizing parallel transportation of vectors around the surface of a sphere of radius R, as shown in Fig. 12.15. The path in the figure starts at the North Pole with a vector tangent to a great circle, then transports the vector along the equator and then returns to the North Pole along another great circle. The initial and final vectors at the North Pole have an opening angle φ between them and the Gauss–Bonnet theorem implies that φ is proportional to the curvature of the surface, $\varphi = \oiint K dA$, and on a sphere this relation reduces to the statement that the area of the geodesic triangle shown in the figure is φR^2.

One can generalize this observation by considering an infinitesimal closed path along which we parallel transport a vector A^μ [2],

$$\Delta A^\mu = \oint \Gamma^\mu_{\alpha\beta} A^\alpha dx^\beta$$

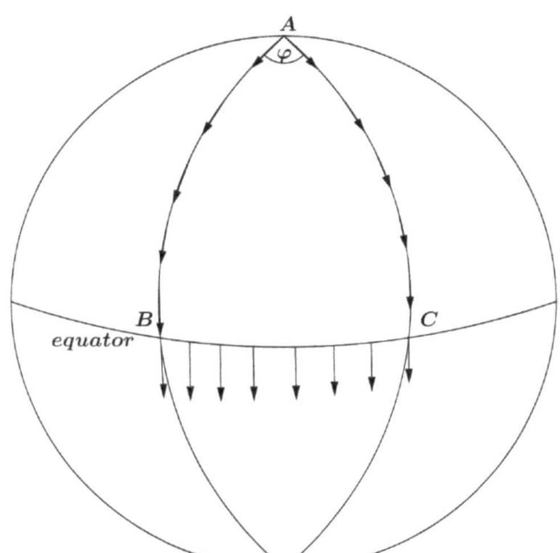

Figure 12.15 Parallel transportation of a vector around a geodesic triangle on the surface of a sphere of radius *R*.

Since ΔA^μ is the difference of two vectors at one point, it is a contravariant vector. Clearly ΔA^μ is nonzero only because space—time is curved. Another way to view this construction is to focus on the path dependence of parallel translation. In general, if we parallel transport a vector A^μ along two separate paths from point 1 to point 2, the results will be different. This observation produces a useful relation for the intrinsic curvature. Let D_μ represent covariant differentiation in the direction μ and let D_ν represent covariant differentiation in the direction ν. Then consider two points p and q, which are separated by two differentials, dx^μ in the μ direction and another, dx^ν in the ν direction. One can travel from p to q in two ways: first, move infinitesimally in the μ direction and then in the ν direction, or, second, move infinitesimally in the ν direction and then in the μ direction. The difference in the results of doing the translations from p to q along the two paths is a measure of the local curvature. One can write,

$$D_\mu D_\nu A^\rho - D_\nu D_\mu A^\rho = -\sum_\sigma R^\rho_{\sigma\mu\nu} A^\sigma$$

which can be written as a "commutator,"

$$(D_\mu D_\nu - D_\nu D_\mu)A^\rho = -\sum_\sigma R^\rho_{\sigma\mu\nu} A^\sigma \tag{12.45}$$

This equation, which defines the Riemann curvature tensor, $R^\rho_{\sigma\mu\nu}$, is visualized in Fig. 12.16.

Since the left-hand side of Eq. (12.45) is the difference of two four-vectors, the relation is a valid tensor equation, which holds in any curvilinear coordinate system. In addition, the fourth(!) rank tensor in Eq. (12.45) $R^\sigma_{\sigma\mu\nu}$, the Riemann curvature tensor, is independent of the vector A_ρ used in the construction.

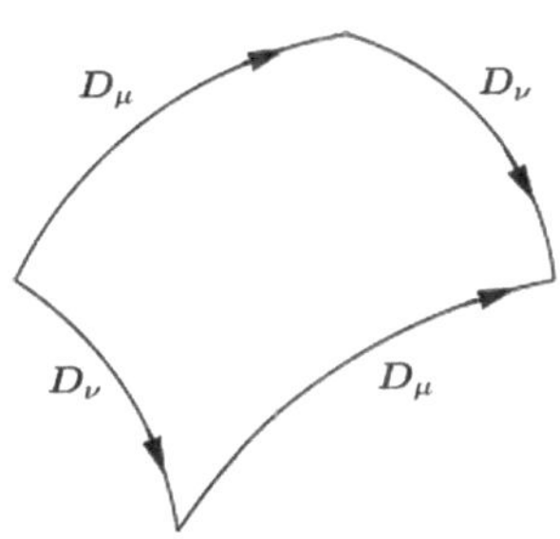

Figure 12.16 A closed path used in the construction of the Riemann curvature tensor.

We will carry through the calculation in Eq. (12.45) and obtain an explicit formula for the Riemann curvature tensor, in terms of the Christoffel symbols and their first derivatives. The result is,

$$R^{\rho}_{\sigma\mu\nu} = \partial_\mu \Gamma^{\rho}_{\nu\sigma} - \partial_\nu \Gamma^{\rho}_{\mu\sigma} + \sum_\alpha \Gamma^{\rho}_{\mu\alpha}\Gamma^{\alpha}_{\nu\sigma} - \sum_\alpha \Gamma^{\rho}_{\nu\alpha}\Gamma^{\alpha}_{\mu\sigma} \tag{12.46}$$

To derive this result, write out the commutator using Eq. (12.37),

$$\begin{aligned} D_\alpha D_\beta V^\rho - D_\beta D_\alpha V^\rho &= \partial_\alpha (D_\beta V^\rho) - \sum_\gamma \Gamma^{\gamma}_{\alpha\beta} D_\gamma V^\rho + \sum_\gamma \Gamma^{\rho}_{\alpha\gamma} D_\beta V^\gamma - (\alpha \leftrightarrow \beta) \\ &= \partial_\alpha \partial_\beta V^\rho + \sum_\gamma \left(\partial_\alpha \Gamma^{\rho}_{\beta\gamma}\right) V^\gamma + \sum_\gamma \Gamma^{\rho}_{\beta\gamma} \partial_\alpha V^\gamma - \sum_\gamma \Gamma^{\gamma}_{\alpha\beta} \partial_\gamma V^\rho \\ &\quad - \sum_{\gamma\sigma} \Gamma^{\gamma}_{\alpha\beta} \Gamma^{\rho}_{\gamma\sigma} V^\sigma + \sum_\gamma \Gamma^{\rho}_{\alpha\gamma} \partial_\beta V^\gamma + \sum_{\gamma\sigma} \Gamma^{\rho}_{\alpha\sigma} \Gamma^{\sigma}_{\beta\gamma} V^\gamma - (\alpha \leftrightarrow \beta) \\ &= \sum_\sigma \left(\partial_\alpha \Gamma^{\rho}_{\beta\sigma} - \partial_\beta \Gamma^{\rho}_{\alpha\sigma} + \sum_\gamma \Gamma^{\rho}_{\alpha\gamma} \Gamma^{\gamma}_{\beta\sigma} - \sum_\gamma \Gamma^{\rho}_{\beta\gamma} \Gamma^{\gamma}_{\alpha\sigma} \right) V^\sigma \end{aligned}$$

where the symmetry of the Christoffel symbol, $\Gamma^{\rho}_{\alpha\beta} = \Gamma^{\rho}_{\beta\alpha}$, and the "$-(\alpha \leftrightarrow \beta)$" terms led to considerable cancellation. Using Eq. (12.44) which expresses the Christoffel symbols in terms of the metric and its derivatives, the Riemann curvature tensor can be written in terms of the metric and its derivatives through second order. The problem set explores these relations, and several problems and Appendix G relates these exercises to the classical differential geometry presented in Section 12.1 above.

The Riemann tensor is very imposing since it has $4 \times 4 \times 4 \times 4 = 256$(!) components. However, it is highly constrained by symmetries. The Riemann tensor symmetry properties can be derived from Eq. (12.46). (Some are clear by inspection, but others require work. They are derived in the problem set.) First, lower the index on the tensor,

$$R_{\rho\sigma\alpha\beta} = \sum_\gamma g_{\rho\gamma} R^{\gamma}_{\sigma\alpha\beta} \tag{12.47}$$

Then the symmetry properties read,

1. $R_{\rho\sigma\alpha\beta} = -R_{\sigma\rho\alpha\beta}$
2. $R_{\rho\sigma\alpha\beta} = -R_{\rho\sigma\beta\alpha}$
3. $R_{\rho\sigma\alpha\beta} = R_{\alpha\beta\rho\sigma}$
4. $R_{\rho\sigma\alpha\beta} + R_{\rho\alpha\beta\sigma} + R_{\rho\beta\sigma\alpha} = 0$
5. $D_\gamma R_{\rho\sigma\alpha\beta} + D_\rho R_{\sigma\gamma\alpha\beta} + D_\sigma R_{\gamma\rho\alpha\beta} = 0$ (Bianchi identity)

Other important curvature tensors of lower rank can be constructed from the Riemann tensor. The second-rank Ricci tensor is

$$R_{\alpha\beta} = \sum_{\gamma} R^{\gamma}_{\alpha\gamma\beta} \equiv \sum_{\gamma\sigma} g^{\gamma\sigma} R_{\sigma\alpha\gamma\beta} \tag{12.48}$$

Properties 1 and 2 imply that it is symmetric, $R_{\alpha\beta} = R_{\beta\alpha}$. Note that the other possible contractions such as $\sum\limits_{\gamma} R^{\gamma}_{\gamma\alpha\beta} = \sum\limits_{\gamma\sigma} g^{\gamma\sigma} R_{\sigma\gamma\alpha\beta}$ or $\sum\limits_{\gamma} R^{\gamma}_{\alpha\beta\gamma} = \sum\limits_{\gamma\sigma} g^{\gamma\sigma} R_{\alpha\beta\gamma\sigma}$ vanish identically because of properties 1 and 3. This observation implies that the Ricci tensor $R_{\alpha\beta}$ is the *only* second-rank tensor that can be constructed from the Riemann tensor.

The Ricci scalar is the fully contracted version,

$$R = \sum_{\gamma} R^{\gamma}_{\gamma} = \sum_{\alpha\beta} g^{\alpha\beta} R_{\alpha\beta} \tag{12.49}$$

In the case of two-dimensional surfaces in three-dimensional Euclidean space, the Ricci scalar is just twice the Gaussian curvature K. This point is demonstrated in the problem set.

In four dimensions the full Riemann tensor $R_{\sigma\alpha\gamma\beta}$ is generally required to specify the curvature(s) of space–time.

12.6 FROM NEWTON'S GRAVITY TO RELATIVISTIC WEAK GRAVITY TO STRONG GRAVITY

Now that differential geometry and curvature have been introduced, we can turn back to physics and consider the differential equation for the gravitational potential in Newton's world,

$$\nabla^2 V(r) = 4\pi G\rho(r) \tag{12.50}$$

where the source of the potential is the mass density $\rho(r)$ and the force on a particle of mass m is $\boldsymbol{F}(r) = -m\nabla V(r)$. We want to recast this equation in the language of general relativity, if possible.

We have seen in several examples discussed in Chapter 11 that the Newtonian gravitational potential modifies the time–time component of the metric: g_{00} becomes $1 + \frac{2V(r)}{c^2}$ in the case of weak gravitational potentials, $2V(r)/c^2 \ll 1$. If there is a particle of mass M at the origin, $V(r) = -\frac{GM}{r}$ is the Newtonian gravitational potential Then the time–time component of the metric becomes,

$$g_{00} = g^{(0)}_{00} + \widetilde{g}_{00} = g^{(0)}_{00} + \frac{2V(r)}{c^2} = 1 - \frac{2GM}{c^2 r} \tag{12.51}$$

where we have introduced $\widetilde{g}_{00}$ to indicate the required modification of the metric of special relativity, $g^{(0)}_{\mu\nu}$.

Now turn back to Eq. (12.50). We want to write this equation in relativistic notation. We have already argued in Section 11.8 that the right-hand side is proportional to the "0-0" component of the energy–momentum tensor: the energy density transforms as the (outer) product of two four-vectors. This was cited in Chapter 11 as one of several critical differences between electromagnetism and gravity. Therefore, the left-hand side should also be a second-rank tensor, and $g_{\mu\nu}$ is a natural candidate. Newton's law of gravitation can be written,

$$\nabla^2 \tilde{g}_{00} = +\frac{8\pi}{c^2} G T_{00} \tag{12.52}$$

A plausible extension of this static equation to time-*dependent* special relativity applications would be,

$$\left(\nabla^2 - \frac{1}{c^2}\frac{\partial^2}{\partial t^2}\right)\tilde{g}_{\mu\nu} = +\frac{8\pi}{c^2} G T_{\mu\nu} \qquad (\textit{tentative guess}) \tag{12.53}$$

We have replaced the Laplacian of the static problem with the invariant wave operator of special relativity. It is convenient to define the wave operator,

$$\Box = \left(\nabla^2 - \frac{1}{c^2}\frac{\partial^2}{\partial t^2}\right) \tag{12.54}$$

Eq. (12.53) incorporates the idea that the metric $g_{\mu\nu}$ becomes a dynamical variable, and the idea that signals should travel at the speed limit, as discussed in Chapters 9 and 11. However, this extension of Newton's law of gravity is only plausible if the deviations $\tilde{g}_{\mu\nu}$ from the Minkowski metric $g^{(0)}_{\mu\nu}$ are numerically small. This is because Eq. (12.53) was built on Lorentz boosts, which are exact symmetries of the Minkowski metric $ds^2 = c^2dt^2 - dr^2$ and *not* general coordinate transformations, which are the symmetries of general relativity. In particular, the wave operator is an invariant operator in special relativity and is not invariant under general coordinate transformations.

In later sections of this chapter, we will apply Eq. (12.53) to the creation and propagation of gravitational waves. Let us consider it more critically in this context.

Sufficiently distant from the source, $T_{\mu\nu}$, the gravity waves predicted by Eq. (12.53) would be tiny fluctuations in an otherwise flat Minkowski space. This element of the problem belongs to special relativity. The decomposition,

$$g_{\mu\nu} = g^{(0)}_{\mu\nu} + \tilde{g}_{\mu\nu}$$

with $\left|\tilde{g}_{\mu\nu}\right| \ll 1$ is critical.

Now consider the left-hand side of Eq. (12.53) again. The "tentative guess" considered there is not unique: there are additional terms that could appear which are consistent with covariance. Those terms should satisfy the three criteria,

1. It must be a second-rank, symmetric tensor,
2. It must be second order in derivatives of $\widetilde{g}_{\mu\nu}$, and
3. It must reduce to Newton's law of gravity in the nonrelativistic limit.

To write its general form consider the four-vectors and tensors that we can use for its construction: ∂_μ, ∂_ν, $g^{(0)}_{\mu\nu}$, and $\widetilde{g}_{\mu\nu}$. With these ingredients we can construct second-rank tensor candidates:

1. $\Box\widetilde{g}_{\mu\nu}$, which we identified above,
2. $\sum\limits_\sigma \partial_\mu\partial^\sigma\widetilde{g}_{\sigma\nu} + \sum\limits_\sigma \partial_\nu\partial^\sigma\widetilde{g}_{\sigma\mu}$,
3. $\partial_\mu\partial_\nu\widetilde{g}^{\,tr}$,
4. $g^{(0)}_{\mu\nu}\Box\widetilde{g}^{\,tr}$,
5. $g^{(0)}_{\mu\nu}\left(\sum\limits_{\sigma\rho}\partial^\sigma\partial^\rho\widetilde{g}_{\sigma\rho}\right)$

where we used the short hand $\widetilde{g}^{\,tr} = \sum\limits_\mu \widetilde{g}^\mu_\mu$. So, we really should consider an equation of the form,

$$a\left(\Box\widetilde{g}_{\mu\nu}\right) - b\left(\sum_\sigma \partial_\mu\partial^\sigma\widetilde{g}_{\sigma\nu} + \sum_\sigma \partial_\nu\partial^\sigma\widetilde{g}_{\sigma\mu}\right) + c(\partial_\mu\partial_\nu\widetilde{g}^{\,tr}) + d\left(g^{(0)}_{\mu\nu}\Box\widetilde{g}^{\,tr}\right) - e\left(g^{(0)}_{\mu\nu}\sum_{\sigma\rho}\partial^\sigma\partial^\rho\widetilde{g}_{\sigma\rho}\right) = +\frac{8\pi}{c^4}GT_{\mu\nu} \tag{12.55}$$

where a, b, c, d, and e are numerical coefficients. The "extra" terms b and c involve time derivatives and do not contribute to the static, nonrelativistic limit, Eq. (12.52), so we have no constraints on them yet.

To make progress, let us consider the decomposition $g_{\mu\nu} = g^{(0)}_{\mu\nu} + \widetilde{g}_{\mu\nu}$ with $\left|\widetilde{g}_{\mu\nu}\right| \ll 1$ more critically. This decomposition picks out just certain frames: we could make a large boost and violate the strong inequality. In other words, our description of the phenomenon does not have full Lorentz invariance. However, there is a large family of frames that respect the decomposition. Suppose we begin in a frame where the decomposition is true and we make only small x-dependent coordinate transformations,

$$x'^\mu = x^\mu + \varepsilon^\mu(x)$$

and suppose that both $\varepsilon^{\mu}(x)$ and its derivatives $\partial_{\nu}\varepsilon^{\mu}(x)$ are very small compared to unity. The metric transforms under coordinate transformations as,

$$g'_{\mu\nu} = \sum_{\sigma\rho} \frac{\partial x^{\sigma}}{\partial x'^{\mu}} \frac{\partial x^{\rho}}{\partial x'^{\nu}} g_{\sigma\rho}$$

and in this case we calculate,

$$\tilde{g}'_{\mu\nu} = \tilde{g}_{\mu\nu}(x) - \partial_{\mu}\varepsilon_{\nu} - \partial_{\nu}\varepsilon_{\mu} + O(\varepsilon^2)$$

Let us use this freedom to our advantage. We copy the strategy of using the Lorenz gauge in electrodynamics as discussed in Appendix F. To motivate this, recall that the energy–momentum tensor is conserved in special relativity, $\sum_{\mu}\partial^{\mu} T_{\mu\nu} = 0$. If we could choose the coordinate system $\varepsilon^{\mu}(x)$ so that $\sum_{\mu}\partial^{\mu}\tilde{g}_{\mu\nu} = 0$, then the wave equation Eq. (12.55) would be simplified. In addition, if the energy–momentum tensors were traceless, $\sum_{\mu} T^{\mu}_{\mu} = 0$, then it would also be advantageous to choose $\varepsilon^{\mu}(x)$ so that the metric fluctuations were also traceless, $\sum_{\mu}\tilde{g}^{\mu}_{\mu} = 0$. We will see below that these conditions are fulfilled in our application to gravitational waves. Then we choose $\partial_{\nu}\varepsilon_{\mu}(x)$ so that $\tilde{g}_{\mu\nu}$ satisfies two subsidiary conditions,

1. $\sum_{\mu}\partial^{\mu}\tilde{g}_{\mu\nu} = 0$, and
2. $\sum_{\mu}\tilde{g}^{\mu}_{\mu} = 0$,

so that $\tilde{g}_{\mu\nu}$ is both conserved and traceless. Problem 12.15 shows how this choice can be made. Now the extra terms *b–e* in the differential equation, Eq. (12.55), for $\tilde{g}_{\mu\nu}$ vanish and the wave equation, Eq. (12.54), remains true! When we solve the wave equation for a given source of gravitational waves, we will check that the two conditions, $\sum_{\mu}\partial^{\mu}\tilde{g}_{\mu\nu} = 0$ and $\sum_{\mu}\tilde{g}^{\mu}_{\mu} = 0$, are valid, so we have not cheated and simplified the real problem at all. "Linearized gravity" is discussed more in the problem set where the student will begin with the full Einstein field equations, to be introduced below, and specialize to weak gravitational effects in a given static Minkowski space–time and prove these assertions. In Problem 12.16 we also consider linearized gravity and the Newtonian limit for static sources of energy–momentum $T_{\mu\nu}$. In this case the energy–momentum tensor is not traceless, and it is not appropriate to enforce subsidiary condition 2. In that case the wave equation Eq. (12.53) is replaced with [2],

$$\Box \overline{h}_{\mu\nu} = \frac{16\pi G}{c^4} T_{\mu\nu} \tag{12.56}$$

where

$$\overline{h}_{\mu\nu} = \widetilde{g}_{\mu\nu} - \frac{1}{2} g^{(0)}_{\mu\nu} h \quad h = \sum_{\alpha} \widetilde{g}^{\alpha}_{\alpha}$$

and $\overline{h}_{\mu\nu}$ satisfies the first subsidiary condition, the Lorenz condition,

$$\sum_{\nu} \partial_\nu \overline{h}^{\mu\nu} = 0$$

So we still have an elegant and simple wave equation, which is useful in other applications. We will see in discussions below that this formulation of linearized gravity is particularly useful, especially for weak gravitational effects outside static sources of mass–energy, as illustrated in the problem set.

Before considering strong gravity, let us confirm that we can retrieve Newtonian physics from this new perspective. Consider the weak field, nonrelativistic case (Newtonian limit) where $T^{00} \to \rho c^2$ and the wave equations Eqs. (12.53) and (12.56) reduce to Newton's law of gravity, $\nabla^2\Phi = 4\pi G\rho$, if we identify $\overline{h}^{00} = 4\Phi/c^2$. Note that all other components of $\overline{h}_{\alpha\beta}$ and $T_{\alpha\beta}$ are negligible here. Then, if we use the definition $\overline{h}_{\mu\nu} = \widetilde{g}_{\mu\nu} - \frac{1}{2} g^{(0)}_{\mu\nu} h$, some algebra (Problem 12.17) reveals that,

$$h = -4\Phi/c^2$$

and,

$$\widetilde{g}^{00} = \widetilde{g}^{11} = \widetilde{g}^{22} = \widetilde{g}^{33} = 2\Phi/c^2$$

So the metric in the weak field case reads,

$$ds^2 = \left(1 + \frac{2\Phi}{c^2}\right) c^2 dt^2 - \left(1 - \frac{2\Phi}{c^2}\right)(dx^2 + dy^2 + dz^2) \tag{12.57}$$

This result is accurate to order $1/c^4$. This coordinate system, which is discussed further in the problem set, is called "isotropic" because of its manifest rotational symmetry in the spatial coordinates. It is generally very handy in weak field applications.

Now let us retrieve Newton's second law in weak, linear gravity. It will follow from the geodesic equation, which predicts how a mass m travels in a force-free space–time described by curvilinear coordinates,

$$\frac{d^2x^\beta}{d^2\tau} + \sum_{\alpha\gamma} \Gamma^\beta_{\gamma\alpha} \frac{dx^\alpha}{d\tau}\frac{dx^\gamma}{d\tau} = 0 \tag{12.58}$$

We need the Christoffel symbols for the metric Eq. (12.57). These are easily calculated using Eq. (12.44). The algebra is minimized if we note that the formulas in linearized, weak field gravity are only accurate to first order in $2\Phi/c^2$, so we can raise and lower indices using the Minkowski metric of the background space–time in Eq. (12.44). Then it is easy to verify (Problem 12.19),

$$\Gamma^0_{00} = \frac{\partial\Phi}{c^2\partial c\tau}, \quad \Gamma^i_{00} = \frac{\partial\Phi}{c^2\partial x^i}$$

We simplify Eq. (12.58) for nonrelativistic kinematics, $p^0 \gg p^i$, and choose the $\beta = 0$ component. The zeroth component of Eq. (12.58) then reduces to,

$$\frac{d}{dt}(E + m\Phi) = 0$$

which is just Newtonian energy conservation to leading order! Similarly, for $\beta = i$, $(i = 1, 2, 3)$, Eq. (12.58) becomes Newton's second law for a particle of mass m in a gravitational field,

$$\frac{d}{dt}\boldsymbol{p} = -m\nabla\Phi$$

to leading order. These are the expected results but obtained from the perspective of general relativity.

This example illustrates that in Einstein's world, gravitation is curvature, expressed through nonzero $\Gamma^\beta_{\gamma\alpha}$, and in Newton's world, gravitation is a universal force, proportional to every body's inertial mass. In the weak field, nonrelativistic regime, these two approaches give identical predictions. Now we will pass to strong gravity where the theories will diverge and the fact that general relativity predicts that energy–momentum warps space–time will generate unique and fascinating predictions.

Our first task is to motivate and present the Einstein field equation to deal with strong gravitational fields where space–time is highly distorted from the Minkowski metric, and we need the full machinery of differential geometry. What are the constructive principles to guide us toward the fundamental differential equations of general relativity? They read as follows:

1. The equation should be invariant to general coordinate transformations: it should be true in all frames of reference.
2. It should express the idea that the energy−momentum density is the source of the curvature of space−time.
3. It should be a second-order differential equation for the space−time metric $g_{\sigma\rho}$.
4. It should embody conservation of energy−momentum locally in space−time.
5. It should reduce to special relativity far from any source of energy−momentum.
6. It should reproduce Newton's theory of gravity in the nonrelativistic, weak field limit.

We have already discussed these principles but have not emphasized item 4 earlier. We have illustrated the energy−momentum tensor for free massive particles, and in a problem set on electrodynamics we constructed the energy−momentum tensor from the electrodynamic field $F_{\mu\nu}$. We want $T_{\mu\nu}$ to express local energy−momentum conservation through a local differential equation so that it will be a true tensor relation in all frames of reference and will be consistent with causality. Recall our discussion of charge conservation that led us to the continuity equation $\sum_{\mu}\partial^{\mu}J_{\mu} = 0$ for the current density J_{μ}. The equation had a simple physical and geometric interpretation: the local charge density ρ in a box changes only when currents $\boldsymbol{J}$ flow into or out of that box. Now we express the conservation of energy−momentum in a similar fashion. In flat Minkowski space−time we have,

$$\sum_{\mu}\partial^{\mu}T_{\mu\nu} = 0 \tag{12.59}$$

and T_{00} is the energy density and T_{0i} is the ith component of momentum density. $T_{\mu\nu}$ is constructed to be symmetric, as was clear from our free particle example and electrodynamics. In a curved space−time Eq. (12.59) becomes

$$\sum_{\mu}D^{\mu}T_{\mu\nu} = 0$$

Now we need the left-hand side of the equation, $? = \frac{8\pi}{c^4} GT_{\mu\nu}$. It should satisfy several criteria,

1. It should be a local, conserved second-rank tensor.
2. It should be constructed out of $g_{\sigma\rho}$, $g^{\sigma\rho}$ and their first and second derivatives in space–time.
3. It should represent the local intrinsic curvature of space–time.

The only candidate tensor with most of these properties is the Ricci tensor $R_{\sigma\rho}$. However, we have not checked whether it is conserved. To begin, we know something about the covariant derivatives of the Riemann tensor $R_{\sigma\rho\nu\mu}$: it satisfies the Bianchi identities of differential geometry, symmetry property 5 discussed in the previous section and derived in Problem 12.11,

$$D_\gamma R_{\rho\sigma\alpha\beta} + D_\rho R_{\sigma\gamma\alpha\beta} + D_\sigma R_{\gamma\rho\alpha\beta} = 0$$

This allows us to calculate the covariant four-divergence of the Ricci tensor,

$$\begin{aligned} 0 &= \sum_{\sigma\gamma} g^{\beta\sigma} g^{\alpha\gamma} (D_\gamma R_{\rho\sigma\alpha\beta} + D_\rho R_{\sigma\gamma\alpha\beta} + D_\sigma R_{\gamma\rho\alpha\beta}) \\ &= \sum_{\gamma} D^\gamma R_{\rho\gamma} - D_\rho R + \sum_{\gamma} D^\gamma R_{\rho\gamma} \end{aligned}$$

So,

$$\sum_{\gamma} D^\gamma R_{\rho\gamma} = \frac{1}{2} D_\rho R \tag{12.60}$$

So, the Ricci tensor is not quite right. However, we can make a conserved, local, second-rank tensor out of $R_{\sigma\rho}$ and R, the Ricci scalar, $R = \sum_{\mu} R^{\mu}_{\mu}$.

The result is the Einstein tensor,

$$G_{\sigma\rho} = R_{\sigma\rho} - \frac{1}{2} R g_{\sigma\rho} \tag{12.61}$$

Eqs. (12.60) and (12.61) imply the desired conservation law,

$$\sum_{\gamma} D^\gamma G_{\rho\gamma} = 0$$

Finally, the Einstein field equation reads,

$$G_{\alpha\beta} = \frac{8\pi G}{c^4} T_{\alpha\beta} \tag{12.62}$$

This is what we have been seeking. It is the centerpiece of the theory. We will investigate its content in the following sections of this chapter.

The Einstein field equation is often written in terms of $R_{\alpha\beta}$ and R. Take the Einstein field equation Eq. (12.62); write it in terms of the Ricci tensor and Ricci scalar and then fully contract the indices,

$$\sum_{\alpha\beta} g^{\alpha\beta} R_{\alpha\beta} - \frac{1}{2}\sum_{\alpha\beta} R g^{\alpha\beta} g_{\alpha\beta} = \frac{8\pi G}{c^4}\sum_{\alpha\beta} g^{\alpha\beta} T_{\alpha\beta}$$

Define $R = \sum_{\gamma} R^{\gamma}{}_{\gamma}$, $T = \sum_{\gamma} T^{\gamma}{}_{\gamma}$ and note that $\delta^{\alpha}{}_{\beta} = \sum_{\gamma} g^{\alpha\gamma} g_{\gamma\beta}$ and $4 = \sum_{\gamma} \delta^{\gamma}{}_{\gamma}$. We learn that,

$$R = -\frac{8\pi G}{c^4} T$$

so the Einstein field equation can be written as,

$$R_{\alpha\beta} = 8\pi G\left(T_{\alpha\beta} - \frac{1}{2} T g_{\alpha\beta}\right) \tag{12.63}$$

We should reflect on Eqs. (12.62) and (12.63). These equations are the heart of general relativity and represent one of the supreme achievements of science. The discussion here is very efficient, tidy, prim, and proper. This is not at all how science works! The creation of general relativity took years of blood sweat and tears, false starts, dead ends and, yes, mistakes. The synthesis of the principles of physics with Riemann's formulation of differential geometry was something utterly new. It represented the first step in the development of modern field theory.

Feynman stated,

> *Einstein arrived at his field equations without the help of a developed field theory, and I must admit that I have no idea how he ever guessed at the final result.....I feel as though he had done it while swimming underwater, blindfolded, and with his hands tied behind his back! [3].*

12.7 THE SCHWARZSCHILD METRIC AND BLACK HOLE

To begin to understand general relativity we need to understand its simplest problem: the space—time outside a spherical, static mass M. This is the first problem one solves in Newton's theory of gravity, and everyone knows the solution for the gravitational potential,

$$\Phi(r) = -\frac{GM}{r} \tag{12.64}$$

as we discussed in our introductory remarks. The force felt by a particle of mass m in the potential Eq. (12.64) is $\boldsymbol{F} = -m\nabla\Phi(r) = -\frac{GMm}{r^2}\hat{\boldsymbol{r}}$: it is a static, radially symmetric force, which obeys the inverse square law.

Can we find the metric in general relativity that supplants this result? The metric should be static and spherically symmetric,

$$ds^2 = U(r)c^2dt^2 - V(r)dr^2 - r^2\left(d\theta^2 + \sin^2\theta d\varphi^2\right) \tag{12.65}$$

in spherical coordinates. The Einstein field equations should determine the functions U and V, which only depend on the radial coordinate r. This metric should apply outside the mass M. If M consists of a mass distribution, its spherical symmetry and static nature in the coordinate system chosen in Eq. (12.65) is required. Finally, the metric must solve the Einstein field equations outside M,

$$R_{\alpha\beta} = 8\pi G\left(T_{\alpha\beta} - \frac{1}{2}g_{\alpha\beta}T\right) = 0 \tag{12.66}$$

and in the nonrelativistic limit we should retrieve Eq. (12.64) and Newton's equation for the motion of a mass m in this environment, $\boldsymbol{F} = -m\nabla\Phi(r) = md^2\boldsymbol{r}/dt^2 = -\frac{GMm}{r^2}\hat{\boldsymbol{r}}$.

Eq. (12.66) is deceptively simple. It hides a set of coupled, second-order, nonlinear differential equations for the components of the metric $U(r)$ and $V(r)$. Why second order? The Einstein field equations were constructed with this attribute in mind because we know from experience with Newton's equations of motion and the equations of electrodynamics that second-order differential equations have physical solutions if they are supplemented with sensible initial conditions. Higher-order differential equations are full of pathologies, including run-away solutions, noncausal effects, etc. Why nonlinear? General relativity is inherently nonlinear. No linear superposition principle here! This fact makes analytic problem solving in general relativity much harder than in electrodynamics. The differential equation will be nonlinear because of the universal character of gravity: any bit of energy—momentum attracts any other, so any patch of curvature attracts any other patch of curvature. In field theory one says that the theory has "intrinsic self-interactions." We discuss these effects further in later sections of this chapter, but we will see them in play immediately.

Let us begin [4]. No tricks, just hard labor. We will illustrate some of the algebra and manipulations involved but will not display all (!) of it. The student should commit themselves to work along here and the problem set will encourage that.

The metric reads,

$$g_{00} = U, \quad g_{11} = -V, \quad g_{22} = -r^2, \quad g_{33} = -r^2 \sin^2 \theta$$

We will need its inverse,

$$g^{00} = \frac{1}{U}, \quad g^{11} = -\frac{1}{V}, \quad g^{22} = -\frac{1}{r^2}, \quad g^{33} = -\frac{1}{r^2 \sin^2 \theta}$$

From these we can calculate the Christoffel symbols,

$$\Gamma^{\mu}_{\nu\sigma} = \frac{1}{2} \sum_{\lambda} g^{\mu\lambda} (\partial_{\sigma} g_{\lambda\nu} + \partial_{\nu} g_{\lambda\sigma} - \partial_{\lambda} g_{\nu\sigma})$$

Many components of $\Gamma^{\mu}_{\nu\sigma}$ are zero because of the static, spherical character of this problem: time derivatives ∂_0 vanish identically and derivatives ∂_{φ} do also and ∂_{θ} is nonzero only if it applies to g_{33} or g^{33}.

We calculate for $i, j = 1, 2, 3$,

$$\Gamma^{0}_{00} = \frac{1}{2} g^{00} (\partial_0 g_{00} + \partial_0 g_{00} - \partial_0 g_{00}) = 0$$

$$\Gamma^{0}_{01} = \frac{1}{2} g^{00} (\partial_1 g_{00} + \partial_0 g_{01} - \partial_0 g_{01}) = \frac{1}{2} g^{00} \partial_1 g_{00} = \frac{\partial_r U}{2U}$$

$$\Gamma^{0}_{ij} = \frac{1}{2} g^{00} (\partial_j g_{0i} + \partial_i g_{0j} - \partial_0 g_{ij}) = 0$$

The full set of nonvanishing Christoffel symbols reads, using a convenient short hand that "primes" indicate differentiation with respect to r, $U' \equiv \partial_r U$, $V' \equiv \partial_r V$,

$$\Gamma^{0}_{01} = \frac{U'}{2U}$$

$$\Gamma^{1}_{00} = \frac{U'}{2V}, \quad \Gamma^{1}_{11} = \frac{V'}{2V}, \quad \Gamma^{1}_{22} = -\frac{r}{V}, \quad \Gamma^{1}_{33} = -\frac{r \sin^2 \theta}{V}$$

$$\Gamma^{2}_{12} = \frac{1}{r}, \quad \Gamma^{2}_{33} = -\sin \theta \cos \theta$$

$$\Gamma^{3}_{13} = \frac{1}{r}, \quad \Gamma^{3}_{23} = \cot \theta$$

The "real" work begins with the Ricci tensor,

$$R_{\mu\nu} = \sum_{\beta} R^{\beta}_{\mu\nu\beta} = \sum_{\beta} \partial_{\nu} \Gamma^{\beta}_{\mu\beta} - \sum_{\beta} \partial_{\beta} \Gamma^{\beta}_{\mu\nu} + \sum_{\alpha\beta} \Gamma^{\alpha}_{\mu\beta} \Gamma^{\beta}_{\alpha\nu} - \sum_{\alpha\beta} \Gamma^{\alpha}_{\mu\nu} \Gamma^{\beta}_{\alpha\beta}$$

Writing this expression out reveals that $R_{\mu\nu} = 0$ identically if $\mu \neq \nu$. This preliminary saves us a lot of extra labor, although its verification, while straight-forward, is tedious.

The nontrivial contributions to R_{00} are,

$$R_{00} = -\partial_1 \Gamma^1_{00} + \Gamma^0_{01}\Gamma^1_{00} - \Gamma^1_{00}\Gamma^1_{11} - \Gamma^1_{00}\Gamma^2_{12} - \Gamma^1_{00}\Gamma^3_{13}$$
$$= -\frac{U''}{2V} + \frac{U'}{4V}\left(\frac{U'}{U} + \frac{V'}{V}\right) - \frac{U'}{rV}$$

Similarly,

$$R_{11} = \frac{U''}{2U} - \frac{U'}{4U}\left(\frac{U'}{U} + \frac{V'}{V}\right) - \frac{V'}{rV}$$
$$R_{22} = \frac{rU'}{2UV} + \frac{1}{V} - \frac{rV'}{2V^2} - 1$$
$$R_{33} = \sin^2\theta R_{22}$$

These are coupled, nonlinear, second-order differential equations. However, notice that if we multiply R_{00} by V/U and add the result to R_{11}, most terms cancel leaving,

$$\frac{U'}{rU} + \frac{V'}{rV} = 0$$

Or more simply,

$$U'V + V'U = 0$$

which implies that the product UV is a constant! To determine the constant, imagine that r becomes large. Then the metric should reduce to the Minkowski metric, so the constant is unity,

$$U(r)V(r) = 1$$

Finally, substitute this relation into $R_{22} = 0$ and find,

$$U + rU' = 1$$

which can be written as,

$$d(rU)/dr = 1$$

which integrates to,

$$rU = r + k \text{ or } U = 1 + k/r$$

where k is a constant. Since $UV = 1$, we learn that $V = (1 + k/r)^{-1}$.

To determine the constant k, imagine the limit of large r. We should obtain the weak gravitational field case that we already analyzed using the equivalence principle,

$$g_{00} \to 1 - \frac{2GM}{c^2 r}$$

So, we read off $k = -\frac{2GM}{c^2}$ and we have the Schwarzschild metric,

$$ds^2 = \left(1 - \frac{2GM}{c^2 r}\right)c^2 dt^2 - \left(1 - \frac{2GM}{c^2 r}\right)^{-1} dr^2 - r^2\left(d\theta^2 + \sin^2\theta d\varphi^2\right) \tag{12.67}$$

The physics in this formula will be the subject of several sections of this chapter. What makes it so exciting? It contains a black hole!

12.8 THE SCHWARZSCHILD BLACK HOLE

The exact Schwarzschild metric reads,

$$ds^2 = \left(1 - \frac{2GM}{c^2 r}\right)c^2 dt^2 - \frac{dr^2}{(1 - 2GM/c^2 r)} - r^2\left(d\theta^2 + \sin^2\theta\ d\varphi^2\right). \tag{12.68}$$

There is an ominous singularity in this expression, for $2GM/rc^2 = 1$, the proper time, $d\tau = \sqrt{1 - 2GM/rc^2}dt$, vanishes, and the proper distance, $dl = dr\Big/\sqrt{1 - 2GM/rc^2}$, diverges. The critical distance,

$$r_{Sch} = \frac{2GM}{c^2},$$

is called the Schwarzschild radius. For a typical celestial body such as the Sun, r_{Sch} is much smaller than the body's actual radius, and these singularities are not physical—the expression for ds^2, Eq. (12.68), is only true *outside* the massive body, just like Newton's gravitational potential, $V(r) = -GM/r$, is only true outside the mass M (r_{Sch} is about 2900 m for the Sun whose physical radius is about 7×10^5 km. For the Earth r_{Sch} is about 0.88 cm!). However, there are astrophysical bodies that have collapsed under their own enormous weight and whose radii are less than their Schwarzschild radii. They are called black holes, and they provide a physical realization of the peculiar, extreme conditions described by the Schwarzschild metric for strong fields.

Consider a Schwarzschild black hole, and suppose that we place a clock, an emitter of light with frequency ν_e, at radius $r > r_{Sch}$. If we reduce r closer and closer toward r_{Sch}, the clock, as measured by an observer who is far away where the gravitational field is negligible, runs slower and slower, as we can read off Eq. (12.68). The redshift becomes infinite in the limit that r coincides with r_{Sch}. In effect, the gravitational field becomes so strong at r_{Sch} that light cannot escape the black hole!

The Schwarzschild metric has singularities at $r = 0$ and $r = 2GM/c^2$. Are either of these singularities properties of the coordinate mesh or are they intrinsic physical properties of the black hole? The Schwarzschild radius is certainly physically significant, but we will learn that it is not a true singular point. One way to see this is to calculate some components of the Riemann tensor, $R_{\alpha\beta\gamma\delta}$, and some curvature invariants and see that they are free of singularities at the Schwarzschild radius. For example, one can calculate the fully contracted "square" of the Riemann tensor,

$$\sum_{\alpha\beta\gamma\delta} R^{\alpha\beta\gamma\delta} R_{\alpha\beta\gamma\delta} = \frac{48G^2M^2}{c^4 r^6}$$

Only the origin is a true singularity. We will find other results below that clarify this finding.

Consider the radial motion of light outside the mass M. It follows the null interval,

$$ds^2 = 0 = \left(1 - \frac{2GM}{c^2 r}\right) c^2 dt^2 - \left(1 - \frac{2GM}{c^2 r}\right)^{-1} dr^2$$

So,

$$\frac{dt}{dr} = \pm \left(1 - \frac{2GM}{c^2 r}\right)^{-1} \tag{12.69}$$

To visualize this equation, plot t against r in a space–time diagram as shown in Fig. 12.17. At large r the slope of the light cone dt/dr approaches ± 1 as in flat Minkowski space–time. However, as the light ray approaches the Schwarzschild radius, a distant observer concludes that the light ray takes longer and longer to make fixed incremental progress toward smaller r.

Imagine a spaceship heading toward r_{Sch}. The captain of the ship signals a distant observer by sending light rays every second according to his watch. The distant observer, however, detects these signals more and more slowly. Since dt/dr becomes unbounded at r_{Sch}, the distant observer never detects the spaceship passing through r_{Sch}. This is a frame-dependent result. What

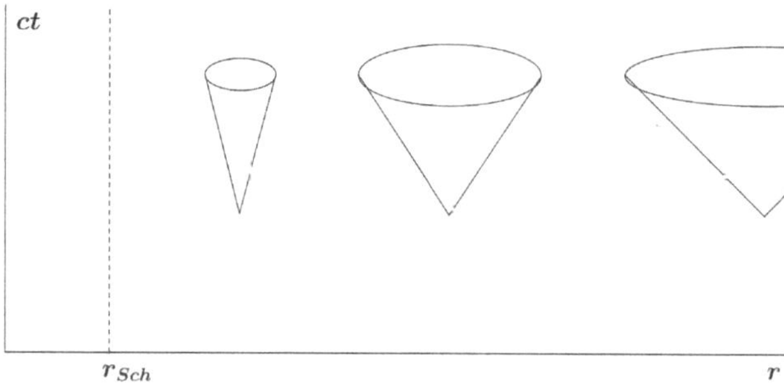

Figure 12.17 Light cones outside but approaching the Schwarzschild radius of a black hole.

happens from the perspective of the captain of the spaceship? To answer this question, let us study the path of the spaceship. If the spaceship is without power it follows a geodesic,

$$\frac{du_\sigma}{d\tau} - \sum_{\mu\rho} \Gamma^\rho_{\mu\sigma} u^\mu u_\rho = 0$$

Here we have written the geodesic equation for the covariant vector u_σ and the Christoffel symbols appear here with the "wrong" sign as discussed in Section 12.3. The second term can be simplified,

$$\sum_{\mu\rho} \Gamma^\rho_{\mu\sigma} u^\mu u_\rho = \sum_{\mu\rho\alpha} \frac{1}{2} g^{\rho\alpha} \left(\partial_\mu g_{\alpha\sigma} + \partial_\sigma g_{\mu\alpha} - \partial_\alpha g_{\mu\sigma} \right) u^\mu u_\rho$$

$$= \sum_{\mu\rho\alpha} \frac{1}{2} \left(\partial_\mu g_{\alpha\sigma} + \partial_\sigma g_{\mu\alpha} - \partial_\alpha g_{\mu\sigma} \right) u^\alpha u^\mu = \frac{1}{2} \sum_{\mu\alpha} \partial_\sigma g_{\mu\alpha} u^\mu u^\alpha$$

where we observed cancellation between the first and third terms in the parentheses when the contraction with the symmetric combination $u^\alpha u^\mu$ was taken.

This form of the geodesic is convenient for finding conserved quantities, which help to solve for the trajectory taken by the spaceship. If there is a direction for which $\partial_\sigma g_{\mu\alpha} = 0$, then u_σ is conserved. Note that if $\sigma = 0$, then $\partial_0 g_{\mu\alpha} = 0$ because the Schwarzschild metric is time independent. In addition, $\partial_\varphi g_{\mu\alpha} = 0$ because the metric has no φ dependence, it is unchanged by rotations around the z axis. Call the first constant $E = u_0$ and the second one, which we will identify as the angular momentum per mass about the z axis, $L = -u_\varphi$. What about motion in the θ direction? If we orient the coordinate system so that $\theta = \pi/2$, then $u_\theta = d\theta/d\tau = 0$ because $du_\theta/d\tau - \frac{1}{2}\partial_\theta g_{\varphi\varphi}u^\varphi u^\varphi = 0$, $g_{\varphi\varphi} = -r^2 \sin^2\theta$, and $\partial_\theta g_{\varphi\varphi} = -2r^2 \sin\theta \cos\theta$, which vanishes at $\theta = \pi/2$. So, if we orient the coordinate system so that initially $\theta = \pi/2$, then θ remains there. So, as expected, from the symmetry of the problem, orbits in the Schwarzschild metric are planar.

We will need the components of the four-velocity with upper indices,

$$u^0 = g^{00}u_0 = \left(1 - \frac{2GM}{c^2 r}\right)^{-1} E, \quad u^r = \frac{dr}{cd\tau}, \quad u^\theta = 0, \quad u^\varphi = g^{\varphi\varphi}u_\varphi = \frac{1}{r^2}L$$

Now we can get the "first integral" of the equations of motion by considering the invariant length of $u^\mu = dx^\mu/d\tau$. This length is conserved along the geodesic, $u\cdot u = \sum_\mu u^\mu u_\mu = c^2$ for the spaceship. Why is $u\cdot u = c^2$? This is a scalar quantity, which can be evaluated in any convenient frame. Choose a free-falling frame where the rules of special relativity apply. In the rest frame of the spaceship the result is immediate. Writing this expression out in terms of the components we have computed,

$$E^2\left(1 - \frac{2GM}{c^2 r}\right)^{-1} - \left(1 - \frac{2GM}{c^2 r}\right)^{-1}\left(\frac{dr}{cd\tau}\right)^2 - \frac{L^2}{r^2} = c^2$$

which we can manipulate into a more standardized form,

$$\frac{1}{2}\left(\frac{dr}{cd\tau}\right)^2 + \frac{1}{2}\left(1 - \frac{2GM}{c^2 r}\right)\left(\frac{L^2}{r^2} + c^2\right) = \frac{1}{2}E^2 \tag{12.70}$$

This expression is useful to solve for orbits in the vicinity of the mass M. Recognize that this is the general relativity version of the Kepler problem of Newtonian mechanics. We will study Eq. (12.70) further in the next section of this chapter, but here we want to specialize to purely radial motion, $L = 0$ with the spaceship starting at rest at some initial radius, $r = R$. Applying these initial conditions to Eq. (12.70), we find E in terms of R, $E^2/c^2 = \left(1 - \frac{2GM}{c^2 R}\right)$. Now Eq. (12.70) becomes,

$$\frac{1}{2}\left(\frac{dr}{cd\tau}\right)^2 + \frac{1}{2}\left(\frac{2GM}{c^2 R} - \frac{2GM}{c^2 r}\right) = 0$$

which predicts the proper time it takes the spaceship to fall into the origin,

$$\tau = \int_0^R \frac{dr}{\left(\frac{2GM}{r} - \frac{2GM}{R}\right)^{1/2}} = \frac{\pi}{2} R \sqrt{\frac{R}{2GM}} \tag{12.71}$$

where we used integral tables in the last step.

We learn two things from these considerations. First, according to the captain of the spaceship a *finite* amount of time passes on her wristwatch before she reaches the origin. The Schwarzschild radius does not play any special role in the operation of her wristwatch. This is in marked contrast with the measurements made by a distant observer who watches the spaceship begin its free fall at $R > r_{Sch}$ and must wait forever for the spaceship to reach the Schwarzschild radius. And second, the result, Eq. (12.71), and the energy equation, Eq. (12.70), bear an uncanny resemblance to the same problem in Newtonian mechanics. In Newton's world, energy conservation reads,

$$E = \frac{1}{2} m v^2 - \frac{GMm}{r} \tag{12.72}$$

where m is the mass of the spaceship. If the spaceship begins at rest at $r = R$, then $E = -\frac{GMm}{R}$, the motion is purely radial, $v = dr/dt$, and the energy conservation equation becomes,

$$\frac{dr}{dt} = \sqrt{2GM} \sqrt{\frac{1}{r} - \frac{1}{R}}$$

which can be flipped over and integrated,

$$t = \int_0^R \frac{dr}{\sqrt{2GM} \sqrt{\frac{1}{r} - \frac{1}{R}}} = \frac{\pi}{2} R \sqrt{\frac{R}{2GM}}$$

So we have the same answer as general relativity with the understanding that Newton's time t is universal, and in relativity the time that was calculated was the proper time of the captain of the spaceship.

Now we face a puzzle! In Newton's world, Eq. (12.72), the velocity v is unrestricted by the speed of light: for fixed E, v becomes as large as you like as r approaches zero. Looking back at our general relativity discussion, we see that $dr/d\tau$ follows the same systematics! Does this mean that the spaceship can travel faster than the speed of light inside a black hole? But notice something special about $dr/d\tau$: it is a "mixed" quantity: r is a coordinate marker, and τ is a proper time. It is not the velocity of an object as measured by an observer in an inertial frame. So we do not have a "manifest" contradiction here, but the matter deserves more thought. That can be found in reference [5].

Now let us return to the Schwarzschild metric and try to understand space—time within the Schwarzschild radius. The coordinates t and r are not suitable for this. First note that g_{tt} and g_{rr} switch signs at r_{Sch}! So, space (radial) becomes timelike and time becomes spacelike (radial) inside the black hole! This suggests that any body, including light rays, inside a black hole inexorably approaches $r = 0$ as a matter of the space—time geometry of the black hole.

Let us follow the historical developments in this field and establish this point. First, the problem with the $t - r$ coordinate system is that progress in the r direction becomes slower and slower with respect to the time t coordinate as r_{Sch} is approached. Let us replace t with a coordinate that moves more slowly in r by solving Eq. (12.69),

$$t = \pm \int \frac{dr}{1 - \dfrac{2GM}{c^2 r}} = \pm \int \left(1 + \frac{1}{1 - \dfrac{c^2 r}{2GM}} \right) dr$$
$$= \pm \left(r + \frac{2GM}{c^2} \ln\left(\frac{c^2 r}{2GM} - 1 \right) \right)$$

Define a new space-like variable, called a "tortoise" coordinate,

$$\widetilde{r} = r + \frac{2GM}{c^2} \ln\left(\frac{c^2 r}{2GM} - 1 \right)$$

and rewrite the Schwarzschild metric in terms of $d\widetilde{r}^2$ instead of dr,

$$ds^2 = \left(1 - \frac{2GM}{c^2 r} \right) \left(dt^2 - d\widetilde{r}^2 \right) - r^2 \left(d\theta^2 + \sin^2 \theta d\varphi^2 \right)$$

Now t and $\widetilde{r}$ are on equal footing, and the only singularity in the metric is at the origin. However, both g_{tt} and $g_{\widetilde{r}\widetilde{r}}$ do vanish at r_{Sch}, and the Schwarzschild radius itself occurs at $\widetilde{r} \to -\infty$. We can do better. Let us borrow from high-energy physics the idea that light-like propagation is best parameterized in light-like coordinates,

$$t_+ = t + \widetilde{r}, \quad t_- = t - \widetilde{r}$$

In this language the "infalling" ray is $t_+ = 0$, and the "outgoing" one is $t_- = 0$. More algebra gives the invariant interval in terms of t_+ and r,

$$ds^2 = \left(1 - \frac{2GM}{c^2 r}\right) dt_+^2 - 2dt_+ dr - r^2\left(d\theta^2 + \sin^2\theta d\varphi^2\right)$$

which is named after its discoverers, the "Eddington–Finkelstein" representation [6]. Again there are no false singularities here and light rays, $ds^2 = 0$, that travel in the r direction satisfy, for "infalling" rays,

$$\frac{dt_+}{dr} = 0$$

And for "outgoing" rays,

$$\frac{dt_+}{dr} = 2\left(1 - \frac{2GM}{c^2 r}\right)^{-1}$$

What do we learn? The light cones vary smoothly at r_{Sch}, and one can plot the paths of light rays and massive particles for all r. Note in addition, that if $r < r_{Sch}$, then $dt_+/dr < 0$, so *all* propagation into the future occurs in the direction of *decreasing* r: so the light cone has "tilted" over. The reader may want to consult references [5,7] for additional graphics. We learn that the surface given by the Schwarzschild radius is a one-way membrane—once a light ray or massive particle passes through it from the outside, it can never return. Such a surface is called an "event horizon." A distant observer cannot perceive anything within r_{Sch}. Note that an event horizon is a global structure, not local: the spaceship passed from $r > r_{Sch}$ to $r < r_{Sch}$ without trauma, but he or she can never retrace his or her steps. Since this behavior is a matter of space–time geometry, it has no compelling analogue in Newton's world. Once the spaceship passes through the event horizon, then $dt_+/dr < 0$, and the spaceship relentlessly falls to the origin even if it is under its own rocket power. In fact, for a "typical" black hole, the spaceship and the astronauts inside it will be ripped apart by gravitational tidal "forces" before it reaches the surface of the black hole. Those forces will stretch and squeeze it mercilessly. We know how to estimate these forces from our discussion of tidal forces in Section 11.3. Recall that if an astronaut is falling into a black hole feet first, that unfortunate person feels a tidal acceleration across his length of,

$$a = \frac{2GMd}{R^3}$$

where R is his distance to the center of the mass M, d is his height, and G is Newton's constant, $6.67 \times 10^{-8} \text{cm}^3/\text{gm s}^2$. Let us suppose that the black hole has a mass of the Sun (1.9×10^{33} g), the astronaut is 2 m tall and R is 100,000 m. Then the acceleration difference across his length is,

$$2 \times (6.67 \times 10^{-8}) \times (1.9 \times 10^{33}) \times (200)/(1.0 \times 10^{7})^3 \approx 51 \times 10^6 \text{ cm}/s^2.$$

Compare this number with the acceleration of gravity at Earth's surface, 980 cm/s^2. Clearly the tidal forces would stretch and squeeze the astronaut out of existence long before reaching the black hole's surface.

There are more than 100 confirmed black holes among the celestial bodies presently under observation. Astrophysicists predict that they are created by the gravitational collapse of very massive stars that run out of fuel late in their life cycles. Binary black holes have also been observed. In 2015, gravitational waves were first observed, and their source proved to be a binary pair of black holes in the process of colliding and merging, as we will discuss in Section 12.13.

An interested student can learn more about black holes in a book [5] written at the same level as this chapter. The book also has a full bibliography and looks at recent astrophysical observations and data as well.

The modern study of black holes, their relation to quantum physics, and unification, a field pioneered by Stephen Hawking, is very stimulating because it places all of our physical laws in an extreme environment and challenges their self-consistency and our understanding of them. When you learn the elements of quantum mechanics and reconsider the properties of black holes, you will learn about Hawking radiation, the fact that black holes are, in fact, hot and radiate electromagnetic waves. They are not really black at all! The unification of gravity and quantum mechanics is an active field of research to which the researchers of the high-energy physics community contribute through their investigations in superstrings, a framework for a theory of everything—electricity and magnetism, radioactivity (weak interactions), nuclear physics (strong interactions), and gravity.

12.9 CIRCULAR ORBITAL MOTION IN THE SCHWARZSCHILD METRIC

There are some striking differences between the Newtonian Kepler problem and orbital motion in the Schwarzschild metric that we should appreciate [5]. We derived in the previous section the crucial orbital equation,

$$\frac{1}{2}\left(\frac{dr}{cd\tau}\right)^2 + \frac{1}{2}\left(1 - \frac{2GM}{c^2 r}\right)\left(\frac{L^2}{r^2} + c^2\right) = \frac{1}{2}E^2 \qquad (12.73)$$

We can write this expression in the form of a one-dimensional equation of motion in the radial coordinate,

$$\frac{1}{2}\left(\frac{dr}{d\tau}\right)^2 + V_{eff}(r) = \frac{1}{2}E^2 \tag{12.74a}$$

where,

$$V_{eff}(r) = \frac{c^2}{2} - \frac{GM}{r} + \frac{L^2}{2r^2} - \frac{GML^2}{c^2 r^3} \tag{12.74b}$$

Compare this equation to the corresponding equation in Newtonian mechanics,

$$\frac{1}{2}\left(\frac{dr}{dt}\right)^2 + V_{eff}^N(r) = E \tag{12.75a}$$

where,

$$V_{eff}^N(r) = -\frac{GM}{r} + \frac{L^2}{2r^2} \tag{12.75b}$$

Special effects in general relativity come from the last term in Eq. (12.74b) which is a relativistic effect, varying as $1/c^2$, that makes the origin more attractive than the centrifugal barrier, $L^2/2r^2$, makes it repulsive! In Fig. 12.18 we compare $V_{eff}^N(r)$ for Newtonian mechanics with $V_{eff}(r)$ of general relativity. The striking difference is that for $L^2 \neq 0$, $V_{eff}(r)$ diverges as $-r^{-3}$ near the origin! This term makes it impossible to have stable circular orbits for small r. We can search for stable circular orbits in $V_{eff}(r)$ by looking for solutions to the two conditions,

$$\frac{dV_{eff}}{dr} = 0, \quad \frac{d^2 V_{eff}}{dr^2} > 0$$

The first requirement gives,

$$\frac{dV_{eff}}{dr} = 0 = \frac{GMr^2/c^2 - L^2 r + 3GML^2/c^2}{r^4}$$

which states how L^2 depends on the radius of the orbit,

$$L^2 = GMr\left(1 - \frac{3GM}{c^2 r}\right)^{-1} \tag{12.76}$$

So, we can have circular orbits only for $r > 3GM/c^2$. In addition, since $dr/d\tau = 0$ at a circular orbit, $E^2 = 2V_{eff}(r)$ there and,

$$E^2 = \left(1 - \frac{2GM}{c^2 r}\right)^2 \left(1 - \frac{3GM}{c^2 r}\right)^{-1} \tag{12.77}$$

there as well. Finally, we can solve the expression Eq. (12.76) for r,

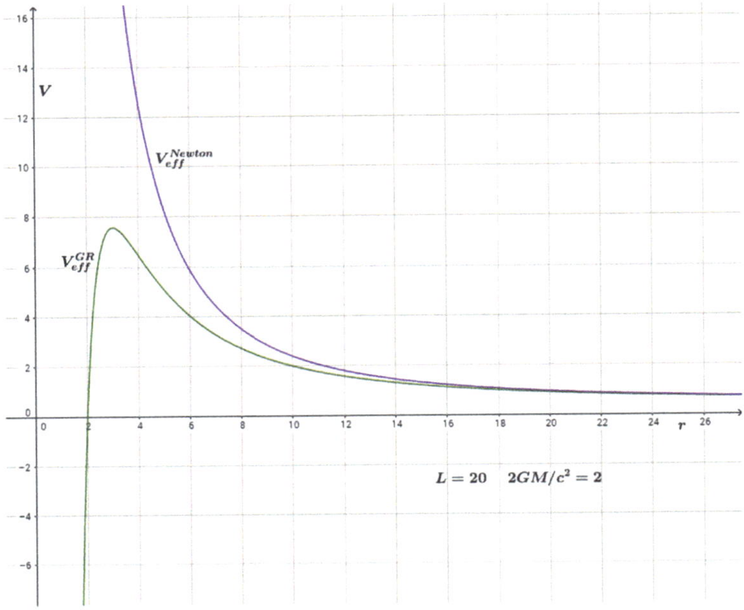

Figure 12.18 Comparison of the general relativistic effective potential and the Newtonian effective potential for the radial coordinate around a black hole. The parameter choice is given in the figure.

$$r_{\pm} = \frac{L^2 \pm \sqrt{L^4 - 12G^2L^2M^2/c^4}}{2GM/c^2} \tag{12.78}$$

We learn that if $L^2 < 12G^2M^2/c^4$ then there are no circular orbits. In addition, one can check that $\frac{d^2V_{eff}}{dr^2} > 0$ for r_+ and $\frac{d^2V_{eff}}{dr^2} < 0$ for r_-. But since L^2 must be greater than $12G^2M^2/c^4$ for stable orbits to exist, it follows from Eq. (12.78) that $r_+ > 6GM/c^2$.

We must understand the term $-\frac{GML^2}{c^2r^3}$ in V_{eff}. It changes the Newtonian effective potential qualitatively near the origin, and the extra attraction overwhelms the centrifugal barrier term, which is so important in many physics problems ranging from planetary motion to quantum mechanics. The modifications to the Newton–Kepler predictions of planetary motion are dramatic even for our solar system where there are no exotic objects such as black holes, and the new term is very, very small compared with the other terms in V_{eff}. One can check that for planetary problems the radial coordinate is always much, much larger than r_{Sch} so we are always working in the "weak" gravity regime where we expect Newtonian predictions to be very accurate. In fact the new term is a tiny perturbation, but it makes distinct and measureable modifications to the predictions of Newtonian mechanics. The reason is interesting. Recall that Newtonian mechanics predicts that the orbits of masses in a $1/r$ potential are perfect ellipses, *closed* periodic orbits. This is a special feature, a "dynamical symmetry," of the $1/r$ potential of Newton's law of gravity. The only other potential that produces *closed*, periodic orbits is the harmonic oscillator, a potential that grows as r^2 and is not relevant here. The point is that if there are any perturbations to the $1/r$ potential of Newton's law of gravity, then the planetary equations lose their special symmetry and would *not* produce closed, periodic ellipses. Instead, the orbits would precess slightly year to year. Since the very slow precession of an orbit accumulates over time, the effect proves to be measurable using data obtained from observations employing conventional telescopes! The new term in V_{eff} did, in fact, address a long-standing discrepancy in planetary science: the other planets that orbit the Sun perturb the orbit of Mercury and also cause it to precess. However, when these Newtonian effects were calculated, only 93% of the observed precession of Mercury was accounted for. The last 7%, 43 s of arc per century, remained a mystery for over 200 years. In fact, the general relativity prediction for the precession rate of Mercury is almost exactly 43 s of arc per century! This was a great success for general relativity in its early days!

But why does general relativity predict the new term, $-\frac{GML^2}{c^2 r^3}$? We can motivate it by writing it in a more suggestive form,

$$\left(-\frac{GM}{r}\right)\left(\frac{L^2/r^2}{c^2}\right) \tag{12.79}$$

Here L^2/r^2 is the energy due to the angular motion of the object and dividing by c^2 gives its mass equivalent. (A word about units here: L was defined as angular momentum per mass. So L^2/r^2 behaves as transverse velocity squared. Multiplying Eq. (12.74) by the particle's mass restores energy units.) The first factor in Eq. (12.79) is the gravitational potential generated by the mass M, so the equation records the fact that all energies attract one another in a universal fashion. Newtonian mechanics misses this term because in Newton's world only masses attract one another, not other forms of energy. In Einstein's world, energy and mass and momentum are all unified into a single four-vector, so the principle of covariance insists that they all contribute equally to gravitational attraction. Since the new term is a long-range force, it falls off as the third power of the distance, it is important in planetary physics where gravitational effects are all weak. But it is also highly singular as $r \to 0$, so in the vicinity of black holes it can dominate over the more familiar centrifugal repulsion potential and produce new effects.

The new term illustrates that in general relativity all forms of energy attract one another. For example, the energy carried by the gravitational field itself attracts all other forms of energy. This makes the theory highly nonlinear, as we have already seen in several applications, most explicitly in the Schwarzschild metric. This fact also shows up in attempts to make a quantum theory of gravity where gravitons, the analogue to photons of electrodynamics, must interact with themselves. These interactions appear to generate new interactions, which are very singular at short distances and lead to pathologies. Although classical general relativity is a theory that is full of triumphs, the search for a quantum version continues unrewarded.

Now let us return to the problem at hand and consider the radial "infalling" of the particle with angular momentum. Consider Eq. (12.73). We want to calculate the proper time for the free radial motion by solving that equation for $d\tau$, generalizing the discussion in Eq. (12.71),

$$\int c d\tau = \int_{r_0}^{r_1} dr \left(E^2 - \left(1 - \frac{2GM}{c^2 r}\right)\left(c^2 + \frac{L^2}{r^2}\right)\right)^{-1/2}$$

The new term, $\frac{GML^2}{c^2 r^3}$, enters the denominator with a positive sign indicating that it tends to *decrease* the proper time needed to fall between r_1 and r_0. The centrifugal barrier, $\frac{L^2}{r^2}$, enters with a minus sign and works in the other direction, as expected. However, if $r < r_{Sch}$, then the new term dominates and *increasing* the angular momentum *decreases* the proper time, the lifetime of the "infalling" object!

12.10 THE SPEED OF LIGHT IN A GRAVITATIONAL FIELD

Although light propagates at the speed limit c in an inertial frame, it need not do so in the space near a mass M. There is a set of experiments that have investigated this point, called radar echoes. The experimenters shot light from Earth to a nearby planet near the Sun and measured the distance to the planet as well as the time of transit of the light pulse. They then compared the total transit time of the light ray to the distance divided by the speed limit c to investigate whether gravity speeds up or slows down the light ray.

Let us work out the prediction of general relativity for these experiments [4]. Suppose the Earth resides at the position (r_e, θ, φ) and a beam of light is transmitted to a planet at (r_p, θ, φ) and back, as shown in Fig. 12.19.

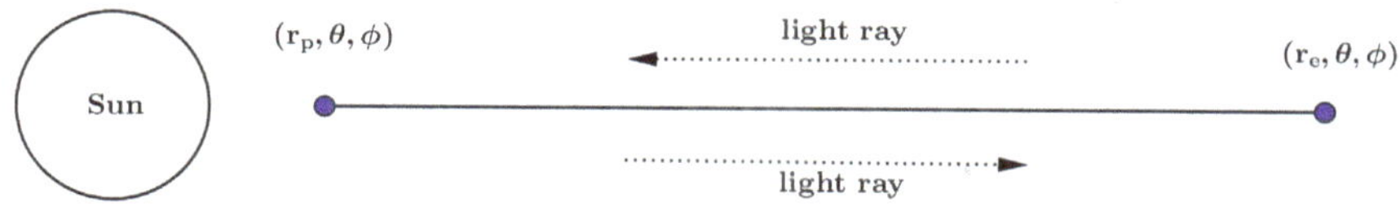

Figure 12.19 Earth resides at the position (r_e, θ, φ), and a beam of light is transmitted to a planet at (r_p, θ, φ) and back. The Sun is to the left of the planet.

The nearby huge mass of the Sun creates a gravitational field that affects the trajectory of light. The gravitational fields produced by the planets are small and negligible by comparison.

First we need to calculate the physical, proper distance between Earth and the reflecting planet. This distance is not $r_e - r_p$, the radial coordinate difference because of the gravitational field itself. In fact, from the Schwarzschild metric, the proper distance dl is related to the coordinate difference by, $dl = dr\big/\sqrt{1 - 2GM/rc^2}$. So, if the coordinate difference is $r_e - r_p$, then the physical, proper distance is

$$\int_{r_p}^{r_e} \frac{dr}{\sqrt{1 - 2GM/rc^2}} = \left[\sqrt{r(r - 2GM/c^2)} \right.$$

$$\left. + \frac{2GM}{c^2} \ln\left(\sqrt{r} + \sqrt{r - 2GM/c^2}\right)\right]_{r_p}^{r_e}.$$

Expanding the right-hand side of this result in powers of the small quantity $2GM/rc^2$, we find

$$\int_{r_p}^{r_e} \frac{dr}{\sqrt{1 - 2GM/rc^2}} \approx r_e - r_p + \frac{GM}{c^2} \ln\left(\frac{r_e}{r_p}\right). \tag{12.80}$$

We see that the physical distance is slightly larger than the coordinate difference $r_e - r_p$.

Finally, when the light propagates along a spoke of the spherical coordinate system, variable r but fixed θ and φ, we set $ds^2 = 0$ in the Schwarzschild metric and find the relation between dr and dt,

$$\left(1 - \frac{2GM}{rc^2}\right) c^2 dt^2 = \left(1 - \frac{2GM}{rc^2}\right)^{-1} dr^2,$$

which gives

$$\frac{dr}{dt} = \pm c\left(1 - \frac{2GM}{rc^2}\right).$$

So, the coordinate time needed for the whole trip, to and fro, is

$$\Delta t = -\frac{1}{c}\int_{r_e}^{r_p}\left(1 - \frac{2GM}{rc^2}\right)^{-1} dr + \frac{1}{c}\int_{r_p}^{re}\left(1 - \frac{2GM}{rc^2}\right)^{-1} dr$$

$$\Delta t = \frac{2}{c}\int_{rp}^{r_e}\left(1 - \frac{2GM}{rc^2}\right)^{-1} dr.$$

The proper time $\Delta\tau$ that passes at r_e where the measurements are made is related to Δt, referring to the Schwarzschild metric,

$$\begin{aligned}
\Delta\tau &= \sqrt{1 - \frac{2GM}{r_e c^2}}\Delta t \\
\Delta\tau &= \frac{2}{c}\sqrt{1 - \frac{2GM}{r_e c^2}}\int_{r_p}^{r_e}\left(1 - \frac{2GM}{rc^2}\right)^{-1} dr \\
\Delta\tau &= \frac{2}{c}\sqrt{1 - \frac{2GM}{r_e c^2}}\left[r_e - r_p + \frac{2GM}{c^2}\ln\left(\frac{r_e - 2GM/c^2}{r_p - 2GM/c^2}\right)\right] \\
\Delta\tau &\approx \frac{2}{c}\left[r_e - r_p - \frac{GM}{c^2}\frac{(r_e - r_p)}{r_e} + \frac{2GM}{c^2}\ln\left(\frac{r_e}{r_p}\right)\right].
\end{aligned} \tag{12.81}$$

The difference between Eq. (12.80) multiplied by $2/c$, and Eq. (12.81) tells us how much the gravitational field influences the time of flight of the light in physical units:

$$\Delta \approx \frac{2GM}{c^3}\left[\ln\left(\frac{r_e}{r_p}\right) - \frac{(r_e - r_p)}{r_e}\right].$$

So, if $r_e \gg r_p$, then Δ is a positive quantity, and the gravitational field has slowed down the light ray compared with naive expectations. The effect, however, is tiny and is very difficult to measure.

Clever experimentalists, however, applied similar calculational methods to the case where the planet lies on the other side of the Sun, as shown in Fig. 12.20. This configuration produces a much larger effect because the light ray must graze by the Sun, where the gravitational field is relatively large. The prediction of general relativity was confirmed to several percent in this case.

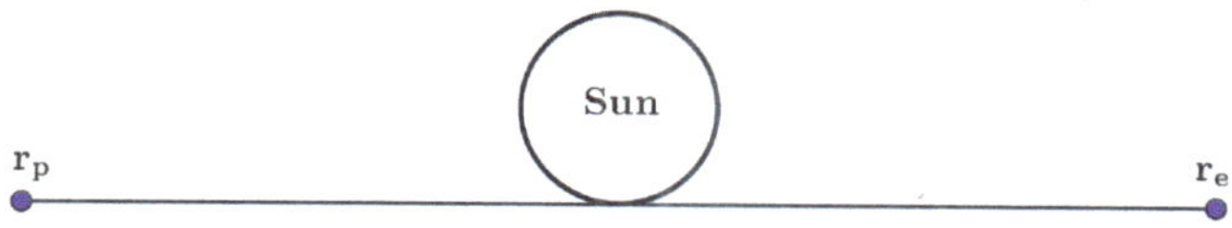

Figure 12.20 A variation of the experiment shown in Fig. 12.19, but now the planet lies on the other side of the Sun, and the light ray grazes the Sun where the gravitational field is relatively large.

12.11 RELATIVISTIC TIDAL FORCES

It is interesting to generalize the discussion of tidal forces from Newtonian mechanics to general relativity. We will see that the Riemann tensor replaces the second derivatives of the gravitational potential in the equation of motion for the difference of the positions of two nearby free-falling objects. This shows us that the curvature of space—time is the general relativistic version of tidal forces. Tidal forces are more fundamental than one might have thought! Let us understand them in general relativity.

Consider the position of a freely falling object $x^\alpha(\tau)$. Its trajectory is given by the geodesic,

$$\frac{d^2x^\alpha}{d\tau^2} + \sum_{\mu\nu} \Gamma^\alpha_{\mu\nu}(x) u^\mu u^\nu = 0 \tag{12.82a}$$

where $u^\mu = dx^\mu/d\tau$. A nearby freely falling object follows the trajectory $\widetilde{x}^\alpha(\tau)$,

$$\frac{d^2\widetilde{x}^\alpha}{d\tau^2} + \sum_{\mu\nu} \Gamma^\alpha_{\mu\nu}(\widetilde{x}) \widetilde{u}^\mu \widetilde{u}^\nu = 0 \tag{12.82b}$$

where we choose the same proper time variable to parameterize this nearby trajectory. Following the discussion of Newtonian tidal forces, the two trajectories are infinitesimally close together at $\tau = 0$ and are parallel as well. We are interested in the time evolution of the difference $\varepsilon^\alpha(\tau) = \widetilde{x}^\alpha(\tau) - x^\alpha(\tau)$. First, consider Eq. (12.82b) and Taylor expand the Christoffel symbol about the position of the first particle $x^\alpha(\tau)$,

$$\Gamma^\alpha_{\mu\nu}(x^\alpha + \varepsilon^\alpha) = \Gamma^\alpha_{\mu\nu}(x^\alpha) + \sum_\alpha \varepsilon^\alpha \partial_\alpha \Gamma^\alpha_{\mu\nu} + O(\varepsilon^2)$$

Substituting into Eq. (12.82b) and keeping terms of first order in ε^α,

$$\frac{d^2(x^\alpha + \varepsilon^\alpha)}{d\tau^2} + \sum_{\mu\nu}\left\{\Gamma^\alpha_{\mu\nu}(x^\alpha) + \sum_\alpha \varepsilon^\alpha \partial_\alpha \Gamma^\alpha_{\mu\nu} + O(\varepsilon^2)\right\}\frac{d(x^\alpha + \varepsilon^\alpha)}{d\tau}\frac{d(x^\alpha + \varepsilon^\alpha)}{d\tau} = 0$$

$$\frac{d^2x^\alpha}{d\tau^2} + \frac{d^2\varepsilon^\alpha}{d\tau^2} + \sum_{\mu\nu}\Gamma^\alpha_{\mu\nu}u^\mu u^\nu + \sum_{\mu\nu}\varepsilon^\alpha\partial_\alpha\Gamma^\alpha_{\mu\nu}u^\mu u^\nu + 2\sum_{\mu\nu}\Gamma^\alpha_{\mu\nu}u^\mu\frac{d\varepsilon^\nu}{d\tau} \approx 0$$

Subtracting Eq. (12.82a) we have,

$$\frac{d^2\varepsilon^\alpha}{d\tau^2} + \sum_{\mu\nu\sigma}\varepsilon^\sigma\partial_\sigma\Gamma^\alpha_{\mu\nu}u^\mu u^\nu + 2\sum_{\mu\nu}\Gamma^\alpha_{\mu\nu}u^\mu\frac{d\varepsilon^\nu}{d\tau} \approx 0 \qquad (12.83)$$

Next we need the evolution of the four-vector $\varepsilon^\alpha(\tau)$ itself. From Section 12.2,

$$\frac{D\varepsilon^\alpha}{D\tau} = \frac{d\varepsilon^\alpha}{d\tau} + \sum_{\mu\nu}\Gamma^\alpha_{\mu\nu}u^\mu\varepsilon^\nu \qquad (12.35)$$

And,

$$\frac{D^2\varepsilon^\alpha}{D\tau^2} = \frac{d}{d\tau}\left(\frac{d\varepsilon^\alpha}{d\tau} + \sum_{\beta\gamma}\Gamma^\alpha_{\beta\gamma}u^\beta\varepsilon^\gamma\right) + \sum_{\delta\eta}\Gamma^\alpha_{\delta\eta}u^\delta\left(\frac{d\varepsilon^\eta}{d\tau} + \sum_{\beta\gamma}\Gamma^\eta_{\beta\gamma}u^\beta\varepsilon^\gamma\right)$$

$$\frac{D^2\varepsilon^\alpha}{D\tau^2} = \frac{d^2\varepsilon^\alpha}{d\tau^2} + 2\sum_{\beta\gamma}\Gamma^\alpha_{\beta\gamma}u^\beta\frac{d\varepsilon^\gamma}{d\tau} + \sum_{\beta\gamma\delta}\partial_\delta\Gamma^\alpha_{\beta\gamma}u^\delta u^\beta\varepsilon^\gamma + \sum_{\beta\gamma}\Gamma^\alpha_{\beta\gamma}\frac{du^\beta}{d\tau}\varepsilon^\gamma$$
$$+ \sum_{\beta\gamma\delta\eta}\Gamma^\alpha_{\delta\eta}\Gamma^\eta_{\beta\gamma}u^\beta u^\delta\varepsilon^\gamma$$

where we used,

$$\frac{d\Gamma^\alpha_{\beta\gamma}}{d\tau} = \sum_\delta\frac{\partial\Gamma^\alpha_{\beta\gamma}}{\partial x^\delta}\frac{dx^\delta}{d\tau} = \sum_\delta\frac{\partial\Gamma^\alpha_{\beta\gamma}}{\partial x^\delta}u^\delta$$

Finally we eliminate $\frac{d^2\varepsilon^\alpha}{d\tau^2}$ from the expression for $\frac{D^2\varepsilon^\alpha}{D\tau^2}$ by using Eq. (12.83), and we use Eq. (12.82a) to eliminate $\frac{du^\beta}{d\tau} = \frac{d^2x^\beta}{d\tau^2}$. Now we have, noting that the terms involving $d\varepsilon^\alpha/d\tau$ cancel,

$$\frac{D^2\varepsilon^\alpha}{D\tau^2} = -\sum_{\beta\gamma\delta}\partial_\delta\Gamma^\alpha_{\beta\gamma}u^\beta u^\gamma\varepsilon^\delta + \sum_{\beta\gamma\delta}\partial_\delta\Gamma^\alpha_{\beta\gamma}u^\beta u^\delta\varepsilon^\gamma - \sum_{\beta\gamma\delta\eta}\Gamma^\alpha_{\beta\gamma}\Gamma^\beta_{\delta\eta}u^\delta u^\eta\varepsilon^\gamma$$
$$+ \sum_{\beta\gamma\delta\eta}\Gamma^\alpha_{\delta\eta}\Gamma^\eta_{\beta\gamma}u^\beta u^\delta\varepsilon^\gamma$$

Relabeling indices,

$$\frac{D^2\varepsilon^\alpha}{D\tau^2} = -\left(\sum_{\beta\gamma\delta}\partial_\gamma\Gamma^\alpha_{\beta\delta} - \sum_{\beta\gamma\delta}\partial_\delta\Gamma^\alpha_{\beta\gamma} + \sum_{\beta\gamma\delta\eta}\Gamma^\alpha_{\gamma\eta}\Gamma^\eta_{\beta\delta} - \sum_{\beta\gamma\xi\eta}\Gamma^\alpha_{\delta\eta}\Gamma^\eta_{\beta\gamma}\right)u^\beta u^\delta \varepsilon^\gamma$$

Recognize the Riemann tensor here,

$$\frac{D^2\varepsilon^\alpha}{D\tau^2} = -\sum_{\beta\gamma\delta} R^\alpha_{\beta\gamma\delta}u^\beta u^\delta \varepsilon^\gamma + O(\varepsilon^2) \tag{12.84}$$

So, we learn that to first order in the small four-vector difference $\varepsilon^\alpha(\tau)$, its evolution is determined by the curvature of space–time. This result applies to strong gravity, which tests the limits of general relativity. Excellent result, Eq. (12.84)! Compare it with Eq. (12.3) and the associated discussion for the analogous result in classical differential geometry, which shows that intrinsic curvature determines the equation of motion of the geodesic deviation.

In Appendix G, we use Eq. (12.84) to analyze the detection of gravitational waves.

Let us compare this result, Eq. (12.84), to Newtonian tidal forces. We should have agreement between general relativity and Newton's theory if we limit our attention to: (1) weak gravitational fields and (2) nonrelativistic motion so $u^\mu = (cdt/d\tau,\ dx/d\tau,\ dy/d\tau,\ dz/d\tau) \approx c(1,\ 0,\ 0,\ 0)$. Then,

$$\frac{D^2\varepsilon^i}{D\tau^2} \approx \frac{d^2\varepsilon^i}{dt^2} \approx -c^2\sum_k R^i_{0k0}\varepsilon^k \tag{12.85}$$

where we used the fact that R^i_{000} vanishes identically, and the sum over k is from 1 to 3. But the Newtonian prediction was,

$$\frac{d^2\boldsymbol{\varepsilon}}{dt^2} = -\nabla(\boldsymbol{\varepsilon}\cdot\nabla\Phi(r))$$

Comparing with Eq. (12.85) we have,

$$\nabla^2\Phi = c^2\sum_k R^k_{0k0} = c^2 R_{00} = c^2\frac{8\pi G}{c^4}\left(T_{00} - \frac{1}{2}g_{00}T\right)$$

where we identified the Ricci tensor, $R_{\alpha\beta} = \sum_\gamma R^\gamma_{\alpha\gamma\beta}$ and used the Einstein field equation, $R_{\alpha\beta} = \frac{8\pi G}{c^4}\left(T_{\alpha\beta} - \frac{1}{2}g_{\alpha\beta}T\right)$. For a static mass density ρ, we saw in Section 11.8 that $T_{00} = \rho c^2$. Then $T = \sum_{\gamma} T^{\gamma}{}_{\gamma} = T^0{}_0 = -\rho c^2$, so $T_{00} - \frac{1}{2}g_{00}T = \rho c^2 - \frac{1}{2}(-1)(-\rho c^2) = \frac{1}{2}\rho c^2$. Collecting everything,

$$\nabla^2\Phi = \frac{1}{2}8\pi G\rho = 4\pi G\rho$$

which is the differential form of Newton's law of gravity. All is well.

12.12 THE DISCOVERY OF GRAVITATIONAL WAVES

Gravitational radiation was predicted by Einstein in 1916. It is a crucial consequence of general relativity because its existence does for gravity what electromagnetic radiation did for electromagnetism: it establishes the dynamical field theoretic character of the gravitational force. As you are discovering in the problem sets, gravity is a much weaker force than the electromagnetic force. This caused the experimental discovery of gravitational waves to take a century of hard work!

However, on February 11, 2016, the LIGO collaboration announced the first observation of gravitational waves in a very exciting Webcast. The critical observation occurred 5 months earlier on September 14, 2015, and several months of data analysis were required to understand the result in sufficient detail to justify publication. The data matched the predictions of general relativity if the source of the waves was the inward spiral and merger of a pair of black holes and subsequent "ringdown" of the resulting single black hole. The observation demonstrated the existence of binary stellar-mass black hole systems and constituted the first observation of a binary black hole merger. The data could be fit extraordinarily well by the hypothesis that two black holes of masses 29 and 36 solar masses merged into one black hole at a distance of about 1.3 billion light years from Earth. During the final fraction of a second of the merger, it released more than 50 times the power of all the stars in the observable universe combined. The signal increased in frequency from 35 to 250 Hz (cycles per second) over 10 cycles (5 orbits of the pair of black holes) as it rose in strength for a period of 0.2 s. The mass of the new black hole was 62 solar masses, so the energy equivalent of three solar masses was emitted as gravitational waves in this catastrophic event. The signal was detected at both sites of the LIGO experiment, one in Hanford, WA and another in Livingston, LA, with a time difference of 7 ms between them. It was inferred that the source, the merging binary black hole system, was 1.3 billion light years away in the general direction of the Magellanic Clouds in the Southern Celestial Hemisphere. What a historic moment!

This achievement was the result of decades of work by the LIGO collaboration, which dated back to 1984, 31 years before fruition. LIGO stands for " Laser Interferometer Gravitational Wave Observatory," and the generation of the detectors that were successful in 2015 constitute the "advanced LIGO" version of the upgraded experiment. In fact, the upgraded detectors (laser interferometers) had been operating for just 2 days before the discovery occurred. Each LIGO detector, which are 1865 miles apart, are L-shaped detectors that have two arms at right angles to each other and are 2.5 miles in length from a central laboratory building. Lasers are beamed down each arm and are bounced back by free-standing suspended mirrors many times to act as extraordinarily precise measuring devices. In fact the lasers allow the length of the arms to be measured to a precision of 10^{-4} the width of a single proton. This accuracy is required to measure the small scale of the effects imparted by a passing gravitational wave. The two LIGO detectors act as a check on one another. It is an amazing engineering feat to detect such tiny signals in an otherwise noisy world!

The basis of each LIGO detector is two perpendicular interferometers, as shown in Fig. 12.21. These devices merge two light rays, which creates an interference pattern. If the peaks of the two light rays overlap, they combine to form a larger peak but when a valley of one light ray overlaps

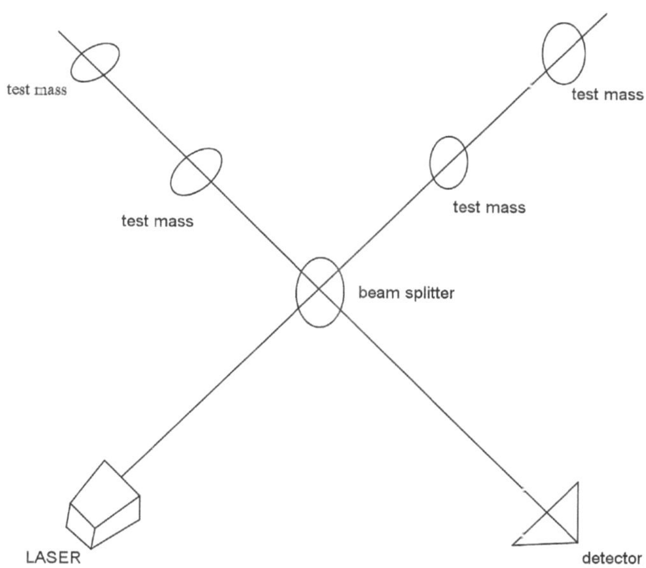

Figure 12.21 A Laser Interferometer Gravitational Wave Observatory (LIGO) detector sketch showing the two perpendicular interferometers.

with the peak of the other ray, the two rays cancel out. The laser beams can be arranged to cancel each other out in ordinary operations. Then when a gravitational wave passes through the facility, it would stretch one arm and compress the perpendicular one (we will derive this crucial property of gravity waves later), the exact cancellation would be destroyed and some light would reach a sensitive photodetector. The responses of the photodetector then measure the character of the gravity wave, its amplitude, and frequency. As mentioned above, the upgraded detector of advanced LIGO is sensitive to amplitudes as small as 10^{-4} the width of a single proton (this requires lasers with sufficient power and accuracy to make coherent beams of light, which can travel more than 100 times up and down an arm of the facility before combining with the laser beam traveling down the other arm) and to be sensitive to gravity waves with frequencies ranging from 75 to 500 Hz. We shall see that this frequency range matches the frequencies of the expected sources. It is amusing to note that these are common acoustic frequencies, the middle A tuning frequency provided by the oboe in symphony orchestras is 440 Hz, so the LIGO detectors act as super-sensitive human ears!

12.13 GRAVITATIONAL RADIATION

Now let us discuss gravitational waves [2] in more detail with an eye toward LIGO applications. We will need to model the sources of the gravity waves and the radiation itself. The sources will consist of two masses under the influence of their gravitational attraction. We will treat this problem with Newtonian mechanics because it is adequate up until the last fraction of a second before the merger of the two black holes. In addition, we will treat the observed gravity wave as a tiny distortion in the Minkowski metric, so Eq. (12.53) will serve us well. These approximations will give us a semiquantitative grasp of the real phenomenon. Of course the experts do better! They treat the merging black holes with full, nonlinear general relativity. Supercomputing is essential here. The subject of numerical general relativity where the field equations are simulated in regions of strong gravitational effects is a mature subject. It is interesting that the simple approximations and idealizations that we will sketch here match the realistic calculations rather well in most cases. Far from the source the experts also use linearized gravity, essentially Eq. (12.56), because the deviation from the Minkowski metric in the vicinity of the LIGO detectors is extraordinarily small.

Before we consider the LIGO experiment in more detail, let us consider the different frames of reference used in our subsequent discussions and analyses. First, there is the frame fixed to the Earth. Since the mechanical structures of LIGO's arms are bound together by electromagnetic forces which are many, many orders of magnitude greater than gravitational forces, the gravity waves detected by the apparatus do not have any effect here. In particular the speed of light relative to the Earth frame, the massive arms of the structure, is the speed limit c because the gravitational fields in the vicinity of the LIGO apparatus are weak. In addition, the gravitational potential due to the Earth's mass is essentially constant over the apparatus and does not play a role in LIGO's analysis. Gravity waves do effect proper lengths and times, such as the proper length between the freely suspended mirrors at the end of each arm of the detector. These are the lengths and times the analysis focuses on.

To begin, suppose that the LIGO detector lies in the x-y plane, and the source is far away in the z direction as shown in Fig. 12.22. Then we

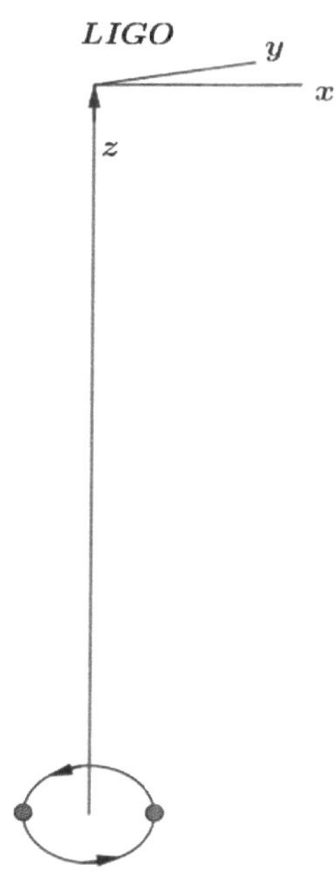

Figure 12.22 An ideal geometric configuration for the detection of gravitation waves emitted from a source on the *z* axis. The Laser Interferometer Gravitational Wave Observatory (LIGO) detector lies in the *x*-*y* plane and the source, two orbiting massive objects, is far away in the *z* direction as shown.

parameterize the traveling gravitational wave $\varepsilon(x^\mu)$ and write the metric in either of two forms,

$$ds^2 = dt^2 - (1-\varepsilon)dx^2 - (1+\varepsilon)dy^2 - dz^2 \quad (\varepsilon \ll 1, +\text{polarization}) \tag{12.86a}$$

or,

$$ds^2 = dt^2 - dx^2 - 2\varepsilon dxdy - dy^2 - dz^2 \quad (\varepsilon \ll 1, \times\text{polarization}) \tag{12.86b}$$

Let us concentrate of the + polarization case expressed in the metric Eq. (12.86a). We are anticipating here that the source of the gravity wave will be an oscillating quadrupole moment of a mass distribution, so if there is a "stretch" in one direction transverse to the propagation direction of the wave, the z axis, then there is a "compression" in the perpendicular direction. The length of an arm of the LIGO detector is about 4×10^3 m, and the wavelength of the gravity wave is $\lambda = c/\nu \approx 3 \times 10^8/10^2 = 3 \times 10^6$ m. Therefore, $\varepsilon(x^\mu)$ can be treated as a constant across the apparatus. Since $\varepsilon(x^\mu)$ satisfies the free-field wave equation in the vicinity of the detector and the gravity wave from the merging black holes is traveling along the z axis, ε is a function of $t - z/c$. When this wave passes and ε is positive, then the physical length of the x-arm of the detector increases and the physical length of the y-arm of the detector decreases according to Eq. (12.86),

$$\Delta x_{arm} = \sqrt{1+\varepsilon}\,\Delta x \approx \left(1+\frac{\varepsilon}{2}\right)\Delta x, \quad \Delta y_{arm} = \sqrt{1-\varepsilon}\,\Delta y \approx \left(1-\frac{\varepsilon}{2}\right)\Delta y \tag{12.87}$$

Therefore, the time difference between light traveling to and fro along the two arms is,

$$\Delta t_{arm} = \left(\frac{1}{2}\varepsilon + \frac{1}{2}\varepsilon\right)2L/c = 2\varepsilon L/c \tag{12.88}$$

For N trips of each laser beam, down and back, along each arm,

$$\Delta t_{arm} = 2N\varepsilon L/c \tag{12.89}$$

The problems discuss these predictions and LIGO's parameters in more detail.

Now let us consider the source of the gravity waves, the right-hand side of Eq. (12.53).

Let us model the source as two masses M_1 and M_2 orbiting one another at a distance r. They move nonrelativistically so Newtonian mechanics applies. An exercise in Newtonian mechanics that is reviewed in the problem set reminds us that the binding energy of the pair is,

$$E_B = -\frac{GM_1M_2}{2r} \tag{12.90}$$

Since the masses are in a constant accelerated state, the system can radiate gravity waves and gradually lose energy. When this happens, the distance r between them will monotonically decrease and they will eventually collide and merge in the case of two black holes. This system becomes a model for realistic systems that could radiate gravity waves and be detected by advanced LIGO. The rate of energy loss to gravity waves requires a detailed analysis of Eq. (12.53) that will be discussed below. The result is that the rate of energy loss to gravity waves is,

$$\frac{dE_B}{dt} = -\frac{32G^4}{5c^5r^5}(M_1M_2)^2(M_1+M_2) \tag{12.91}$$

By combining Eqs. (12.90) and (12.91), one derives the rate at which r decreases due to the energy loss,

$$\frac{dr}{dt} = -\frac{64G^3}{5c^5r^3}M_1M_2(M_1+M_2) \tag{12.92}$$

These expressions have excellent experimental support from the detection of binary neutron stars.

These results are pursued in more detail in the problem sets.

Finally we quote the result for the gravitational wave radiated from this system,

$$\varepsilon(t,z) = -\frac{4G^2M_1M_2}{rzc^4}\cos(2\pi\nu(t-z/c)) \tag{12.93}$$

where ν is twice the frequency of the orbiting binary, ν_{orbit}. We will show below that the quadrupole moment of the mass distribution is responsible for the radiation and the relation $\nu = 2\nu_{orbit}$ follows just as it does for quadrupole radiation in electromagnetism. The Newtonian prediction for ν_{orbit} follows from Kepler's law,

$$\nu_{orbit} = \frac{1}{2\pi}\left(\frac{G(M_1+M_2)}{r^3}\right)^{1/2} \tag{12.94}$$

These relations are important because LIGO is sensitive to the range of frequencies 50—500 Hz, and this matches binaries that are the candidates for LIGO sources. This will be discussed further below.

Now let us return to the wave equation Eq. (12.53) and derive some of the features of gravity waves emitted by a binary system. We will be satisfied with a semiquantitative discussion. If the reader worked through Appendix F, she knows the form of the solution to Eq. (12.53). It has the same form as the four-vector potential radiated by a general charge distribution. In this case the gravity wave piece of the metric, the solution to Eq. (12.53), reads,

$$\tilde{g}_{\mu\nu}(t,x,y,z) = \frac{4G}{c^4}\iiint T_{\mu\nu}(x',y',z',t-R/c)\frac{dx'\,dy'\,dz'}{R} \tag{12.95}$$

where $R^2 = (x-x')^2 + (y-y')^2 + (z-z')^2 = c(t-t')$. The temporal offset, $t-t'$ accounts for the fact that the bits of the energy–momentum tensor that effect the gravity wave at time t originated a time earlier that accounts for the time needed for the signal to a travel a distance R at the speed limit c. This result was discussed further in Appendix F. In our application the observation point is very(!) far from the source, so we can pull the factor of R^{-1} outside the integral,

$$\tilde{g}_{\mu\nu}(t,x,y,z) = \frac{4G}{c^4 R_0}\iiint T_{\mu\nu}(x',y',z',t-R/c)dx'\,dy'\,dz' \tag{12.96}$$

where R_0 is the distance from the observation point to the origin of the coordinate system, which is chosen at the center of mass of the radiating system. We model the radiating system as two masses $M_1$and M_2 orbiting one another in the x-y plane, as shown in Fig. 12.23. We treat the two-body system with Newtonian mechanics, which proves to be adequate until the last fraction of a second when the two black holes merge. Furthermore, if we model the system with a circular orbit, then the two bodies simply rotate around their common center of mass with a frequency $\nu_{orbit} = \omega/2\pi$, which is given by Kepler's law, as reviewed in the problem set. The orbits of the two masses are given by,

$$x_1^{(1)}(t) = r_1\cos\omega t, \quad x_2^{(1)}(t) = r_1\sin\omega t, \quad x_3^{(1)}(t) = 0 \tag{12.97a}$$

and similarly, for the second particle,

$$x_1^{(2)}(t) = -r_2\cos\omega t, \quad x_2^{(2)}(t) = -r_2\sin\omega t, \quad x_3^{(2)}(t) = 0 \tag{12.97b}$$

The energy–momentum density Eq. (12.96) is proportional to the sum of two terms, one for each of the two particles, one proportional to $\delta\left(x' - x_1^{(1)}(t)\right)\delta\left(y' - x_2^{(1)}(t)\right)\delta(z')$ and one proportional to

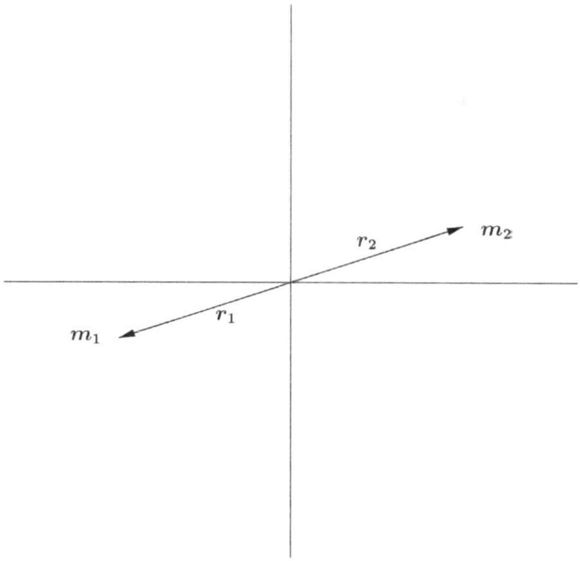

Figure 12.23 The binary system of masses $M_1$and M_2 orbiting one another and radiating gravitational waves.

$\delta\left(x' - x_1^{(2)}(t)\right)\delta\left(y' - z_2^{(2)}(t)\right)\delta(z')$. The source of the gravity waves is the energy—momentum tensor $T_{\mu\nu}$, discussed in Section 11.8, whose Cartesian coordinates behave as,

$$T_{ij} = \left(m_1 v_i^{(1)} v_j^{(1)} + m_2 v_i^{(2)} v_j^{(2)}\right) \tag{12.98}$$

where $v_i^{(1)} = dx_i^{(1)}/dt$ for mass 1 and $v_j^{(2)} = dx_j^{(2)}/dt$ for mass 2. So, differentiating Eq. (12.97) to make the velocities $v_i^{(1)}$ and $v_j^{(2)}$, the nonzero spatial components of $T_{\mu\nu}$ are,

$$T_{11} = I\omega^2 \cos^2 \omega t = \frac{1}{2} I\omega^2(1 + \cos 2\,\omega t)$$

$$T_{12} = T_{21} = I\omega^2 \cos \omega t \sin \omega t = \frac{1}{2} I\omega^2 \sin 2\,\omega t$$

$$T_{22} = I\omega^2 \sin^2 \omega t = \frac{1}{2} I\omega^2(1 - \cos 2\,\omega t)$$

where $I = M_1 r_1^2 + M_2 r_2^2$ is the moment of inertia of the two-body system. LIGO is only sensitive to the time-dependent pieces of $T_{\mu\nu}$ which read,

$$\begin{aligned} T_{11} &= +\frac{1}{2} I\omega^2 \cos 2\,\omega t \\ T_{12} &= T_{21} = I\omega^2 \sin 2\,\omega t \\ T_{22} &= -\frac{1}{2} I\omega^2 \cos 2\,\omega t \end{aligned} \tag{12.99}$$

Since $\widetilde{g}_{ij}(t)$ is proportional to $T_{ij}(t - z/c)$ where z, as shown in Fig. 12.22, is the distance to the source,

$$\widetilde{g}_{ij}(t) = \frac{2G}{c^4 z} T_{ij}(t - z/c) \tag{12.100}$$

We see that (1) $\widetilde{g}_{11}(t) = -\widetilde{g}_{22}(t)$, as anticipated in Eq. (12.86a), and (2) $\omega_{gravity} = 2\omega_{orbit}$, the frequency of the gravity wave is determined by that of the quadrupole moment, which is twice that of the orbiting masses. Note also that $\widetilde{g}_{ij}$ falls as the inverse of the distance to the source.

The reader is encouraged to follow the operation of advanced LIGO and see what additional progress it makes in observational astronomy. A space-based version of LIGO called "LISA" (Laser Interferometer Space Antenna) is also in the planning and prototyping stage. It should be sensitive to much lower frequency waves and should observe the merging of superheavy black holes that astronomers believe exist at the centers of galaxies.

The discussion here is only qualitative because we have applied the wave equation Eq. (12.53) near the merging black holes where space–time is actually strongly curved. The full Einstein field equations must be used to make quantitative predictions.

Appendix G.3 contains a discussion of gravity wave detection using the geodesic deviation ("tidal forces") equations applied to LIGO's mirrors. It uses the mathematics of linearized gravity developed in the problem set.

After observing four sources of gravitational waves (colliding black holes) over a 2-year period, on August 17, 2017, LIGO detected the inspiraling and collision of two neutron stars. The neutron stars, each approximately 12 miles in diameter and consisting of neutrons packed together at a density two to three times that of ordinary nuclear matter and each weighing 20%–30% more than the Sun, emitted gravitational waves as they orbited around each other and then collided and exploded, emitting a burst of gamma rays perpendicular to their orbital plane. LIGO detectors, one in Washington state, another in Louisiana, and a third in Italy, detected the gravitational waves and directed optical telescopes toward the galaxy NGC 4993 in the Constellation Hydra, located 130 million light years away from earth. NASA's Fermi-GLAST satellite independently detected the burst of gamma rays within 2 s of the detection of the gravitational waves. Subsequent observations by land-based telescopes detected light at optical and radio frequencies from the exploding neutron stars, called a "kilonova." A kilonova is similar but much smaller than a "supernova" explosion, which produced each neutron star in the first place, but 1000 times as powerful as a "nova" event, a star showing a sudden large increase in brightness and then slowly returning to its original state over a few months.

This observation has been hailed as a watershed event in the history of astrophysics. For the first time we have observed a catastrophic natural event where both light and gravitational waves were copiously produced. The almost-simultaneous production (the explosion lags the burst of gravitational waves by a few seconds) of these waves and the huge distance the waves travel to us allow astronomers to infer, with great accuracy, that both waves travel at the speed limit! Studies of the emitted light, whose frequencies range from gamma rays to radio waves, will lead to a better understanding of the dynamics of neutron stars. Early indications are that the hypothesis that kilonovas are the source of heavy elements in the universe, gold and platinum, for example, has been verified. Earthly observations of more kilonovas should produce better estimates of the Hubble constant, the rate of expansion of the universe, and may shed light on the nature of dark energy, which is thought to be responsible for the accelerating expansion of the universe. Dark energy will be introduced in the next and final section of this book.

12.14 CONTRASTING SPECIAL AND GENERAL RELATIVITY: THE COSMOLOGICAL CONSTANT AND DARK ENERGY

Let us continue our discussion started in Chapter 11 where we considered some of the major differences between electromagnetism and relativity, and now contrast special relativity with general relativity.

Special relativity is characterized by having global symmetries. There are 10 in all

1. translations in four space–time dimensions,
2. rotations about the three spatial axes, and
3. boosts (Lorentz transformations) along any of the three spatial axes.

This collection of 10 symmetries constitute the "Poincare group." General relativity takes these global symmetries and makes them local, i.e., space–time dependent. In special relativity the 10 symmetry operations have the same action over all of space–time. In general relativity the degree of symmetry is much higher. One requires that the theory be invariant to space–time-dependent coordinate transformations. The theory must be invariant to coordinate transformations, which might be appreciable only in the vicinity of some space–time event and negligible elsewhere. This was the guiding principle in the construction of Einstein's field equations.

In special relativity we took great care to set up a coordinate system of meter sticks and clocks over all of space–time so that measurements could be compared between frames in a local fashion. We found that clocks that were synchronized in one frame were not synchronized in moving frames, and we discovered time dilation, Lorentz contraction, and the relativity of simultaneity. We never embarked on an analogous activity in general relativity because, as we found out when considering parallel translation, comparing measurements at nonzero distances in general relativity is not generally possible. The curvature of space–time makes any attempt in this direction path dependent and hence nonuniversal. In the context of differential geometry, this is a consequence of the fact that tangent planes at different points on a curved surface, or freely falling frames in general relativity, do not coincide and do not give the basis of global comparisons of time and spatial measurements.

We have seen that the implementation of the equivalence principle can be quite subtle. The equivalence principle means that in small enough regions of space–time, the laws of physics reduce to those of special relativity. In a freely falling coordinate system it becomes impossible to detect the gravitational field by strictly local experiments. In a local inertial reference frame the metric at a point P is given by the Minkowski metric $g^{(0)}_{\sigma\rho}(P)$, and the first derivatives of the metric must vanish, $\partial_\alpha g_{\sigma\rho}(P) = 0$ in the falling frame. This condition, $\partial_\alpha g_{\sigma\rho}(P) = 0$, means that there are no forces in the freely falling frame at P. This is the essence of the equivalence principle. One must consider second derivatives of the metric to find nonzero elements in the freely falling frame. These quantities produce the curvature of space–time.

The equivalence principle also makes general relativity intrinsically nonlinear. The idea is that gravity couples universally to energy—momentum. But the gravitational field itself carries energy—momentum, so it must couple to itself. We saw the nonlinearity in the Einstein field equations and the Schwarzschild metric. Even simpler, the expression for the Riemann tensor has terms that are linear as well as quadratic in the Christoffel symbols. The Christoffel symbols themselves involve the products of the inverse of the metric, the metric and its first derivatives. It is the intrinsic nonlinearity of these equations, which makes it so challenging to find exact solutions to general relativity. The theory does *not* enjoy the principle of linear superposition that plays such an important role in classical electrodynamics.

Modern quantum field theories of elementary particles are also based on local symmetry principles and universality. These theories are formulated in Minkowski space—time, but they have "internal" symmetry operations that are local symmetries, and these symmetries dictate interactions and conservation laws. The theory of strong interactions "quantum chromodynamics" is based on the premise that the three colors of quarks become a local symmetry principle. The theory contains quark fields as well as gluon fields, which also carry color. The gluons are analogous to the electromagnetic field of electrodynamics, but they carry the color quantum number themselves in contrast to the electromagnetic fields and its quantum, the photon, which are neutral. The fact that the gluons in quantum chromodynamics carry color implies that they interact among themselves, which makes the theory intrinsically nonlinear. This property of quantum chromodynamics is believed to underlie its attribute of quark confinement. Certainly the gauge theories of elementary particles and general relativity have more in common than appear at first. We saw in our introduction of the Riemann tensor, which underlies the Einstein curvature tensor that enters the field equation for general relativity, that closed loops are the essential geometric objects used to formulate the theory. Closed loops are also the basic geometric object underlying the construction of gauge theories of elementary particle physics. In all cases we are formulating theories with local symmetry groups and products of operators around closed loops capture the character of the theories precisely. In general relativity the local symmetry group is that of Poincare, in quantum chromodynamics it is rotations in local color space of quarks and gluons. Covariant derivatives must be constructed and used in each theory to express coordinate-free physical differences. Of course the language of quantum field theory is needed for elementary particle physics. Unfortunately, a quantum formulation of gravity eludes physics, and it is not even known if the two foundations of modern physics are mutually compatible. Perhaps string theory will shed light here.

Another particularly significant difference between field theories based on a flat Minkowski metric and general relativity concerns the nature of the scale of energy. In all these theories except general relativity, the absolute scale of energy is not significant—only differences of energy matter. For example, in Newton's world, forces are given by the gradient of potentials so the zero point of the potential energy is irrelevant. All of this changes in general relativity because the source of the Einstein tensor is the energy-momentum tensor, not its derivatives! The zero point of energy is established by the energy density of the vacuum. In a quantum field theory, this is very perplexing. In quantum field theory, quantum fluctuations in the vacuum contribute to the vacuum energy. These contributions occur over all length scales from zero to infinity. It appears that to understand and calculate the vacuum energy one must understand physics at arbitrarily high energies! Without some unknown principle or symmetry, this knowledge is utterly beyond our reach. Given these obstacles, the only way forward appears to be phenomenology.

This brings us to an introductory discussion of the "cosmological constant." If the vacuum energy must be included in the Einstein field equations then it should have the form,

$$T_{\mu\sigma}^{(Vac)} = \rho_{Vac} g_{\mu\sigma} \tag{12.101}$$

And Einstein's equation becomes,

$$G_{\mu\sigma} = \frac{8\pi G}{c^4}\left(T_{\mu\sigma}^{(M)} + \rho_{Vac} g_{\mu\sigma}\right)$$

where $T_{\mu\sigma}^{(M)}$ is the matter field contribution to the energy—momentum tensor that we have discussed and illustrated earlier. This equation is usually written,

$$G_{\mu\sigma} - \Lambda g_{\mu\sigma} = \frac{8\pi G}{c^4} T_{\mu\sigma}^{(M)} \tag{12.102}$$

where Λ is the "cosmological constant" and $\Lambda = 8\pi G\rho_{Vac}$. Eq. (12.101) has been coined "dark energy."

We shall see that the cosmological constant produces an acceleration of the expansion of the universe. The observation of the spectra of light emitted by certain supernova indicates that,

$$\left|\rho_{Vac}^{obs}\right| \approx (10^{-12}\ \text{GeV}) \sim 10^{-8}\text{erg/cm}^3 \tag{12.103}$$

This is a tiny energy density that is numerically negligible on the scale of galaxies, etc. but has large cosmological effects. It is interesting to ask if quantum field theory can provide a prediction for Λ. In some future quantum field theory of gravity, one might use dimensional analysis to make an "estimate." The dimensional numbers in the theory are the Planck constant $\hbar$, the speed of light c, and Newton's gravitational constant G. These constants can be combined to produce a "Planck energy,"

$$E_P = \left(\frac{\hbar c^5}{G}\right)^{1/2} \approx 1.22 \times 10^{19}\ \text{GeV} \approx 1.95 \times 10^{16}\ \text{erg}$$

If this "natural" energy scale sets the scale for the energy density of the vacuum, we would "predict,"

$$\rho_{Vac} \sim (10^{19}\ \text{GeV})^4 \approx 10^{112}\ \text{erg/cm}^3$$

So, our "back of the envelop" estimate of ρ_{Vac} is wrong by 120 orders of magnitude! A cosmological constant of this order would overwhelm the right-hand side of Einstein's equation and vitiate all of its great successes. Something is terribly(!) wrong with these estimates. Many suggestions have been made to circumvent this disaster, but none of them are truly satisfactory and a resolution is an open challenge! Perhaps a full-fledged quantum field theory of gravity would provide a solution. Breakthroughs in string theory might be required as well. For the time being, we proceed phenomenologically and include the nonzero estimate Eq. (12.103) in cosmological calculations of the expansion of the universe.

Without reviewing modern developments in applications of general relativity to cosmology, the large-scale structure of the universe, we can argue in elementary terms that Λ leads to an accelerating expansion of the universe. Consider the Newtonian limit of Eq. (12.102). Retracing our steps in Section 12.6, we easily find that Λ modifies Newton's law of gravity,

$$\nabla^2 V = 4\pi G\rho - \Lambda c^2 \qquad (12.104)$$

We again see that Λ had better be very small on terrestrial scales to have escaped observation! Using Eq. (12.104) we calculate the gravitational acceleration a distance r outside a mass M,

$$\boldsymbol{g} = -\nabla V(r) = -\frac{GM}{r^2}\hat{\boldsymbol{r}} + c^2 \Lambda\, \mathrm{r}\, \hat{\boldsymbol{r}}$$

We find a universal acceleration outward that grows(!) with distance.

The reader is encouraged to consult the literature on this timely subject and develop a more sophisticated understanding of the theoretical and experimental issues in this challenging field.

PROBLEMS

12.1 Consider a circle of radius R drawn in the x-z plane with its center at $x=0$, $z=R$.

a. Show that in the vicinity of the origin $z=\frac{1}{2R}x^2$ + corrections, where "corrections" are of order x^4/R^3.

b. Consider a curve $z=f(x)$, which is tangent to the x axis at $x=z=0$. Show that $f(x)=\frac{1}{2}\kappa x^2$ + corrections, near the origin and κ is its curvature.

12.2 Consider a surface in three dimensions parameterized as $z=f(x, y)$. Orient the coordinate system so that the x-y plane is the surface's tangent plane at $x=y=z=0$ and choose the x and y axes to coincide with the principal directions.

a. Show that $z=\frac{1}{2}(\kappa_1 x^2+\kappa_2 y^2)$ + higher orders, for infinitesimal x and y.

b. Consider a hyperbolic paraboloid, $z=x^2-2\,y^2$. Plot it and recognize it as a "saddle." Note that $\kappa_1=2$, $\kappa_2=-4$, $K=-8$ at the origin.

c. Consider an elliptic paraboloid $z=x^2+2y^2$. Plot it. Note that $\kappa_1=2$, $\kappa_2=4$, $K=8$ at the origin.

d. Consider a sphere $(z-R)^2+x^2+y^2=R^2$. Plot it. Note that $\kappa_1=R^{-1}$, $\kappa_2=R^{-1}$, $K=R^{-2}$ everywhere on the surface.

12.3 Let us get an explicit formula for the Gaussian curvature K in terms of the metric and its derivatives. To avoid a wealth of subscripts let us use "old-fashioned" notation for the first and second fundamental forms,

$$ds^2=d\mathbf{x}\cdot d\mathbf{x}=Edu^2+2Fdudv+Gdv^2$$

$$-d\mathbf{x}\cdot d\mathbf{N}=edu^2+2fdudv+gdv^2$$

We will use subscripts to denote derivatives with respect to the coordinate mesh (u, v). So, the tangent plane at a point P is spanned by the vectors $\boldsymbol{x}_u=\partial\mathbf{x}/\partial u$ and $\boldsymbol{x}_v=\partial\mathbf{x}/\partial v$. An infinitesimal change in the normal vector at P is,

$$d\mathbf{N}=\mathbf{N}_u du+\mathbf{N}_v dv$$

a. Show that

$$e = -\boldsymbol{x}_u \cdot \boldsymbol{N}_u, \quad 2f = -(\boldsymbol{x}_u \cdot \boldsymbol{N}_v + \boldsymbol{x}_v \cdot \boldsymbol{N}_u), \quad g = -\boldsymbol{x}_v \cdot \boldsymbol{N}_v$$

b. Since $\boldsymbol{x}_u \cdot \boldsymbol{N} = \boldsymbol{x}_v \cdot \boldsymbol{N} = 0$, show that $\boldsymbol{x}_u \cdot \boldsymbol{N}_v = \boldsymbol{x}_v \cdot \boldsymbol{N}_u$, so

$$f = -\boldsymbol{x}_u \cdot \boldsymbol{N}_v = -\boldsymbol{x}_v \cdot \boldsymbol{N}_u$$

And,

$$e = \boldsymbol{x}_{uu} \cdot \boldsymbol{N}, \quad f = \boldsymbol{x}_{uv} \cdot \boldsymbol{N}, \quad g = \boldsymbol{x}_{vv} \cdot \boldsymbol{N}$$

c. Use $\boldsymbol{N} = \frac{(\boldsymbol{x}_u \times \boldsymbol{x}_v)}{|\boldsymbol{x}_u \times \boldsymbol{x}_v|} = \frac{(\boldsymbol{x}_u \times \boldsymbol{x}_v)}{\sqrt{EG - F^2}}$ to show that,

$$e = (EG - F^2)^{1/2} \det \begin{vmatrix} x_{uu} & y_{uu} & z_{uu} \\ x_u & y_u & z_u \\ x_v & y_v & z_v \end{vmatrix}$$

and corresponding expressions for f and g.

d. Recall that the Gaussian curvature is the ratio of the determinants of the second and first fundamental forms,

$$K = \frac{eg - f^2}{EG - F^2}$$

On a large sheet of paper, write out this expression in terms of the three determinants in the numerator obtained in part c. Use the identities $\det(AB) = \det A \det B$ and $\det A = \det A^T$, where "T" means transpose, to write the expression for K as the difference of two determinants,

$$K = \frac{1}{(EG - F^2)^2} \left\{ \det \begin{vmatrix} \boldsymbol{x}_{uu} \cdot \boldsymbol{x}_{vv} & \boldsymbol{x}_{uu} \cdot \boldsymbol{x}_u & \boldsymbol{x}_{uu} \cdot \boldsymbol{x}_v \\ \boldsymbol{x}_u \cdot \boldsymbol{x}_{vv} & \boldsymbol{x}_u \cdot \boldsymbol{x}_u & \boldsymbol{x}_u \cdot \boldsymbol{x}_v \\ \boldsymbol{x}_v \cdot \boldsymbol{x}_{vv} & \boldsymbol{x}_v \cdot \boldsymbol{x}_u & \boldsymbol{x}_v \cdot \boldsymbol{x}_v \end{vmatrix} \right.$$

$$\left. - \det \begin{vmatrix} \boldsymbol{x}_{uv} \cdot \boldsymbol{x}_{uv} & \boldsymbol{x}_{uv} \cdot \boldsymbol{x}_u & \boldsymbol{x}_{uv} \cdot \boldsymbol{x}_v \\ \boldsymbol{x}_u \cdot \boldsymbol{x}_{uv} & \boldsymbol{x}_u \cdot \boldsymbol{x}_u & \boldsymbol{x}_u \cdot \boldsymbol{x}_v \\ \boldsymbol{x}_v \cdot \boldsymbol{x}_{uv} & \boldsymbol{x}_v \cdot \boldsymbol{x}_u & \boldsymbol{x}_v \cdot \boldsymbol{x}_v \end{vmatrix} \right\}$$

e. Recognize that all the entries in these determinants are components of the metric, E, G, and F, and their derivatives except $\boldsymbol{x}_{uu} \cdot \boldsymbol{x}_{vv}$ and $\boldsymbol{x}_{uv} \cdot \boldsymbol{x}_{uv}$. Use the definitions,

$$E = \boldsymbol{x}_u \cdot \boldsymbol{x}_u, \quad F = \boldsymbol{x}_u \cdot \boldsymbol{x}_v, \quad G = \boldsymbol{x}_v \cdot \boldsymbol{x}_v$$

to show that,

$$-\frac{1}{2}E_{vv} + F_{uv} - \frac{1}{2}G_{uu} = \boldsymbol{x}_{uu}\cdot\boldsymbol{x}_{vv} - \boldsymbol{x}_{uv}\cdot\boldsymbol{x}_{uv}$$

Why is this useful? Imagine expanding the determinants in part d: $\boldsymbol{x}_{uu}\cdot\boldsymbol{x}_{vv}$ multiplies $EG - F^2$ in the first one and $\boldsymbol{x}_{uv}\cdot\boldsymbol{x}_{uv}$ multiplies $EG - F^2$ in the second one, so K only depends on the difference, $\boldsymbol{x}_{uu}\cdot\boldsymbol{x}_{vv} - \boldsymbol{x}_{uv}\cdot\boldsymbol{x}_{uv}$.

f. Identify the entries in the determinant in part d. and use the observation in part e. to obtain,

$$K = \frac{1}{(EG - F^2)^2}\left\{\det\begin{vmatrix} -\frac{1}{2}E_{vv} + F_{uv} - \frac{1}{2}G_{uu} & \frac{1}{2}E_u & -\frac{1}{2}E_v \\ F_v - \frac{1}{2}G_u & E & F \\ \frac{1}{2}G_v & F & G \end{vmatrix}\right.$$

$$\left. -\det\begin{vmatrix} 0 & \frac{1}{2}E_v & \frac{1}{2}G_u \\ -\frac{1}{2}E_v & E & F \\ \frac{1}{2}G_u & F & G \end{vmatrix}\right\}$$

g. Choose to parameterize the surface with an orthogonal mesh $(u,\ v)$ so that $\boldsymbol{x}_u\cdot\boldsymbol{x}_v = F = 0$ everywhere. Then the expression in part f. simplifies. Show that,

$$K = -\frac{1}{\sqrt{EG}}\left[\frac{\partial}{\partial u}\left(\frac{1}{\sqrt{E}}\frac{\partial\sqrt{G}}{\partial u}\right) + \frac{\partial}{\partial v}\left(\frac{1}{\sqrt{G}}\frac{\partial\sqrt{E}}{\partial v}\right)\right]$$

You have derive by brute force(!) Gauss' Theorema Egregium: the Gaussian curvature is an intrinsic property of the surface; it only depends on the metric and its derivatives, quantities that do not rely on the existence of an embedding space.

12.4 Now let us prove the relation $\boldsymbol{N}_u \times \boldsymbol{N}_v = K\boldsymbol{x}_u \times \boldsymbol{x}_v$, which was used in the text to derive the geometric interpretation of the Gaussian curvature K, Eq. (12.17).

a. Show that $\boldsymbol{N}_u$ and $\boldsymbol{N}_v$ are linearly independent and lie in the tangent plane at P. (Hint: To show that the vectors $\boldsymbol{N}_u$ and $\boldsymbol{N}_v$ lie in the tangent plane differentiate $\widehat{\boldsymbol{N}}\cdot\widehat{\boldsymbol{N}} = 1$ with respect to u and/or v.)

b. Argue that at any point P on the surface $\boldsymbol{N}_u \times \boldsymbol{N}_v$ points in the direction $\widehat{\boldsymbol{N}} = (\boldsymbol{x}_u \times \boldsymbol{x}_v)/\sqrt{\det(g)}$, so there is a scalar c such that $\boldsymbol{N}_u \times \boldsymbol{N}_v = c\boldsymbol{x}_u \times \boldsymbol{x}_v$.

c. To determine c, first show that,

$$(\boldsymbol{x}_u \times \boldsymbol{x}_v)\cdot(\boldsymbol{N}_u \times \boldsymbol{N}_v) = eg - f^2 = \det(b)$$

and we have used the notation of the previous problem, parts a and b. Hint: Apply the vector identity used in the text and appendices, $(\boldsymbol{a}\times\boldsymbol{b})\cdot(\boldsymbol{c}\times\boldsymbol{d}) = (\boldsymbol{a}\cdot\boldsymbol{c})(\boldsymbol{b}\cdot\boldsymbol{d}) - (\boldsymbol{a}\cdot\boldsymbol{d})(\boldsymbol{b}\cdot\boldsymbol{c})$

d. Combine the result of part c with $\boldsymbol{N}_u \times \boldsymbol{N}_v = c\boldsymbol{x}_u \times \boldsymbol{x}_v$, and the result established in the text, $(\boldsymbol{x}_u \times \boldsymbol{x}_v)\cdot(\boldsymbol{x}_u \times \boldsymbol{x}_v) = \det\ (g)$, to establish that,

$$c = \frac{\det(b)}{\det(g)}$$

which is the Gaussian curvature K!

12.5 In the weak gravitational Newtonian limit, we saw that the gravitational potential $V(r)$ contributed to the zero-zero component of the metric, $g_{00} \approx 1 + \frac{2V}{c^2}$. Let us check that $\frac{2V}{c^2}$ is actually small in many applications:

a. Show that $\frac{2V}{c^2} \approx 10^{-9}$ on the surface of the Earth.

b. Show that $\frac{2V}{c^2} \approx 10^{-6}$ on the surface of the Sun.

c. Show that $\frac{2V}{c^2} \approx 10^{-4}$ on the surface of a white dwarf star.

12.6 Let us provide some of the steps summarized in Appendix G, which derives the equation of motion of a particle in curvilinear coordinates and obtains an explicit formula for the Christoffel symbols.

a. In Appendix G, show that Eq. (G.5) follows from Eq. (G.4).

b. In Appendix G, verify the validity of Eq. (G.9) and (G.10). Show that the Christoffel symbol is given by,

$$\Gamma^k_{ij} = \frac{1}{2}\sum_l g^{kl}\left(\frac{\partial g_{il}}{\partial x_j} + \frac{\partial g_{lj}}{\partial x_i} - \frac{\partial g_{ij}}{\partial x_l}\right)$$

as claimed there.

12.7 Consider an ordinary sphere of radius R. Recall that the invariant interval on the surface of the sphere is $ds^2 = R^2 d\theta^2 + R^2 \sin^2\theta d\varphi^2$ where θ and φ are the polar and azimuthal angles of the usual polar coordinate system.

a. In matrix notation, show that the metric and its inverse are,

$$g_{ij} = \begin{pmatrix} R^2 & 0 \\ 0 & R^2\sin^2\theta \end{pmatrix}$$

and,

$$g^{ij} = \begin{pmatrix} \frac{1}{R^2} & 0 \\ 0 & \frac{1}{R^2\sin^2\theta} \end{pmatrix}$$

b. The Christoffel symbols with all lower indices are given by, as found in Appendix G,

$$\Gamma_{mjk} = \frac{1}{2}\left(\partial_k g_{mj} + \partial_j g_{mk} - \partial_m g_{jk}\right)$$

Show that the only nonzero Christoffel symbols for the surface of a sphere are,

$$\Gamma_{\varphi\varphi\theta} = R^2 \sin\theta\cos\theta, \quad \Gamma_{\theta\varphi\varphi} = -R^2\sin\theta\cos\theta$$

$$\Gamma^{\varphi}{}_{\varphi\theta} = g^{\varphi\varphi}\Gamma_{\varphi\varphi\theta} = \frac{\cos\theta}{\sin\theta}, \quad \Gamma^{\theta}{}_{\varphi\varphi} = g^{\theta\theta}\Gamma_{\theta\varphi\varphi} = -\sin\theta\cos\theta$$

c. The Riemann curvature tensor is,

$$R^i{}_{jkm} = \partial_k\Gamma^i{}_{jm} - \partial_m\Gamma^i{}_{jk} + \sum_n \Gamma^i{}_{nk}\Gamma^n{}_{jm} - \sum_n \Gamma^i{}_{nm}\Gamma^n{}_{jk}$$

Show that,

$$R^{\theta}{}_{\varphi\theta\varphi} = \sin^2\theta$$

and all the nonzero components of the tensor can be obtained from this one.

d. Show that the Ricci tensor for the surface of the sphere is,

$$R_{jm} = \sum_i R^i{}_{jim} = \frac{1}{R^2} g_{jm}$$

e. Show that the Ricci scalar, the trace of R_{jm}, is,

$$\sum_i R^i{}_i = \frac{2}{R^2}$$

so the "curvature" grows larger as the sphere becomes smaller.

f. The geodesic equation on the sphere is,

$$\frac{d^2x^m}{dt^2} + \sum_{ij} \Gamma^m{}_{ij} \frac{dx^i}{dt}\frac{dx^j}{dt} = 0$$

Show that,

$$\frac{d^2\theta}{dt^2} - \sin\theta\cos\theta\left(\frac{d\varphi}{dt}\right)^2 = 0$$

$$\frac{d^2\varphi}{dt^2} + 2\cot\theta\frac{d\theta}{dt}\frac{d\varphi}{dt} = 0$$

Identify the Coriolis and centripetal "forces" in these equations

12.8 Christoffel symbols, plane polar coordinates, and Newtonian mechanics.

Consider a nonrelativistic particle in a gravitational potential generated by a mass M at the origin,

$$\frac{d^2\boldsymbol{r}}{dt^2} = -\frac{GM}{r^2}\hat{\boldsymbol{r}}$$

This is a problem with a central potential so the motion of the particle in the potential will be planar. Let us use plane polar coordinates so,

$$dx^2 + dy^2 = dr^2 + r^2 d\theta^2$$

In plane polar coordinates $\frac{dr}{dt} \equiv u^i = \left(\frac{dr}{dt}, \frac{d\theta}{dt}\right)$, $g_{11} = g^{11} = 1$, $g_{22} = r^2$, $g^{22} = 1/r^2$ and the equation of motion written in (r, θ) coordinates reads,

$$\frac{du^i}{dt} + \sum_{kl} \Gamma^i_{kl} u^k u^l = -\frac{GM}{r^2}\delta^i_1$$

a. Calculate the Christoffel symbols from the metric in plane polar coordinates, and verify that $\Gamma^1_{22} = -r$ and $\Gamma^2_{12} = \Gamma^2_{21} = 1/r$ with all others vanishing.

b. Show that the equations of motion read,

$$\frac{d^2r}{dt^2} - r\left(\frac{d\theta}{dt}\right)^2 = -\frac{GM}{r^2}$$

$$\frac{d^2\theta}{dt^2} + \frac{2}{r}\frac{dr}{dt}\frac{d\theta}{dt} = 0$$

c. Show that the second equation gives the conservation law for the orbital angular momentum, $r^2\frac{d\theta}{dt} = L$. Substitute this into the first equation to convert it to a one-dimensional differential equation for the radial coordinate.

d. Argue that the terms coming from the Christoffel symbols are the centripetal and Coriolis forces discussed in Section 11.1.

12.9 Calculating Christoffel symbols from spatial dependence of coordinate system.

Let us calculate the Christoffel symbols for plane polar coordinates using the definition in the text,

$$\partial_\beta \boldsymbol{e}_\alpha = \sum_\gamma \Gamma^\gamma_{\alpha\beta}\boldsymbol{e}_\gamma$$

In Cartesian coordinates $\boldsymbol{r} = r\cos\theta\,\boldsymbol{i} + r\sin\theta\,\boldsymbol{j}$ and the coordinate vectors of plane polar coordinates are,

$$\frac{\partial \boldsymbol{r}}{\partial r} = \boldsymbol{e}_r = \cos\theta\,\boldsymbol{i} + \sin\theta\,\boldsymbol{j} \qquad \frac{\partial \boldsymbol{r}}{\partial \theta} = \boldsymbol{e}_\theta = -r\sin\theta\,\boldsymbol{i} + r\cos\theta\,\boldsymbol{j}$$

a. Show that

$$\frac{\partial \boldsymbol{e}_r}{\partial r} = 0, \qquad \frac{\partial \boldsymbol{e}_r}{\partial \theta} = \frac{1}{r}\boldsymbol{e}_\theta, \qquad \frac{\partial \boldsymbol{e}_\theta}{\partial r} = \frac{1}{r}\boldsymbol{e}_\theta, \qquad \frac{\partial \boldsymbol{e}_\theta}{\partial \theta} = -r\boldsymbol{e}_r$$

b. Show that the results in part a give the Christoffel symbols,

$$\Gamma^\theta{}_{r\theta} = \Gamma^\theta{}_{\theta r} = \frac{1}{r}, \qquad \Gamma^r_{\theta\theta} = -r$$

and the other Christoffel symbols vanish.

12.10 Plane polar coordinates, and the covariant derivative.

Let us do one more exercise with plane polar coordinates. We want to illustrate that D_μ, the covariant derivative, measures the *real* variation in vector fields. For example, the Cartesian unit vectors $\hat{\boldsymbol{i}}$ and $\hat{\boldsymbol{j}}$ are constant vectors in the plane. However, when written in terms of $\boldsymbol{e}_r$ and $\boldsymbol{e}_\theta$, a basis that varies in magnitude and direction over the plane, this obvious fact is obscured. Let us check that the formalism is smart by doing a "really stupid" calculation: calculate the components $D\hat{i}$ and $D\hat{j}$ in plane polar coordinates and verify that, indeed, $\hat{i}$ and $\hat{j}$ are constant vectors, $D\hat{i} = D\hat{j} = 0$!

a. In Problem 12.9 you calculated $\boldsymbol{e}_r$ and $\boldsymbol{e}_\theta$ in terms of $\hat{i}$ and $\hat{j}$. Invert these expressions and find,

$$\hat{i} = \cos\theta\, \boldsymbol{e}_r - \frac{1}{r}\sin\theta\, \boldsymbol{e}_\theta, \quad \hat{j} = \sin\theta\, \boldsymbol{e}_r + \frac{1}{r}\cos\theta\, \boldsymbol{e}_\theta$$

b. Verify $D_\theta i^r = D_\theta i^\theta = D_r i^r = D_r i^\theta = 0$ and $D_\theta j^r = D_\theta j^\theta = D_r j^r = D_r j^\theta = 0$. First note that $i^r = \cos\theta$ and $i^\theta = -\frac{1}{r}\sin\theta$ from part a. Then using the Christoffel symbols from Problem 12.8, $\Gamma^\theta_{r\theta} = \Gamma^\theta_{\theta r} = \frac{1}{r}$ and $\Gamma^\gamma_{\theta\theta} = -r$, verify,

$$D_\theta i^\theta = \partial_\theta i^\theta + \Gamma^\theta_{r\theta} i^r = -\frac{1}{r}\cos\theta + \frac{1}{r}\cos\theta = 0$$

Check the seven other cases.

The Christoffel term in the covariant derivative expression removes the variation of the curvilinear coordinates so the covariant derivative produces the *real* rate of change of the vector it acts on.

12.11 Symmetries of the Riemann tensor.

An elegant approach to deriving the symmetries of the Riemann tensor is to begin in a locally inertial frame at a point P. If we can prove tensor relations there, they will be generally true, in any coordinate system.

a. Recall that $\Gamma^\alpha_{\gamma\delta} = 0$ at P and use Eq. (12.44) to find the derivative of $\Gamma^\alpha_{\gamma\delta}$ at P,

$$\partial_\sigma \Gamma^\alpha_{\mu\nu} = \frac{1}{2}\sum_\beta g^{\alpha\beta}\left(\partial_\nu\partial_\sigma g_{\beta\mu} + \partial_\sigma\partial_\mu g_{\beta\nu} - \partial_\beta\partial_\sigma g_{\mu\nu}\right)$$

b. Use Eq. (12.46) to write the Riemann tensor at P in terms of the metric and its second derivatives,

$$R^\alpha_{\beta\mu\nu} = \frac{1}{2}\sum_\sigma g^{\alpha\sigma}\left(\partial_\beta\partial_\mu g_{\sigma\nu} - \partial_\beta\partial_\nu g_{\sigma\mu} + \partial_\sigma\partial_\nu g_{\beta\mu} - \partial_\sigma\partial_\mu g_{\beta\nu}\right)$$

$$R_{\alpha\beta\mu\nu} = \sum_\lambda g_{\alpha\lambda} R^\lambda_{\beta\mu\nu} = \frac{1}{2}\left(\partial_\beta\partial_\mu g_{\alpha\nu} - \partial_\beta\partial_\nu g_{\alpha\mu} + \partial_\alpha\partial_\nu g_{\beta\mu} - \partial_\alpha\partial_\mu g_{\beta\nu}\right)$$

c. Use the result in part b to verify the symmetry properties listed in the text,

$$R_{\alpha\beta\mu\nu} = -R_{\beta\alpha\mu\nu} = -R_{\alpha\beta\nu\mu} = R_{\mu\nu\alpha\beta}$$
$$R_{\alpha\beta\mu\nu} + R_{\alpha\nu\beta\mu} + R_{\alpha\mu\nu\beta} = 0$$

Since these relations are tensor equations, if they are true in one coordinate system, they are true in all coordinate systems! We used a locally inertial frame to minimize our labors.

d. Use the symmetries to argue that the number of independent components of the Riemann tensor is "only" 20. (Without these symmetries the tensor would have had $4 \times 4 \times 4 \times 4 = 256$ independent components.)

e. Starting from Eq. (12.46), show that in a locally inertial frame at point P,

$$\partial_\lambda R_{\alpha\beta\mu\nu} = \frac{1}{2}\left(\partial_\beta\partial_\mu\partial_\lambda g_{\alpha\nu} - \partial_\beta\partial_\nu\partial_\lambda g_{\alpha\mu} + \partial_\alpha\partial_\nu\partial_\lambda g_{\beta\mu} - \partial_\alpha\partial_\mu\partial_\lambda g_{\beta\nu}\right)$$

f. Use the result of part e to show that,

$$\partial_\lambda R_{\alpha\beta\mu\nu} + \partial_\nu R_{\alpha\beta\lambda\mu} + \partial_\mu R_{\alpha\beta\nu\lambda} = 0$$

This expression can be written in terms of the covariant derivative since the covariant derivative and the partial derivatives coincide in a locally inertial frame,

$$D_\lambda R_{\alpha\beta\mu\nu} + D_\nu R_{\alpha\beta\lambda\mu} + D_\mu R_{\alpha\beta\nu\lambda} = 0$$

But this is a tensor relation, so it is true in all coordinate systems. It is the Bianchi identity referenced and used in the text.

g. Recall the definition of the Ricci tensor,

$$R_{\alpha\beta} = \sum_\lambda R^\lambda_{\ \alpha\lambda\beta}$$

Note that the Ricci tensor is symmetric, $R_{\alpha\beta} = R_{\beta\alpha}$. Use the symmetry properties of $R_{\alpha\beta\mu\nu}$ to argue that other contractions, such as $\sum_\lambda R^\lambda_{\ \alpha\beta\lambda}$, either vanish identically or reduce to $\pm R_{\alpha\beta}$. So, the Ricci tensor is the unique second-rank tensor that can be made from the Riemann tensor.

12.12 The symmetry properties of the Riemann tensor can be found from another method, which does not depend on using an inertial frame.

a. Show from Eq. (12.46) for the Riemann tensor and Eq. (12.44) for the Christoffel symbol that a general expression for $R_{\alpha\beta\gamma\delta}$ reads,

$$R_{\alpha\beta\gamma\delta} = \frac{1}{2}\left(\partial_\delta\partial_\alpha g_{\beta\gamma} - \partial_\delta\partial_\beta g_{\alpha\gamma} + \partial_\gamma\partial_\beta g_{\alpha\delta} - \partial_\gamma\partial_\alpha g_{\beta\delta}\right)$$

$$-\sum_{\rho\sigma} g^{\rho\sigma}\left(\Gamma_{\rho\alpha\gamma}\Gamma_{\sigma\beta\delta} - \Gamma_{\rho\alpha\delta}\Gamma_{\sigma\beta\gamma}\right)$$

b. Prove the symmetry properties 1–3 listed in the text after Eq. (12.47) using the result of part a.

12.13 The Poincare Disc and Non-Euclidean Geometry

We have seen that the surface of a sphere has a constant positive Gaussian curvature. The plane has vanishing Gaussian curvature (a cylinder does also since it is obtained from a plane by bending). What about a surface of constant *negative* curvature? There is a famous theorem (D. Hilbert) that states that a regular complete two-dimensional surface of constant negative curvature cannot be embedded in three dimensions.

However, Poincarè pointed out that one could invent a two-dimensional space with a metric so that $K = -1$ everywhere! It is the Poincarè disc.

Consider plane polar coordinates (r, θ) with $r < 1$ and a metric,

$$ds^2 = \frac{4\left(dr^2 + r^2 d\theta^2\right)}{\left(1 - r^2\right)^2}$$

a. Calculate the Gaussian curvature K, using the formula from Problem 12.3 or Appendix G.2, and verify that it is a constant -1 over the entire Poincarè disc.

b. Let us compare the radius of a circle to its circumference in such a space. Calculate the radius R of a line segment from $(0, 0)$ to $(r, 0)$,

$$R = \int_0^r \frac{2dr'}{1 - r'^2} = \ln\left(\frac{1 + r}{1 - r}\right)$$

So $r = (e^R - 1)/(e^R + 1)$. We learn that $R \to \infty$ as $r \to 1$ and the physical space is infinite!

c. Calculate the circumference of a circle of physical radius R and centered at the origin,

$$C=\int_0^{2\pi}\frac{2rd\theta}{1-r^2}=2\pi\frac{2r}{1-r^2}=2\pi\left(\frac{e^R-e^{-R}}{2}\right)=2\pi\sinh R$$

d. Use the geometric definition of Gaussian curvature, Eq. (12.2), and verify that $K=-1$ at the origin.

e. To develop a deeper perspective on this problem, read about Escher tilings, non-Euclidean geometry of Bolyai and Lobachevsky, and its formulation in the complex plane and the unit disc.

12.14 The Schwarzschild metric in isotropic coordinates.

We want to write the Schwarzschild metric in a coordinate system where the spatial part of the ds^2 is flat, Euclidean three-space,

$$ds^2=A(r)dt^2-B(r)\left(dx^2+dy^2+dz^2\right)$$

We write this expression in spherical coordinates to compare with the results in Section 12.7,

$$ds^2=A(r)dt^2-\lambda^2(\rho)\left(d\rho^2+\rho^2d\theta^2+\rho^2\sin^2\theta d\varphi^2\right)$$

Comparing with the Schwarzschild metric obtained and studied in the text,

$$ds^2=\left(1-\frac{2GM}{c^2r}\right)c^2dt^2-\left(1-\frac{2GM}{c^2r}\right)^{-1}dr^2-r^2\left(d\theta^2+\sin^2\theta d\varphi^2\right) \quad (12.67)$$

we see that,

$$A(r)=\left(1-\frac{2GM}{c^2r}\right),\quad r^2=\lambda^2\rho^2,\quad \left(1-\frac{2GM}{c^2r}\right)^{-1}dr^2=\lambda^2d\rho^2$$

a. Integrate the expression $\left(1-\frac{2GM}{c^2r}\right)^{-1}dr^2=\lambda^2d\rho^2$ and require that $r\sim\rho$ as either variable becomes large, and find

$$r=\rho\left(1+\frac{GM}{2c^2\rho}\right)^2$$

so that

$$\lambda^2=\left(1+\frac{GM}{2c^2\rho}\right)^4$$

b. Show that the Schwarzschild metric reads in isotropic coordinates,

$$ds^2 = \frac{\left(1 - \frac{GM}{2c^2r}\right)^2}{\left(1 - \frac{GM}{2c^2r}\right)^2} c^2dt^2 - \left(1 + \frac{GM}{2c^2r}\right)^4 \left(dr^2 + r^2d\theta^2 + r^2\sin^2\theta d\varphi^2\right)$$

where we have relabeled ρ as r in the last step. This metric can be written,

$$ds^2 = \frac{\left(1 - \frac{GM}{2c^2r}\right)^2}{\left(1 + \frac{GM}{2c^2r}\right)^2} c^2dt^2 - \left(1 + \frac{GM}{2c^2r}\right)^4 (dx^2 + dy^2 + dz^2)$$

as promised, where $r^2 = x^2 + y^2 + z^2$.

Note that in the weak gravity case, we can write this result to linear accuracy in the "small" parameter $\frac{GM}{2c^2r} \ll 1$,

$$ds^2 \approx \left(1 - \frac{2GM}{c^2r}\right)c^2dt^2 - \left(1 + \frac{2GM}{c^2r}\right)(dx^2 + dy^2 + dz^2)$$

We have obtained this result, accurate to $O(c^{-4})$, in several different ways in Chapters 11 and 12 and the problem sets. It is important in cosmology.

12.15 Linearized gravity, the divergenceless, traceless "gauge."

In the text, we obtained the transformation property for small fluctuations about a flat Minkowski metric,

$$\tilde{g}'_{\mu\nu} = \tilde{g}_{\mu\nu}(x) - \partial_\mu\varepsilon_\nu - \partial_\nu\varepsilon_\mu + O(\varepsilon^2)$$

In the general relativity literature it is conventional to define,

$$\bar{h}_{\mu\nu} = \tilde{g}_{\mu\nu} - \frac{1}{2}g^{(0)}_{\mu\nu}h, \qquad h = \sum_\alpha \tilde{g}^\alpha_\alpha$$

a. Show that the transformation law for $\bar{h}_{\mu\nu}$ is,

$$\bar{h}'_{\mu\nu} = \bar{h}_{\mu\nu} - \partial_\mu\varepsilon_\nu - \partial_\nu\varepsilon_\mu + g^{(0)}_{\mu\nu}\sum_\alpha \varepsilon^\alpha_\alpha$$

and show that,

$$\sum_\nu \partial_\nu \overline{h}'^{\mu\nu} = \sum_\nu \partial_\nu \overline{h}^{\mu\nu} - \sum_\nu \partial^\nu \partial_\nu \varepsilon^\mu$$

So if we choose ε^μ to solve the differential equation, $\sum_\nu \partial_\nu \overline{h}^{\mu\nu} - \sum_\nu \partial^\nu \partial_\nu \varepsilon^\mu = 0$, then $\overline{h}'^{\mu\nu}$ is divergenceless, $\sum_\nu \partial_\nu \overline{h}'^{\mu\nu} = 0$.

b. There is still some freedom in our choose of ε^μ: we can always add a solution to the wave equation, $\sum_\nu \partial^\nu \partial_\nu \omega^\mu = 0$. Now,

$$\widetilde{g}''_{\mu\nu} = \widetilde{g}'_{\mu\nu} - \partial_\mu \omega_\nu - \partial_\nu \omega_\mu$$

Show that

$$\sum_\mu \widetilde{g}''^{\mu}_{\mu} = \sum_\mu \widetilde{g}'^{\mu}_{\mu} - 2\sum_\mu \partial^\mu \omega_\mu$$

So if ω_μ solves the differential equation $\sum_\mu \partial^\mu \omega_\mu = \frac{1}{2}\sum_\mu \widetilde{g}'^{\mu}_{\mu}$, then $\sum_\mu \widetilde{g}''^{\mu}_{\mu} = 0$.

We learn that we can always choose a divergenceless, traceless "gauge" where,

$$\sum_\nu \partial_\nu \widetilde{g}^{\mu\nu} = 0, \quad \sum_\mu \widetilde{g}^{\mu}_{\mu} = 0$$

In this gauge

$$h = 0, \quad \overline{h}_{\mu\nu} = \widetilde{g}_{\mu\nu}$$

12.16 Linearized gravity near a source.

Consider small fluctuations around the Minkowski metric,

$$g_{\mu\nu} = g^{(0)}_{\mu\nu} + \widetilde{g}_{\mu\nu}$$

a. Show that the Riemann tensor is, to first order in $\widetilde{g}_{\mu\nu}$,

$$R_{\alpha\beta\gamma\delta} = \frac{1}{2}\left(\partial_\beta \partial_\gamma \widetilde{g}_{\alpha\delta} + \partial_\alpha \partial_\delta \widetilde{g}_{\beta\gamma} - \partial_\beta \partial_\delta \widetilde{g}_{\alpha\gamma} - \partial_\alpha \partial_\gamma \widetilde{g}_{\beta\delta}\right)$$

b. Show that $R_{\alpha\beta\gamma\delta}$ is "gauge" invariant, i.e., under a transformation $\widetilde{g}'_{\mu\nu} = \widetilde{g}_{\mu\nu}(x) - \partial_\mu \varepsilon_\nu - \partial_\nu \varepsilon_\mu$, $R_{\alpha\beta\gamma\delta}$ is unchanged. Note that this is another proof that the physics is invariant to small coordinate transformations in the case of linearized gravity.

c. Introduce the field,

$$\overline{h}_{\mu\nu} = \widetilde{g}_{\mu\nu} - \frac{1}{2} g^{(0)}_{\mu\nu} h \quad h = \sum_{\alpha} \widetilde{g}^{\alpha}_{\alpha}$$

as in Problem 12.15. Show that this equation can be written in the form,

$$\widetilde{g}_{\mu\nu} = \overline{h}_{\mu\nu} - \frac{1}{2} g^{(0)}_{\mu\nu} \overline{h} \quad \overline{h} = \sum_{\alpha} \overline{h}^{\alpha}_{\alpha} = -h$$

d. Show that to first order in $\widetilde{g}_{\mu\nu}$ the Einstein tensor reads,

$$G_{\alpha\beta} = -\frac{1}{2}\left(\sum_{\mu} \partial^{\mu}\partial_{\mu}\overline{h}_{\alpha\beta} + g^{0}_{\alpha\beta}\sum_{\mu\nu}\partial^{\mu}\partial^{\nu}\overline{h}_{\mu\nu} - \sum_{\mu}\partial^{\mu}\partial_{\beta}\overline{h}_{\alpha\mu} - \sum_{\mu}\partial^{\mu}\partial_{\alpha}\overline{h}_{\beta\mu}\right)$$

e. If we impose the Lorenz gauge to $\overline{h}_{\mu\nu}$, $\sum_{\nu}\partial_{\nu}\overline{h}^{\mu\nu} = 0$, show that,

$$R_{\alpha\beta} = G_{\alpha\beta} = -\frac{1}{2}\sum_{\gamma}\partial^{\gamma}\partial_{\gamma}\overline{h}_{\alpha\beta}$$

So, the linearized Einstein field equation reduces to the wave equation for the metric fluctuation $\overline{h}_{\mu\nu}$ in a flat background Minkowski metric,

$$\left(\nabla^2 - \frac{1}{c^2}\frac{\partial^2}{\partial t^2}\right)\overline{h}_{\mu\nu} = \frac{16\pi G}{c^4} T_{\mu\nu}$$

Note that the Lorenz gauge choice here makes the equation compatible with energy–momentum conservation, $\sum_{\nu}\partial_{\nu} T^{\mu\nu} = 0$.

12.17 Linearized gravity, the Newtonian limit, and isotropic coordinates.

Let us apply what we learned in Problem 12.14 to the Newtonian limit of general relativity.

Consider the weak field case (Newtonian limit) where $T^{00} \to \rho c^2$ and the wave equation of part e of Problem 12.16 reduces to Newton's law of gravity, $\nabla^2\Phi = 4\pi G\rho$, if we identify $\overline{h}^{00} = 4\Phi/c^2$, with all other components of $\overline{h}_{\alpha\beta}$ and $T_{\alpha\beta}$ negligible. Show that in this limit,

$$h = -4\Phi/c^2$$

and use the definition, $\overline{h}_{\mu\nu} = \widetilde{g}_{\mu\nu} - \frac{1}{2} g^{(0)}_{\mu\nu} h$, to find,

$$\widetilde{g}^{00} = \widetilde{g}^{11} = \widetilde{g}^{22} = \widetilde{g}^{33} = 2\Phi/c^2$$

So the metric in the static, weak field case reads,

$$ds^2 = \left(1 + \frac{2\Phi}{c^2}\right)c^2dt^2 - \left(1 - \frac{2\Phi}{c^2}\right)(dx^2 + dy^2 + dz^2)$$

This result is accurate to order $1/c^4$. This coordinate system, introduced in Problem 12.14, is called "isotropic" because of its manifest rotational symmetry in the spatial coordinates.

12.18 Weak gravity and isotropic coordinates.

In Chapter 11 we argued, on the basis of the equivalence principle, that the metric in a weak $2GM/c^2r \ll 1$ gravitational field was approximately,

$$ds^2 = \left(1 - \frac{2GM}{c^2r}\right)c^2dt^2 - \left(1 + \frac{2GM}{c^2r}\right)dr^2 - r^2\left(d\theta^2 + \sin^2\theta d\varphi^2\right) + O\left(\frac{1}{c^4}\right)$$

We can write this in isotropic form if we make a small transformation to the radial coordinate,

$$r = \rho + \frac{GM}{c^2}$$

Show that the metric becomes,

$$ds^2 = \left(1 - \frac{2GM}{c^2\rho}\right)c^2dt^2 - \left(1 + \frac{2GM}{c^2\rho}\right)\left(d\rho^2 + \rho^2d\theta^2 + \rho^2\sin^2\theta d\varphi^2\right) + O\left(\frac{1}{c^4}\right)$$

If we define $\rho^2 = r'^2 = x'^2 + y'^2 + z'^2$, $x' = \rho\sin\theta\cos\varphi$, $y' = \rho\sin\theta\cos\varphi$, and $z' = \rho\cos\theta$,

$$ds^2 = \left(1 - \frac{2GM}{c^2r}\right)c^2dt^2 - \left(1 + \frac{2GM}{c^2r}\right)(dx^2 + dy^2 + dz^2) + O\left(\frac{1}{c^4}\right)$$

where we dropped the primes for clarity.

12.19 Newtonian mechanics in a weakly curved isotropic metric.

Consider the metric of Problem 12.17,

$$ds^2 = \left(1 + \frac{2\Phi}{c^2}\right)c^2dt^2 - \left(1 - \frac{2\Phi}{c^2}\right)(dx^2 + dy^2 + dz^2)$$

where $2\Phi/c^2 \ll 1$.

a. Calculate the nonvanishing Christoffel symbols for this metric and verify,

$$\Gamma^0_{00} = \frac{\partial \Phi}{c^2 \partial c\tau}, \quad \Gamma^i_{00} = \frac{\partial \Phi}{c^2 \partial x^i}$$

b. Consider a particle of mass m. Its equation of motion is given by the geodesic,

$$\frac{d^2 x^\beta}{d^2 \tau} + \sum_{\alpha\gamma} \Gamma^\beta_{\gamma\alpha} \frac{dx^\alpha}{d\tau} \frac{dx^\gamma}{d\tau} = 0$$

Simplify this expression for nonrelativistic kinematics, $p^0 \gg p^i$, and show that for the index $\beta = 0$; it becomes Newtonian energy conservation to leading order,

$$\frac{d}{dt}(E + m\Phi) = 0$$

And for $\beta = i$, $(i = 1, 2, 3)$, it becomes Newton's second law for a particle in a gravitational field,

$$\frac{d}{dt}\boldsymbol{p} = -m\nabla\Phi$$

to leading order.

This example illustrates that in Einstein's world, gravitation is curvature, expressed through nonzero $\Gamma^\beta_{\gamma\alpha}$, and in Newton's world, gravitation is a universal force, proportional to every body's inertial mass.

12.20 Newtonian mechanics and circular orbits.

In our illustrations of sources of gravity waves, we will use Newtonian nonrelativistic mechanics to model the system and make estimates. This naïve approach works surprisingly well up until the last fraction of a second when the constituent stars collide.

Consider a two-body system in their center of mass frame as shown in Fig. 12.23. The system rotates with an angular velocity ω with a fixed distance between the stars. Newton's second law reads,

$$m_1 \frac{v_1^2}{r_1} = m_2 \frac{v_2^2}{r_2} = G \frac{m_1 m_2}{(r_1 + r_2)^2}$$

and the condition for the center of mass reads $m_1 r_1 = m_2 r_2$ and $\omega = v_1/r_1 = v_2/r_2$.

a. Derive Kepler's law for circular orbits,

$$\omega = \sqrt{\frac{G(m_1 + m_2)}{r^3}}$$

where $r = r_1 + r_2$.

b. Show that the total energy of the system in the center of mass frame, $\frac{1}{2}m_1 v_1^2 + \frac{1}{2}m_2 v_2^2 - \frac{Gm_1 m_2}{r}$ is,

$$E_B = -\frac{Gm_1 m_2}{2r}$$

where the subscript "B" indicates "binding" energy.

c. Use this expression for the binding energy and the formula for the power loss due to the radiation of gravity waves,

$$\frac{dE_B}{dt} = -\frac{32G^4}{5c^5 r^5} m_1^2 m_2^2 (m_1 + m_2)$$

to derive the rate at which the distance between the pair decreases,

$$\frac{dr}{dt} = -\frac{64G^3}{5c^5 r^3} m_1 m_2 (m_1 + m_2)$$

12.21 Indirect evidence for gravitational waves ("Hulse—Taylor" binary).

On July 2, 1974, R.A. Hulse and J.H. Taylor detected signals from a pulsar using the Arecibo radio telescope. After many observations and data analyses, they inferred that they were observing a binary system consisting of a pulsar and another neutron star, which were orbiting one another with a period of 7.75 h. The pulsar period was measured to be 59 ms, 17 times per second, which allowed the observers to infer precise results for the characteristics of the binary.

a. The mass of the pulsar was 1.44 solar masses, and the mass of its companion was 1.39 solar masses.

b. The semimajor axis of the pair is, $a = 2.3418$ light—seconds (702,053 km).

c. The semiminor axis of the pair is, $b = 1.8426$ light—seconds.

d. The rate of decrease of the orbital period was 76.5 μs per Earth year.

Let us model this system with a simpler one we can easily analyze, a circular orbit where the distance between the stars is $r = a + b = 4.185$ light—seconds. Let us apply Newtonian mechanics to this binary as in the problems above and use those formulas to learn more about it,

a. What is the power radiated in gravity waves by this idealized model of the binary? (A more realistic estimate predicts approximately 7.35×10^{24} W, which is 1.9% of the power radiated as light by the Sun.)

b. For one orbit, calculate the approximate change in r. (Hint: The fractional change in r is small so there is no need to do an integration. A more realistic calculation predicts a rate of decrease of the semimajor axis of 3.5 m per Earth year.)

c. What is the orbital velocity of the stars relative to the center of mass? (A more realistic calculation predicts an orbital velocity of the stars at the point of closest approach of ~450 km/s and at the point of greatest separation, ~110 km/s)

d. Could advanced LIGO detect the gravity waves from this system? Evaluate Eq. (12.93) assuming the present best estimate of the distance to the binary of 20,000 light-years. Evaluate the frequency of the gravity waves.

12.22 Comparison of radiation rates for electromagnetism and gravity.

Consider a hydrogen atom in its ground state. According to classical physics, it should radiate electromagnetic waves and lose energy at the rate (Larmor formula),

$$\frac{dE}{dt} = -\frac{2kq^2a^2}{3c^3}$$

where q is the charge of the electron and a is its acceleration. This formula applies as long as $v/c \ll 1$.

a. Look up the size of the first Bohr radius, and estimate the acceleration a classically. You may assume Newtonian mechanics, the Coulomb force law, and a circular orbit so Problem 12.20 applies.

b. Evaluate $\frac{dE}{dt}$ from the Larmor formula and calculate the rate at which classical physics predicts the size of the Bohr orbit should decrease.

c. Derive the classical prediction for the lifetime of the Bohr orbit.

d. Repeat this exercise for gravitational radiation from the hydrogen atom. Compare the rate of electromagnetic radiation with that of gravitational radiation numerically.

12.23 More on radiation: electromagnetic and gravitational.

a. Suppose an electron decelerates at a constant rate a from an initial velocity v_0 to zero. Assume that $v_0/c \ll 1$ so that nonrelativistic kinematics applies. What fraction of its initial kinetic energy is radiated away? Use the Larmor formula given in Problem 12.22.

b. Repeat part a. for gravitational radiation. Replace the charged particle with a particle of mass m but no charge. (Hint. You can view the particle from a freely falling frame in a constant gravitational field. Keep the equivalence principle in mind. Be careful!)

c. In light of part b. reconsider part a. Consider a charged particle at rest on the surface of the Earth. It experiences a constant acceleration of 9.8 m/s^2 but it remains at rest in this noninertial frame. Does it radiate? Next, view the electron from the perspective of a freely falling observer. The observer's local frame is inertial so special relativity should apply. Does the charged particle radiate in his frame? (This "simple" problem is actually very challenging, and very famous. See Ref. [8] for a discussion and a list of references.)

12.24 Schwarzschild radius in Newtonian mechanics.

a. Show that the Newtonian escape velocity from a mass M at distance r is $v_{escape} = \sqrt{2GM/r} = c\sqrt{r_{Sch}/r}$.

If we impose the speed limit of special relativity here, we see that the particle ejected in this fashion could not escape to infinity if initially $r < r_{Sch}$.

b. Calculate numerically the acceleration of gravity at the Schwarzschild radius. Compare with g on the Earth.

c. Continuing the line of thinking in part a, note that a particle *could* escape by simply accelerating from $r < r_{Sch}$ to $r > r_{Sch}$in Newton's world. How much energy is required to move a mass m from just below the Schwarzschild radius to infinity in Newton's world? Is quantum physics and/or relativistic mechanics required in a proper description of particle dynamics in the vicinity of the Schwarzschild radius?

12.25 Newton's turntable.

Let us return to our first discussion in Chapter 11 where we discussed "apparent" forces on a turntable in Newton's world. Let us unleash the formalism of general relativity and get those results systematically. Talk about cracking a nut with a sledgehammer!

Anyway, we start with Cartesian coordinates $x^\alpha = (t, x^1, x^2)$ and want to transform to general coordinates $\varepsilon^\alpha = (t, \varepsilon^1, \varepsilon^2)$. Note that the first variable in *both* coordinate systems is Newton's universal time, t. In the Cartesian coordinate system we have Newton's second law,

$$\frac{d^2x^i}{dt^2} = F^i$$

where the index i ranges over 1 and 2.

The metric in Newton's world is purely spatial,

$$ds^2 = \sum_i \left(dx^i\right)^2 = \sum_{imn} \frac{\partial x^i}{\partial \varepsilon^m}\frac{\partial x^i}{\partial \varepsilon^n} d\varepsilon^m d\varepsilon^n = \sum_{mn} g_{mn} d\varepsilon^m d\varepsilon^n$$

And,

$$g_{mn} = \sum_i \frac{\partial x^i}{\partial \varepsilon^m}\frac{\partial x^i}{\partial \varepsilon^n}$$

Here the indices m and n run over 1 and 2.

a. Repeat the derivation for geodesics in the text or, better, in Appendix G, and find Newton's second law in the general coordinate system,

$$\frac{d^2\varepsilon^i}{dt^2} + \sum_{jk\alpha\beta} g^{ij}\frac{\partial x^k}{\partial \varepsilon^j}\frac{\partial^2 x^k}{\partial \varepsilon^\alpha \partial \varepsilon^\beta}\frac{d\varepsilon^\alpha}{dt}\frac{d\varepsilon^\beta}{dt} = f^i$$

where α and β run over 0, 1, and 2, and the indices i and j run over 1 and 2 only. (We could have identified a Christoffel symbol here, following earlier discussions.)

Here f^I is the force in the second coordinate system,

$$f^i = \sum_j \frac{\partial \varepsilon^i}{\partial x^j} F^j$$

Now let us apply this to the turntable of Chapter 11,

$$x = r\cos(\theta + \omega t) \quad y = r\sin(\theta + \omega t)$$

b. Find the metric in the second coordinate system, (t, r, θ),

$$g_{ij} = \begin{pmatrix} 1 & 0 \\ 0 & r^2 \end{pmatrix} \quad g^{ij} = \begin{pmatrix} 1 & 0 \\ 0 & r^{-2} \end{pmatrix}$$

c. Now for some algebra! We need all the partial derivatives of x and y with respect to t, r, and θ.

Let us use some abbreviations: $c = \cos(\theta + \omega t)$ and $s = \sin(\theta + \omega t)$.

Verify,

$$\frac{\partial x}{\partial t} = -\omega rs \quad \frac{\partial x}{\partial r} = c \quad \frac{\partial x}{\partial \theta} = -rs$$

and the analogous expressions for y, following from the substitution, $x \to y$, $s \to -c$, and $c \to s$,

$$\frac{\partial y}{\partial t} = \omega rc \quad \frac{\partial y}{\partial r} = s \quad \frac{\partial y}{\partial \theta} = rc$$

Similarly,

$$\frac{\partial^2 x}{\partial t^2} = -\omega^2 rc \quad \frac{\partial^2 x}{\partial r \partial t} = -\omega s \quad \frac{\partial^2 x}{\partial \theta \partial t} = -\omega rc$$
$$\frac{\partial^2 x}{\partial t \partial r} = -\omega s \quad \frac{\partial^2 x}{\partial r^2} = 0 \quad \frac{\partial^2 x}{\partial \theta \partial r} = -s$$
$$\frac{\partial^2 x}{\partial t \partial \theta} = -\omega rc \quad \frac{\partial^2 x}{\partial r \partial \theta} = -s \quad \frac{\partial^2 x}{\partial \theta^2} = -rc$$

and the analogous expressions for y following from the substitution, $x \to y$, $s \to -c$, and $c \to s$,

$$\frac{\partial^2 y}{\partial t^2} = -\omega^2 rs \quad \frac{\partial^2 y}{\partial r \partial t} = \omega c \quad \frac{\partial^2 y}{\partial \theta \partial t} = -\omega rs$$
$$\frac{\partial^2 y}{\partial t \partial r} = \omega c \quad \frac{\partial^2 y}{\partial r^2} = 0 \quad \frac{\partial^2 y}{\partial \theta \partial r} = c$$
$$\frac{\partial^2 y}{\partial t \partial \theta} = -\omega rs \quad \frac{\partial^2 y}{\partial r \partial \theta} = c \quad \frac{\partial^2 y}{\partial \theta^2} = -rs$$

d. Since $r^2 = x^2 + y^2$, show that,

$$\frac{\partial r}{\partial x} = \frac{x}{r} = c \quad \frac{\partial r}{\partial y} = \frac{y}{r} = s$$

and since $x = r\cos(\theta + \omega t)$, and $y = r\sin(\theta + \omega t)$, show that,

$$\frac{\partial \theta}{\partial x} = -\frac{s}{r} \quad \frac{\partial \theta}{\partial y} = \frac{c}{r}$$

So the transformation law for the external force is,

$$f^1 = cF^1 + sF^2 \quad f^2 = -\frac{s}{r}F^1 + \frac{c}{r}F^2$$

e. Finally, using these bits we can write out the equation of motion in part a. Verify,

$$\frac{d^2r}{dt^2} - \omega^2 r - 2\omega r\frac{d\theta}{dt} - r\left(\frac{d\theta}{dt}\right)^2 = \frac{d^2r}{dt^2} - r\left(\omega + \frac{d\theta}{dt}\right)^2 = f^1$$

$$\frac{d^2\theta}{dt^2} + \frac{2\omega}{r}\frac{dr}{dt} + \frac{2}{r}\frac{dr}{dt}\frac{d\theta}{dt} = \frac{d^2\theta}{dt^2} + \frac{2}{r}\frac{dr}{dt}\left(\omega + \frac{d\theta}{dt}\right) = f^2$$

where we see the centripetal and Coriolis forces yet again.

12.26 Show that if a particle contributes $T^{\alpha\beta} = \rho_0 u^\alpha u^\beta$ to the energy–momentum tensor, then the conservation condition, $\sum_\alpha D_\alpha T^{\alpha\beta} = 0$ implies that the particle travels on a geodesic.

REFERENCES

[1] E. Kreyszig, Differential Geometry, Dover Publications, Inc., New York, 1991.
[2] L.D. Landau, E.M. Lifshitz, The Classical Theory of Fields, Pergamon Press, Oxford, 1962.
[3] R.P. Feynman, F.B. Morinigo, W.G. Wagner, Feynman Lectures on Gravitation, Addison-Wesley Publishing Company, Reading, Massachusetts, 1995.
[4] J. Foster, J.D. Nightingale, A Short Course in General Relativity, Springer-Verlag, New York, 1995.
[5] E.F. Taylor, J.A. Wheeler, Exploring Black Holes, Addison Wesley Longman, New York, 2000.
[6] W. Rindler, Essential Relativity, Springer-Verlag, Berlin, 1971.
[7] E.F. Taylor, J.A. Wheeler, Spacetime Physics, W. H. Freeman, New York, 1992.
[8] C. de Almeida, A. Saa, The radiation of a uniformly accelerated charge is beyond the horizon: a simple derivation, Am. J. Phys (December 2, 2005) arXiv:physics/0506049v5

APPENDIX A

Physical Constants, Data, and Conversion Factors

1. Speed of light in empty space: $c = 2.9979\ldots \cdot 10^{8}$ m/s
2. Gravitational constant: $G = 6.673 \cdot 10^{-11}$ m^3/kg-s^2
3. Planck's constant: $h = 6.626 \cdot 10^{-31}$ kg-m^2/s
4. Electron's charge: $e = 1.602 \cdot 10^{-19}$ coulombs
5. Electron's rest mass: $m = 9.109 \cdot 10^{-31}$ kg $= 0.511$ MeV/c^2
6. Proton's rest mass: $m = 1.673 \cdot 10^{-27}$ kg $= 938$ MeV/c^2
7. Mass of Earth: $M = 5.9742 \cdot 10^{24}$ kg
8. Radius of Earth: $R = 6.371 \cdot 10^{6}$ m
9. Mass of the Sun: $M = 1.989 \cdot 10^{30}$ kg
10. Radius of the Sun: $R = 6.960 \cdot 10^{8}$ m
11. Distance of Earth to the Sun: $D = 1.50 \cdot 10^{11}$ m
12. Orbital speed of Earth around the Sun: $v = 2.98 \cdot 10^{4}$ m/s
13. Electronvolt conversion: 1 eV $= 1.602 \cdot 10^{-19}$ J
14. Kilometer conversion: 1 km $= 0.6214$ miles
15. Length of a year: 1 year $= 3.156 \cdot 10^{7}$ s

APPENDIX B

Solutions to Selected Problems

B.1 CHAPTER 2 PROBLEMS

Problem 2-1

a. In the spaceship frame, events 1 and 2 do not occur at the same space point; that is, event 2 occurs on Earth. However, both events 1 and 2 occur at the same place in the Earth frame, so it is a proper time interval in the Earth frame.

b. Following the same reasoning as in part (a), the time interval between events 2 and 3 is not a proper time interval in either frame.

c. The time interval between events 1 and 3 is a proper time interval in the spaceship frame, but not in the Earth frame.

d. Because the time between events 1 and 2 is a proper time interval in the Earth frame, all that the spaceship sees is a dilated time value,

$$t_2' = \gamma\, t_e = \frac{10}{\sqrt{1 - \frac{v^2}{c^2}}} \text{ min} = 12.5 \text{ min}.$$

e. The velocity of Earth according to the spaceship is $0.6c$. Additionally, the time between events 2 and 1, according to the spaceship, is 12.5 min. as found in part (d). So the distance of Earth at event 2, according to the spaceship, is

$$l_2' = 12.5 \cdot 60 \cdot 0.6c = 1.35 \cdot 10^{11} \text{m}.$$

f. The time between events 3 and 2 is (according to the spaceship)

$$t_3' - t_2' = \frac{l_2'}{c} = 7.5 \text{ min}.$$

And we know the time of event 2 according to the spaceship. So the time of event 3 is

$$t_3' = 7.5 + 12.5 = 20 \text{ min}.$$

g. From Earth's perspective, when Earth emits the pulse (event 2), the spaceship is at a distance

$$l_2 = 10 \text{ min}\cdot 0.6c = 1.08\cdot 10^{11}\text{m}.$$

When the pulse reaches the spaceship, the spaceship has moved an additional distance. Let the time for the pulse to travel to the spaceship be called Δt, where

$$\Delta t = t_3 - t_2$$

and

$$c\Delta t = 1.08\cdot 10^{11} + v\Delta t$$

$$\Delta t = \frac{1.08\cdot 10^{11}}{2.9979\cdot 10^{8}\cdot 0.4} = 900 \text{ s} = 15 \text{ min}.$$

So the time of event 3 according to Earth is

$$t_3 = t_2 + \Delta t = 25 \text{ min}.$$

h. We know that the time interval between events 3 and 1 is a proper time in the spaceship frame (part c). So the time interval between events 3 and 1 in the Earth frame should just be the dilated value of the time interval in the rocket frame:

$$t_3 - t_1 = \gamma\left(t_3' - t_1'\right).$$

Now $t_1 = t_1' = 0$, so we should have

$$t_3 = \gamma\, t_3'.$$

Let us see if this is true: $t_3 = 25$ min, while $t_3' = 20$ min.

$$t_3 = \gamma\, t_3' = \frac{1}{\sqrt{1 - \dfrac{v^2}{c^2}}} 20 = \frac{5}{4}\cdot 20 = 25 \text{ min}.$$

Hence our results are consistent.

Problem 2-2

a. The relative velocity v_r is the velocity of rocket B measured by A. The distance traveled in A's frame is 100 m (the length of rocket A in its own frame), and the time taken, as measured by A's clocks, is $1.5\cdot 10^{-6}$ s. So,

$$v_r = \frac{100}{1.5 \cdot 10^{-6}} = 6.67 \cdot 10^7 \text{ m/s}.$$

b. The time measurement depends only on the relative velocity, and we know that the relative velocity of B with respect to A equals the relative velocity of A with respect to B. Hence the time (shown by the clocks on B) for the front end of A to pass the entire length of B is also $1.5 \cdot 10^{-6}$ s.

c. An observer sitting in the front end of B sees a contracted length of rocket A. This length contraction is given by

$$l' = \frac{1}{\gamma} l = \sqrt{1 - (v_r/c)^2}\, l = 0.9749.100 = 97.49 \text{ m}.$$

The time taken is therefore

$$t = \frac{l'}{v_r} = \frac{l}{v_r \gamma} = \frac{1.5 \cdot 10^{-6}}{\gamma} = 0.9749 \cdot 1.5 \cdot 10^{-6} = 1.46 \cdot 10^{-6} \text{ s}.$$

This time is not expected to agree with the time in part (b) because the events occur at two different points in A's frame, namely the front and back ends. In part (b) only the front end is being timed.

Problem 2-3

a. Given

$$N(t) = N_0 2^{-t/T},$$

the time for one-third of the pions to decay (in the rest frame of the pions) is given by

$$\frac{2}{3} = 2^{-t'}/T$$

$$t' = T \frac{\ln(3/2)}{\ln(2)} = 1.05 \cdot 10^{-8} \text{ s}.$$

The length traveled by the pions in the lab frame is 35 m. So the pions see a contracted length in their own rest frame. This is given by

$$l' = \frac{1}{\gamma} = 35\sqrt{1 - v^2/c^2} \text{ m}.$$

Additionally, we have

$$v = \frac{l'}{t'} = \frac{1}{1.05 \cdot 10^{-8}} 35\sqrt{1 - v^2/c^2}.$$

We solve the equation by squaring both sides

$$v^2 = \frac{10^{16}}{1.1025} 1225\left(1 - \frac{v^2}{c^2}\right)$$

$$v^2\left(1 + \frac{1111.11 \cdot 10^{16}}{9.10^{16}}\right) = 1111.11 \cdot 10^{16}$$

$$v^2 = 8.93 \cdot 10^{16},$$

which gives us

$$v = 2.985 \cdot 10^8 \text{m/s},$$

which is just slightly less than the speed of light.

b. $l' = \frac{l}{\gamma} = 35\sqrt{1 - v^2/c^2}\ \text{m} = 2.27\ \text{m}.$

Problem 2-5

a. According to the rocket frame, the signal travels from the nose of the rocket to the tail. This is just the length of the rocket in its own frame. Moreover, the speed of light is invariant, so the time taken in the rocket frame is given by

$$t' = \frac{l_0}{c}.$$

b. At time $t = 0$, an observer in frame S measures the tail of the spaceship to be a distance l_0/γ to the left of A, according to Lorentz contraction. The tail of the spaceship is moving at velocity v to the right, so the time t when light reaches the tail is given by

$$ct = l_0/\gamma - vt.$$

Solving for t, we find

$$t = l_0/[c\gamma(1 + v/c)] = \frac{l_0}{c}\sqrt{\frac{1 - v/c}{1 + v/c}}.$$

c. An observer in frame S sees a contracted length of the spaceship. So the time for the whole ship to pass point A is given by

$$t_2 = \frac{1}{v}\frac{l_0}{\gamma}.$$

B.2 CHAPTER 3 PROBLEMS

Problem 3-1

The events are illustrated in Figs. B.1–B.4.

The figures can be used to read off the values:

1. $x' = 0.58$, $ct' = 0.58$
2. $x' = -1.15$, $ct' = 2.3$
3. $x = 1.7$, $ct = 1.7$
4. $x = 1.15$, $ct = 2.3$

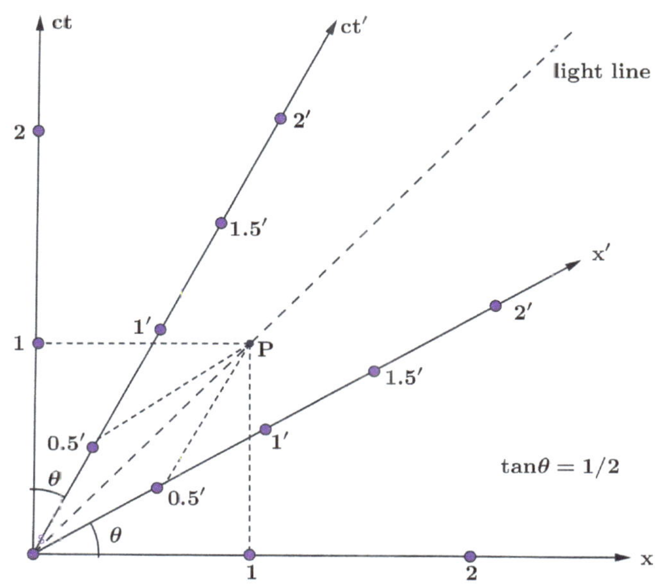

Figure B.1 Minkowski diagram for the event in Problem 3.1 part b.i. $x = 1$, $ct = 1$.

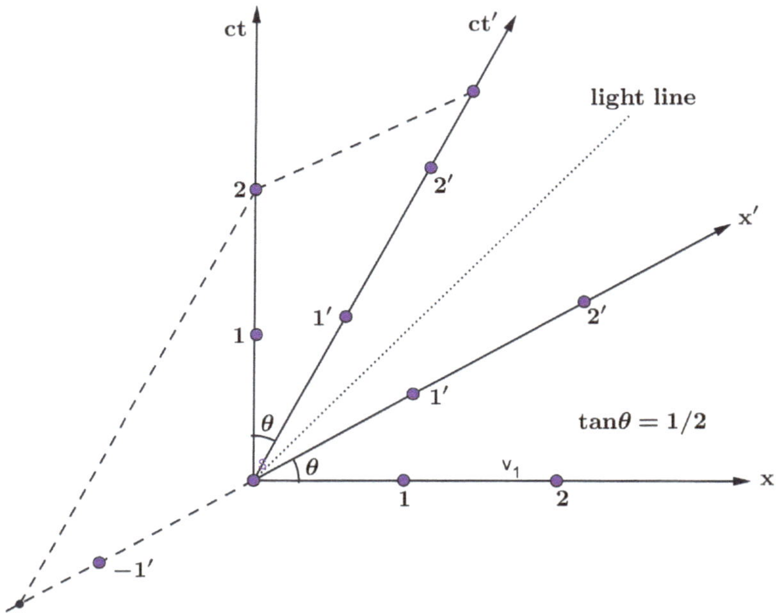

Figure B.2 Minkowski diagram for the event in Problem 3.1 part b.ii. $x = 0$, $ct = 2$.

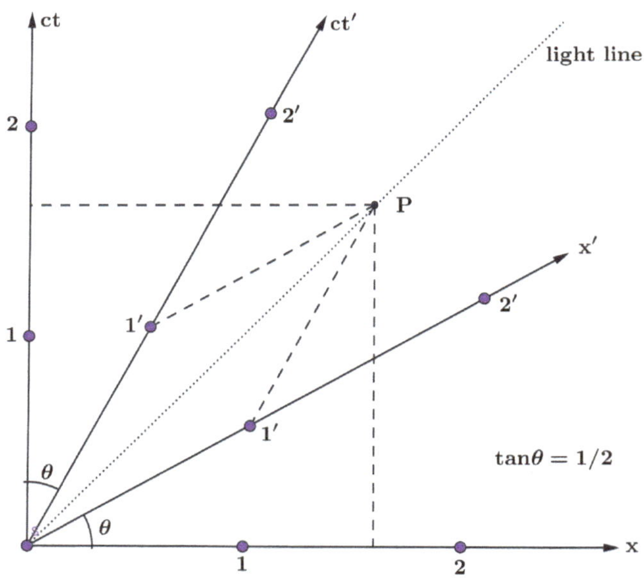

Figure B.3 Minkowski diagram for the event in Problem 3.1 part b.iii. $x' = 1$, $ct' = 1$.

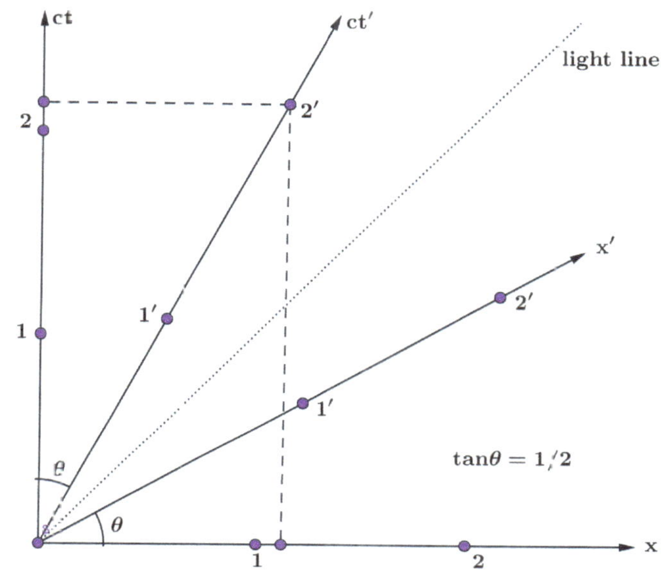

Figure B.4 Minkowski diagram for the event in Problem 3.1 part b.iv. $x' = 0$, $ct' = 2$.

Problem 3-7

The Minkowski diagram for the events is shown in Fig. B.5.

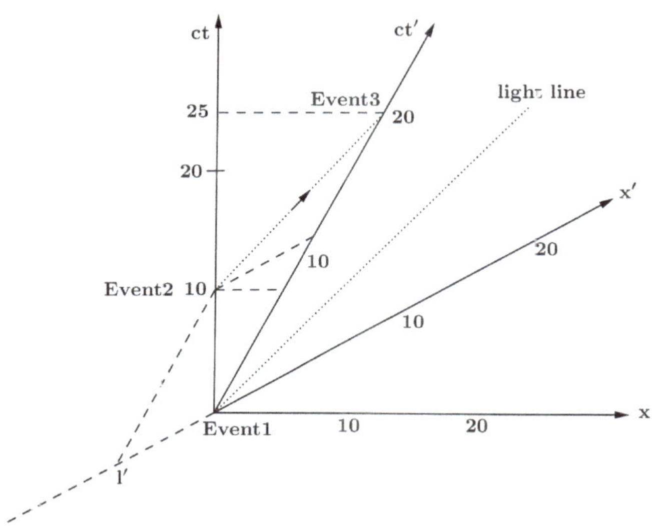

Figure B.5 Minkowski diagram for Problem 2.1.

From the graph, we can see that (a) the time interval between events 1 and 2 is a proper time in the Earth frame, (b) the time interval between events 2 and 3 is not a proper time interval in either frame, while (c) the time interval between events 1 and 3 is a proper time in the spaceship frame. (For an interval to be a proper time, both events should lie on the time axis of some frame.)

We also get (d) $t_2' = 12.5$ min, while (e) the distance of Earth from the spaceship (as measured by the spaceship) at event 2 is $1.35 \cdot 10^{11}$ m.

Additionally, (f) the time of event 3, as seen by the spaceship, is 20 min, while (g) the time of the same event seen by Earth is 25 min.

Thus (h) our results from the diagram match with those of Chapter 2.

B.3 CHAPTER 4 PROBLEMS

Problem 4-1

Let us use invariant intervals to solve this problem. In frame S, the square of the interval is

$$(\Delta x)^2 - (c\Delta t)^2 = 0^2 - (5c)^2 = -25c^2,$$

where $\Delta x = 0$ because the events occur at the same place in S.

The value of the invariant interval is the same when measured in any inertial frame, so an observer in S' has

$$\begin{aligned} -25c^2 &= (\Delta x')^2 - (c\Delta t')^2 \\ &= (\Delta x')^2 - (7c)^2 \\ (\Delta x')^2 &= 24c^2 \\ (\Delta x') &= \sqrt{24}c \text{ m}, \end{aligned}$$

which is the answer we need.

Problem 4-2

In frame S,

$$(\Delta x)^2 - (c\Delta t)^2 = (2000)^2 - 0^2 = 4 \cdot 10^6.$$

In S',

$$4\cdot 10^6 = (\Delta x')^2 - (c\Delta t')^2$$

$$= (4000)^2 - (c\Delta t')^2$$

$$(c\Delta t')^2 = 12\cdot 10^6$$

$$\Delta t' = \sqrt{\frac{12}{9}\cdot 10^{-10}}$$

$$= \sqrt{\frac{4}{3}}\cdot 10^{-5} \text{ s.}$$

Problem 4-3

Use the definition $\beta \equiv v/c = 4/5$ to make the notation efficient. The Lorentz transformations give

$$x = \frac{1}{\sqrt{1-\beta^2}}x' + \frac{\beta}{\sqrt{1-\beta^2}}ct'$$

$$= \frac{1}{0.6}\cdot 100 + \frac{0.8}{0.6}\cdot 3\cdot 10^8\cdot 9\cdot 10^{-8}$$

$$= 202.67 \text{ m}$$

and

$$t = \frac{\beta}{\sqrt{1-\beta^2}}\frac{x'}{c} + \frac{1}{\sqrt{1-\beta^2}}t'$$

$$= \frac{0.8}{0.6}\cdot\frac{100}{3\cdot 10^8} + \frac{1}{0.6}\cdot 9\cdot 10^{-8}$$

$$= 59.44\cdot 10^{-8} \text{ s.}$$

Problem 4-4

Use the Lorentz transformation,

$$t' = \frac{1}{\sqrt{1-v^2/c^2}}(t - vx/c^2).$$

a. Equating the t' for the two events, we get

$$t_1 - \frac{vx_1}{c^2} = t_2 - \frac{vx_2}{c^2}$$

$$\frac{v}{c^2}(2L - L) = \frac{L}{2c} - \frac{L}{c}$$

$$v = -\frac{c}{2}.$$

b.

$$t' = \frac{1}{\sqrt{3/4}}\left(\frac{L}{c} + \frac{L}{2c}\right)$$

$$= \frac{\sqrt{3}L}{c}.$$

Problem 4-5

$$\gamma = \frac{1}{\sqrt{1 - \frac{v^2}{c^2}}} = \frac{1}{\sqrt{1 - 0.8^2}} = \frac{5}{3}.$$

a.

$$t' = \gamma\left(t - \frac{vx}{c^2}\right)$$

$$= \frac{5}{3}\left(5 \cdot 10^{-7} - \frac{400}{5c}\right)$$

$$= \frac{5}{3} \cdot 10^{-7}\left(5 - \frac{8}{3}\right)$$

$$= 1.48 \cdot 10^{-7} \text{ s.}$$

$$x' = \frac{5}{3}\left(100 - \frac{4c}{5} \cdot 5 \cdot 10^{-7}\right)$$

$$= -\frac{100}{3} \text{ m.}$$

b.

$$t_2' = \frac{5}{3}\left(7\cdot 10^{-7} - \frac{200}{5c}\right)$$
$$= 11.67\cdot 10^{-7} \text{ s}$$

and the time difference is

$$\Delta t' = t_2' - t' = 10.19\cdot 10^{-7} \text{ s}.$$

Problem 4-7

a. To find the velocity of the particle in frame S, we use the relativistic addition formula,

$$u = \frac{0.6c - c/3}{1 - 0.6/3} = \frac{c}{3}.$$

b. Use the invariance of the space−time interval,

$$(\Delta x)^2 - c^2(\Delta t)^2 = (\Delta x')^2 - c^2(\Delta t')^2.$$

We know $\Delta t' = 3\cdot 10^{-7} - 10^{-7}$ s, and

$$\Delta x' = u'\Delta t' = \frac{-c}{3}\cdot 2\cdot 10^{-7}.$$

Additionally, we know that $\Delta t = \Delta x/u$, which gives us

$$(\Delta x)^2\left(1 - \frac{c^2}{u^2}\right) = 4.10^{-14}\left(\frac{c^2}{9} - c^2\right),$$

which gives, using u from part (a),

$$(\Delta x) = \frac{2c}{3}\cdot 10^{-7} = 20 \text{ m}.$$

We could have also done the problem equally easily using Lorentz transformations.

Problem 4-9

Let v_S be the velocity of light in the frame S. This is given by the relativistic addition formula

$$v_S = \frac{v_m + v}{1 + \dfrac{v_m v}{c^2}}.$$

We expand the denominator in a Taylor series,

$$\begin{aligned} v_S &= (v_m + v)\left(1 + \frac{v_m v}{c^2}\right)^{-1} \\ &= (v_m + v)\left(1 - \frac{v_m v}{c^2} + \cdots\right), \end{aligned}$$

where we have expanded only to leading order in v/c. We expand the bracket, again dropping the term with v^2/c^2:

$$v_S = v_m + v\left(1 - \frac{v_m^2}{c^2}\right).$$

Problem 4-11

a. Let Earth be at the origin and the star be somewhere in the first quadrant (i.e., it has positive x and positive y coordinates). Then the light signal that is emitted from the star toward Earth has negative components. If Earth and the star did not have a relative velocity, these components would be

$$\begin{aligned} v_x &= -c \cos\theta \\ v_y &= -c \sin\theta. \end{aligned}$$

Now suppose that the star (frame S) is moving away from Earth (frame S') with velocity v in the x direction. We therefore know the x component of velocity in S, and we know the x component of velocity of S' with respect to S. So, to find the x component of velocity in S', we use the relativistic addition formula. Additionally, we note that because the speed of light is invariant, the only way the x component can change is if the angle θ changes. In S', the x component of light is therefore given by

$$v_x' = -c \cos\theta'.$$

So,

$$-c\cos\theta' = \frac{-c\cos\theta - v}{1 - \dfrac{(v)(-c\cos\theta)}{c^2}}.$$

Dividing through by $-c$, we get the desired result:

$$\cos\theta' = \frac{\cos\theta + \frac{v}{c}}{1 + \frac{v\cos\theta}{c}}.$$

b. As in Problem 4.9, we expand the denominator to first order in v/c,

$$\cos\theta' \approx \left(\cos\theta + \frac{v}{c}\right)\left(1 - \frac{v\cos\theta}{c}\right)$$

$$\cos\theta' \approx \cos\theta + \frac{v}{c} - \frac{v}{c}\cos^2\theta$$

$$\approx \cos\theta + \frac{v}{c}\sin^2\theta.$$

c. We now have $\theta \approx \theta'$. We have from part (b)

$$\cos\theta' - \cos\theta = \frac{v}{c}\sin^2\theta.$$

We use a trigonometric identity

$$\cos\theta' - \cos\theta = 2\left(\sin\frac{\theta - \theta'}{2}\right)\left(\sin\frac{\theta + \theta'}{2}\right).$$

We substitute $\alpha = \theta' - \theta$, and use the approximation $\sin\beta \approx \beta$ for $|\beta| \ll 1$. We also use the approximate equality of θ and θ' in the second sine term in the product to get

$$-2\frac{\alpha}{2}\left(\sin\frac{2\theta}{2}\right) \approx \frac{v}{c}\sin^2\theta$$

$$\alpha \approx -\frac{v}{c}\sin\theta.$$

Problem 4-14

a. We just use the relativistic addition formulae

$$v_x = \frac{v'_x + v}{1 + \left(\frac{v'_x v}{c^2}\right)}$$

$$v_y = \frac{v'_y}{\gamma\left(1 + \frac{(v'_x v)}{c^2}\right)} = \frac{v'_y}{\gamma}.$$

b. v_y and v'_y are different, even though the y coordinates remain unchanged, because time dilates.

Problem 4-17

a. An observer at rest in frame S' measures the position of the meter stick at a particular time t' in his frame. But, by the relativity of simultaneity, a clock at -0.5 m lags the time on a clock at $+0.5$ m when viewed from S'. Therefore an observer in S' states that the right end of the rod reaches $y = 0$ before the left end of the rod. The rod is, therefore, seen to tilt up to the right, as shown in Fig. B.6. Curious!

b. First find the time and position in frame S' when the right end of the meter stick crosses the x axis. This event occurs in the S frame at $t_1 = 0$ and $x_1 = 0.5$ m, $y_1 = 0$. Using Lorentz transformations gives

$$x'_1 = \gamma(x_1 - vt_1) = 0.5\gamma$$

$$t'_1 = \gamma\left(t_1 - \frac{vx_1}{c^2}\right) = -\frac{0.5\gamma v}{c^2}$$

$$y'_1 = 0.$$

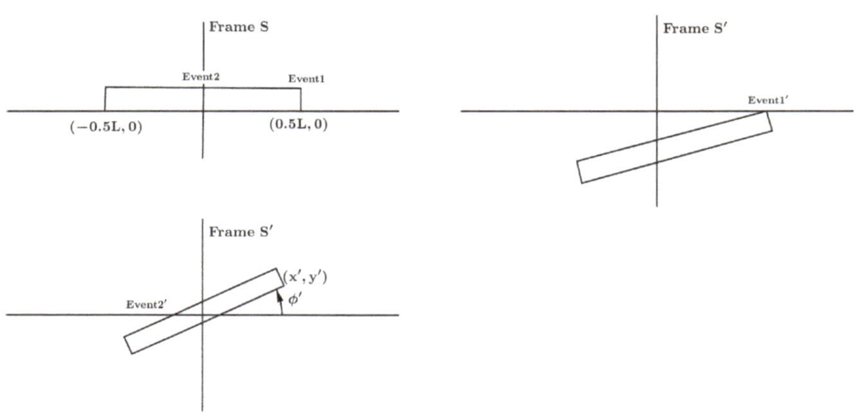

Figure B.6 Visualization of the events 1 and 2 in frames S and S' for Problem 4.18, part b.

Now we look at another event, that of the midpoint of the meter stick crossing the x axis. In the S frame, this happens at $x_2 = y_2 = 0$, $t_2 = 0$. In the S' frame, we have

$$x'_2 = y'_2 = t'_2 = 0.$$

Denote the right-end coordinates at time t_2 to be (x', y').

From Problem 4.14, we compute the velocities in the x and y directions in the S' frame, $v'_x = -v$ and $v'_y = v_y/\gamma$,

$$\begin{aligned} x' &= x'_1 + v'_x(t'_2 - t'_1) \\ &= 0.5\gamma - v\frac{0.5\gamma v}{c^2} = 0.5\gamma(1 - v^2/c^2) = \frac{0.5}{\gamma} \\ y' &= y'\,1 + v'_y(t'_2 - t'_1) \\ &= 0 + \frac{v_y}{\gamma}\frac{0.5\gamma v}{c^2} = \frac{vv_y}{c^2}0.5. \end{aligned}$$

And we have

$$\tan\phi' = \frac{y'}{x'} = \frac{vv_y\gamma}{c^2}.$$

B.4 CHAPTER 5 PROBLEMS

Problem 5-2

a. Let the distance of the observer from point 2 be x. The first pulse, therefore, has to travel a distance of

$$l_1 = d + x = vt + x$$

and the second pulse has to travel a distance of x meters. The observer, therefore, receives the second pulse at a time

$$t_2 = \frac{x}{c} + t$$

and we have

$$t_1 = \frac{l_1}{c} = \frac{x}{c} + t\frac{v}{c}.$$

So the interval between the pulses is

$$T = t_2 - t_1 = t\left(1 - \frac{v}{c}\right).$$

b.

$$\frac{d}{T} = \frac{v}{1 - v/c}$$

follows from part (a). This is the apparent speed the observer sees.

c. If an observer is a distance y to the left of point 1, the two pulses will be received at times

$$t_1 = \frac{y}{c}$$

and

$$t_2 = \frac{y + vt}{c} + t.$$

The time interval is, therefore,

$$T' = t\left(1 + \frac{v}{c}\right)$$

and the apparent speed is

$$\frac{d}{T'} = \frac{v}{1 + (v/c)}.$$

As v approaches c, d/T' approaches $c/2$.

B.5 CHAPTER 6 PROBLEMS

Problem 6-1

The rest mass energy of the proton is 938 MeV. We use the formula for relativistic energy to get

$$E = \frac{mc^2}{\sqrt{1 - v^2/c^2}}$$

$$10^{19}\text{eV} = \frac{938 \cdot 10^6 \text{eV}}{\sqrt{1 - v^2/c^2}}.$$

Solving for v, we get

$$\gamma = \frac{10^{19}}{938 \cdot 10^{6}} = 1.067 \cdot 10^{10}$$

$$\frac{v}{c} = \sqrt{1 - (938 \cdot 10^{-13})^2} \approx 1 - \frac{(938 \cdot 10^{-13})^2}{2}.$$

So, the proton is traveling very close to the speed limit. The fractional difference is only $-4.40 \cdot 10^{-21}$.

a. The time it takes to traverse the galaxy in the galaxy frame is given by

$$t_{gal} = \frac{10^5 c \text{ year}}{v} = \frac{10^5 \text{ year}}{1 - (938 \cdot 10^{-13})^2/2} \approx 10^5 \text{ year}$$

b. The event of the proton traversing the galaxy is a proper time interval in the proton's frame because the two events, the proton starting off and the proton reaching the edge of the galaxy, occur at the same space point in the proton's frame. Therefore the time seen by the galaxy between these two events is just the dilated proton time for the same events. But we know the galaxy time from part (a), so we have

$$t_{gal} = \gamma t_{proton}$$

$$t_{proton} = \frac{t_{gal}}{\gamma}$$

$$= \frac{10^5 \text{ year}}{1 - (938 \cdot 10^{-13})^2/2} \frac{1}{1.067 \cdot 10^{10}}$$

$$\approx 9.38 \cdot 10^{-6} \text{years} \approx 4.93 \text{ min.}$$

Problem 6-2

The rest mass energy of an electron is 0.51 MeV. The energy acquired in the potential drop is 10^5 eV. So, the relativistic energy, γmc^2, is

$$10^5 + 0.51 \cdot 10^6 = \frac{0.51 \cdot 10^6}{\sqrt{1 - \frac{v^2}{c^2}}}$$

$$\sqrt{1 - \frac{v^2}{c^2}} = \frac{5.1}{6.1}$$

$$\frac{v}{c} = 0.55.$$

a. The time to travel 10 m in the lab frame is

$$t_{lab} = \frac{10}{v} = \frac{10}{0.55 \cdot 3 \cdot 10^8} = 6.06 \cdot 10^{-8} \text{ s.}$$

b. Following the same reasoning as in Problem 6.1, we have Lorentz contraction,

$$d_{el} = \frac{d_{lab}}{\gamma} = \frac{5.1}{6.1} \cdot 10 \text{ m} = 8.36 \text{ m.}$$

Problem 6-4

The equations are:

a. Newtonian kinematics,

$$E = \frac{p^2}{2m} + mc^2,$$

where we included the rest mass energy, mc^2, for easy comparison to the relativistic formulas.

b. Relativistic kinematics for a particle with rest mass m,

$$E = +\sqrt{p^2c^2 + m^2c^4}.$$

c. Relativistic kinematics for a massless particle (i.e., photon),

$$E = |p|c.$$

The graphs are shown in Fig. B.7. Curves 1 and 2 agree near $E = mc^2$, whereas curves 2 and 3 agree at large momentum values.

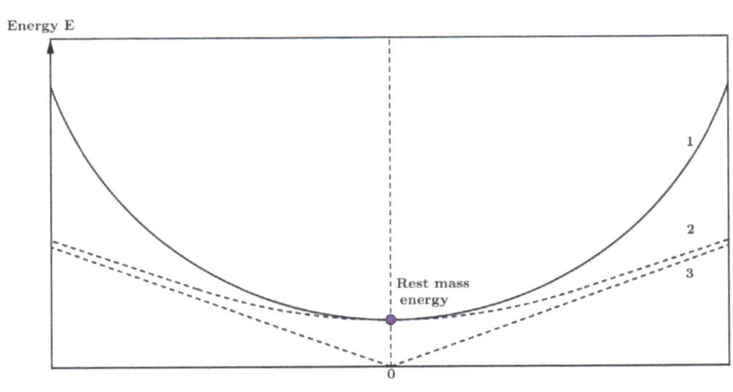

Figure B.7 Energy–momentum plots for the three relations described in parts 1, 2, and 3 of Problem 6.4.

Problem 6-5

From the Lorentz transformation, we have

$$E' = \gamma(E - vp_1).$$

If we specialize to the case of the photon, then $p_1 = E/c$, and we get

$$\begin{aligned} E' &= \gamma\left(E - v\frac{E}{c}\right) \\ &= \frac{1}{\sqrt{1-\frac{v^2}{c^2}}}E\left(1-\frac{v}{c}\right) \\ &= \sqrt{\frac{1-v/c}{1+v/c}}E. \end{aligned}$$

Problem 6-10

The electrostatic potential energy is given by

$$U = \frac{3kQ^2}{5r}.$$

Here Q is the total charge, which is

$$Q = 1.6\cdot 10^{-19}\cdot n,$$

where n is the number of electrons in 1 g of electrons:

$$n = \frac{1\cdot 10^{-3}}{9.1\cdot 10^{-31}} = 1.1\cdot 10^{27}.$$

So, $Q = -1.76\cdot 10^8$ C and

$$U = \frac{3\cdot 9\cdot 10^9\cdot 3.10\cdot 10^{16}}{5\cdot .1} = 1.67\cdot 10^{27}.$$

Equating this to the relativistic rest mass energy, we get

$$mc^2 = U.$$

So,

$$m = 1.86\cdot 10^{10}\ \text{kg},$$

which is amazingly large!

Problem 6-11

Consider the equation requiring that the position of the center of mass of the system does not change. In view of the approximations stated in this problem, this is

$$\Delta\bar{x} = 0 = mL + M\Delta x,$$

where m is the mass equivalent of light, M the mass of the box (neglecting the loss of mass during light transit), and Δx is the amount moved by the box. Now, if we include the fact that the box loses mass during light transit, the second term becomes $(M - m)\Delta x$, and if we include the fact that the box has moved (and hence the light travels less distance before being absorbed), the first term becomes $m(L + \Delta x)$. The signs have been taken according to the conventions in the book. So our equation now reads

$$0 = m(L + \Delta x) + (M - m)\Delta x,$$

which gives us exactly the same equation as the one that was derived using the approximations.

Problem 6-12

The relativistic kinetic energy is given by $E - mc^2$, where E is the total energy

$$E - mc^2 = 2mc^2$$
$$E = 3mc^2.$$

Using the formula for relativistic energy, we get

$$E = \frac{mc^2}{\sqrt{1 - \frac{v^2}{c^2}}}$$

$$3mc^2 = \frac{mc^2}{\sqrt{1 - \frac{v^2}{c^2}}}$$

$$\sqrt{1 - \frac{v^2}{c^2}} = \frac{1}{3}$$

$$v = 0.94c.$$

The momentum is given by

$$p = \frac{mv}{\sqrt{1 - \frac{v^2}{c^2}}} = 3mv = 2.82mc.$$

If the relativistic kinetic energy is five times the rest mass then, $E = 6mc^2$, and we have

$$\sqrt{1 - \frac{v^2}{c^2}} = \frac{1}{6}$$

$$v = 0.986c$$

$$p = 6mv = 5.916mc.$$

Problem 6-13

Since $v/c = 0.99$, the particle's $\gamma = 1\Big/\sqrt{1-(0.99)^2} = 7.09$. For a potential drop of X:

$$X\ \text{eV} = \frac{mc^2}{\sqrt{1 - \frac{v^2}{c^2}}} - mc^2 = 6.09 \cdot 0.51 \cdot 10^6\ \text{eV}$$

$$X = 3.11 \cdot 10^6\ \text{V}.$$

The proton's rest mass is 938 MeV, so the value of X becomes

$$X = 6.09 \cdot 938 \cdot 10^6 = 5712 \cdot 10^6\ \text{V}.$$

Problem 6-15

Let the incident photon energy be E_0 and the reflected energy be E_f. Let the final velocity of the rocket be v. Energy–momentum conservation reads,

$$\frac{E_0}{c} = \gamma m_0 v - \frac{E_f}{c}$$

$$E_0 + m_0c^2 = E_f + \gamma m_0 c^2.$$

These give

$$E_0 + E_f = \gamma m_0 vc$$

$$E_0 - E_f = \gamma m_0 c^2 - m_0 c^2.$$

Thus

$$2E_0 = \gamma m_0 vc + \gamma m_0 c^2 - m_0 c^2$$

$$E_0 = \frac{1}{2} m_0 c^2 \left[\gamma\left(1 + \frac{v}{c}\right) - 1\right].$$

The mass equivalent is

$$\frac{E_0}{c^2} = \frac{1}{2} m_0 \left[\gamma\left(1 + \frac{v}{c}\right) - 1\right].$$

Problem 6-17

Let the photons have a total energy E in the initial rest frame of the rocket. Energy and momentum conservation read

$$M_i c^2 = E + \gamma M_f c^2$$

$$0 = \gamma M_f v - \frac{E}{c}.$$

This gives

$$\gamma M_f v = \frac{E}{c}$$

$$M_i c = \frac{E}{c} + \gamma M_f c.$$

Putting the first equation into the second, we get

$$M_i c = \gamma M_f (c + v)$$

$$\frac{M_i}{M_f} = \frac{1 + \frac{v}{c}}{\sqrt{1 - \frac{v^2}{c^2}}}$$

$$= \left(\frac{1 + \frac{v}{c}}{1 - \frac{v}{c}}\right)^{1/2}.$$

Problem 6.18

The energy of each lump prior to the collision is

$$\frac{mc^2}{\sqrt{1-(3/5)^2}} = \frac{5}{4}mc^2$$

Relativistic energy conservation then reads,

$$\frac{5}{4}mc^2 + \frac{5}{4}mc^2 = Mc^2$$

So,

$$M = \frac{5}{2}m$$

Since $M > 2m$, the kinetic energy has been converted to mass.

Problem 6.19

Relativistic energy and momentum are conserved in the decay,

$$\boldsymbol{p}_\nu = -\boldsymbol{p}_\mu$$

$$E_\mu + E_\nu = m_\pi c^2$$

Since $E_\nu = |\boldsymbol{p}_\nu|c$, we have from momentum conservation $E_\nu = |\boldsymbol{p}_\mu|c$. But $\boldsymbol{p}_\mu^2 c^2 = E_\mu^2 - m_\mu^2 c^2$, so from energy conservation,

$$E_\mu + \sqrt{E_\mu^2 - m_\mu^2 c^2} = m_\pi c^2$$

Solve the quadratic equation,

$$E_\mu = \left(m_\pi^2 + m_\mu^2\right)c^2 \Big/ m_\pi$$

Next, the magnitude of the velocity of the muon is,

$$v_\mu = |p_\mu|c^2/E_\mu = c\left(1 - m_\mu^2 c^4 \Big/ E_\mu^2\right)^{1/2} = c\sqrt{1 - \frac{4m_\pi^2 m_\mu^2}{\left(m_\pi^2 + m_\mu^2\right)^2}}$$

after some algebra.

Problem 6-21

In the rest frame of the K^o meson, the two π mesons are emitted in opposite directions, so that momentum is conserved. We are observing from the lab frame, in which the K^o meson is moving, so the problem is one of relativistic velocity addition. The speed of the meson that is moving in the same direction as the K meson in the lab frame is

$$\begin{aligned} v_{\max} &= \frac{v_1 + v_r}{1 + v_1 v_r / c^2} \\ &= \frac{0.9c + 0.85c}{1 + 0.765} \\ &= 0.992c \end{aligned}$$

and the minimum speed is that of the other pion,

$$\begin{aligned} v_{\min} &= \frac{v_2 + v_r}{1 + v_2 v_r / c^2} \\ &= \frac{0.9c + 0.85c}{1 - 0.765} \\ &= 0.212c. \end{aligned}$$

Problem 6-22

The velocity of B observed by A is, using relativistic addition of velocities,

$$v_r = \frac{2v}{1 + v^2/c^2}.$$

The energy is given by

$$E = \frac{M_0 c^2}{\sqrt{1 - v_r^2/c^2}}$$

$$\begin{aligned} 1 - \frac{v_r^2}{c^2} &= 1 - \frac{4v^2}{c^2\left(1 + v^2/c^2\right)^2} \\ &= \frac{\left(1 + v^2/c^2\right)^2 c^2 - 4v^2}{\left(1 + v^2/c^2\right)^2 c^2} \\ &= \frac{c^2\left(1 - v^2/c^2\right)^2}{c^2\left(1 + v^2/c^2\right)^2}. \end{aligned}$$

So we get

$$E = M_0c^2\frac{1 + v^2/c^2}{1 - v^2/c^2}.$$

Problem 6-23

a. We know that for a relativistic system, energies add, momenta add (vectorially), but rest masses do not add. So the total energy of the two photons is $E_{total} = 200 + 400 = 600$ MeV. The momentum of a photon is given by $p = E/c$, and the total momentum of the system is

$$p_{total} = \sqrt{(400/c)^2 + (200/c)^2} = 200\sqrt{5}\ \text{MeV/c}$$

at an angle $\theta = 63.43$ degrees above the x axis.

b. If a single particle had the total energy and total momentum of the system, it would have a rest mass m given by

$$E_{total}^2 - p_{total}^2c^2 = m^2c^4$$
$$600^2 - 5\cdot 200^2 = m^2c^4,$$

which gives

$$m = 400\ \text{Mev/c}^2.$$

The single particle would travel along the direction making an angle $\theta = 63.43$ degrees above the x axis.

To find the speed, we have

$$\frac{v}{c^2} = \frac{p_{total}}{E_{total}}$$
$$\frac{v}{c} = \frac{p_{total}c}{E_{total}}$$
$$v = \frac{\sqrt{5}}{3}c.$$

Problem 6-25

The total energy is

$$E = m_oc^2 + 3m_oc^2 + 2m_oc^2 = 6m_oc^2.$$

The total momentum is carried solely by the first particle, and this is

$$p^2c^2 = E_1^2 - m_o^2c^4,$$

where $E_1 = m_oc^2 + 3m_oc^2 = 4m_oc^2$.

$$p^2 = 15m_o^2c^2.$$

Now we consider the composite particle.

$$E^2 - p^2c^2 = M_o^2c^4.$$

This gives

$$M_o^2 = 36m_o^2 - 15m_o^2 = 21m_o^2.$$

So, $M_o = \sqrt{21m_o}$.

Problem 6-26

a. Using momentum and energy conservation relations,

$$\frac{E}{c} = \gamma mv$$

$$E + m_oc^2 = \gamma mc^2.$$

Performing the algebra, we get the velocity,

$$\frac{v}{c} = \frac{E}{\gamma mc^2} = \frac{E}{E + m_oc^2}.$$

Next, we calculate $1/\gamma$ in terms of m_o and E:

$$\frac{1}{\gamma} = \sqrt{1 - \frac{v^2}{c^2}} = \frac{\left(2Em_oc^2 + m_o^2c^4\right)^{1/2}}{E + m_oc^2}.$$

Now we get m, the mass of the composite particle:

$$m = \frac{1}{\gamma}\left(m_o + \frac{E}{c^2}\right) = \sqrt{m_o^2 + \frac{2Em_o}{c^2}}.$$

b. The incident particle has a velocity of $v_i = 4c/5$ m/s, and a γ_i given by

$$\gamma_i = \left(\sqrt{1 - 16/25}\right)^{-1} = 5/3.$$

Employing the conservation relations, we have

$$\frac{5}{3}\cdot\frac{4}{5}m_oc = \frac{4}{3}m_oc = \gamma mv$$

$$\frac{5}{3}m_oc^2 + m_oc^2 = \gamma mc^2.$$

Algebra gives

$$\frac{4}{3}m_oc = \gamma mv$$

$$\frac{8}{3}m_o = \gamma m.$$

So,

$$v = \frac{c}{2}$$

$$\gamma = \frac{1}{\sqrt{1 - 1/4}} = \frac{2}{\sqrt{3}},$$

which gives

$$\frac{8}{3}m_o = \frac{2}{\sqrt{3}}m$$

$$m = \frac{4}{\sqrt{3}}m_o.$$

Problem 6-28

a. We have the kinetic energy,

$$K = (\gamma - 1)m_oc^2.$$

The values given are $m_o = 135$ MeV/c^2, $K = 1$ GeV, so

$$10^9 = (\gamma - 1)\cdot 135\cdot 10^6$$

$$\gamma - 1 = 1000/135$$

$$\gamma = \frac{1}{\sqrt{1 - \frac{v^2}{c^2}}} = 8.4074$$

$$v = 2.977.10^8 \text{ m/s}.$$

Let E_1 and E_2 be the energies of the photons, where the photon with E_2 travels in the direction of the incident π meson. We have

$$\gamma m_o v = \frac{E_2}{c} - \frac{E_1}{c}$$
$$\gamma m_o c^2 = E_1 + E_2.$$

We get

$$E_2 + E_1 = \gamma m_o c^2$$
$$E_2 - E_1 = \gamma m_o c v,$$

which gives

$$E_2 = \frac{1}{2}\gamma m_o c(c + v) = 1.131 \ \text{GeV}$$
$$E_1 = \frac{1}{2}\gamma m_o c(c - v) = 3.973 \ \text{MeV}.$$

b. Let θ be the angle between one photon and the incident direction. Then the angle included between the two photons is 2θ. We have

$$\gamma m_o c^2 = 2E$$
$$\gamma m_o v = \frac{2E}{c}\cos\theta.$$

So,

$$\gamma m_o v = \gamma m_o c \cos\theta$$
$$\cos\theta = \frac{v}{c}$$
$$\theta = 6.769 \text{ degrees}$$

and the included angle is $2\theta = 13.539$ degrees.

Problem 6-29

a. We have one photon going forward and one backward. The conservation relations read

$$E_{\bar{p}} + E_p = E_1 + E_2$$
$$1000 + 2 \cdot 938 = E_1 + E_2$$

and

$$p_{\bar{p}} = \frac{1}{c}(E_1 - E_2),$$

where $p_{\bar{p}}$ can be found simply:

$$p_{\bar{p}}^2 c^2 = E_{\bar{p}}^2 - m^2 c^4$$
$$p_{\bar{p}}^2 c^2 = 1938^2 - 938^2$$
$$p_{\bar{p}} c = 1695.9 \text{ MeV}.$$

The two equations give

$$E_1 = 2286 \text{ MeV}$$
$$E_2 = 590 \text{ MeV}.$$

b. We now find the velocity of the antiproton,

$$1938 \text{ MeV} = \frac{938 \text{ Mev}}{\sqrt{1 - \frac{v^2}{c^2}}},$$

which gives

$$v = 0.875c.$$

So the energy of the photons measured in this frame are

$$E_1' = \gamma(E_1 - vp_1) = \gamma\left(E_1 - \frac{v}{c}E_1\right)$$

$$E_1' = \sqrt{\frac{1 - v/c}{1 + v/c}}E_1 = \sqrt{\frac{1 - 0.875}{1 + 0.875}}2286 = 590 \text{ MeV}$$

$$E_2' = \gamma(E_2 - vp_2) = \gamma\left(E_2 + \frac{v}{c}E_2\right)$$

$$E_2' = \sqrt{\frac{1 + v/c}{1 - v/c}}E_1 = \sqrt{\frac{1 + 0.875}{1 - 0.875}}590 = 2286 \text{ MeV}.$$

So, in this frame the energies of the photons switch—which we could have anticipated without calculation!

Problem 6-31

If Q is the energy of the photon, we have

$$Q + \gamma_o m_o c^2 = Q + \gamma_f m_o c^2$$

$$\frac{Q}{c} - \gamma_o m v_o = -\frac{Q}{c} + \gamma_f m_o v_f,$$

where v_f and γ_f refer to the electron after the collision. The first equation gives $\gamma_o = \gamma_f$, which implies $v_o = v_f$. So,

$$\frac{2Q}{c} = 2\gamma_o m_o v_o$$

$$\frac{Q}{c} = \frac{m v_o}{\sqrt{1 - v_o^2/c^2}}$$

$$v_o^2 = \frac{Q^2}{m^2 c^2}\left(1 - \frac{v_o^2}{c^2}\right)$$

$$v_o^2\left(1 + \frac{Q^2}{m_o^2 c^4}\right) = \frac{Q^2}{m_o^2 c^2}$$

$$v_o = \frac{Qc}{\sqrt{Q^2 + m_o^2 c^4}},$$

which is the required answer.

Problem 6-33

Let the momentum of the K meson be p. Because one of the created pions is at rest, the second pion's momentum must be p. The conservation relations read

$$E_K = 137 + E_\pi$$

$$p_K = p_\pi,$$

where the subscript π refers to the moving pion. Squaring the momentum relation equation, multiplying by c^2, and subtracting from the square of the energy conservation equation, we get

$$E_K^2 - p_K^2 c^2 = 137^2 + 2\cdot 137\cdot E_\pi + E_\pi^2 - p_\pi^2 c^2.$$

The left-hand side and the last two terms of the right-hand side are clearly the expressions for rest mass energies $m_K^2 c^4$ and $m_\pi^2 c^4$, which we know, so we get

$$494^2 = 137^2 + 274E_\pi + 137^2 = 37538 + 274E_\pi$$

$$E_\pi = 753.6 \text{ MeV}.$$

The energy of the original K meson is given by

$$E_K = 137 + E_\pi = 890.6 \text{ MeV}.$$

Problem 6-34

The configuration that has the minimum final energy is that in which all the three created particles move in the same direction with equal velocities. (Viewed from the center of momentum frame, this means that all the particles in the final state are created at rest there.) In such a case, the conservation relations read in the lab frame (where E is the gamma-ray energy),

$$E + 0.51 \text{ Mev} = 3E_e,$$

where E_e is the energy of each of the created particles.

$$p_e = \frac{E}{3c}$$

because momenta are shared equally. Combining these equations,

$$E_e^2 = 0.51^2 + p_e^2 c^2 = 0.51^2 + \frac{E^2}{9} = \left(\frac{E + 0.51}{5}\right)^2.$$

We can solve for E:

$$9\cdot\left(0.51^2 + \frac{E^2}{9}\right) = E^2 + 0.51^2 + 1.02E$$

$$E = \frac{8\cdot 0.51^2}{1.02} = 2.04 \text{ MeV}.$$

Problem 6-35

Let us use invariant methods to solve this problem. Let p be the energy–momentum four-vector $p = \left(E/c, \overrightarrow{p}\right)$. The invariant length of p is $p^2 = E^2/c^2 - \overrightarrow{p}^2 = m_o^2c^2$, where we identified the rest mass m_o in the special case that p refers to a single particle. Note that p^2 is the same in all frames, so we could evaluate it in any frame to solve a practical problem.

In this problem, energy–momentum conservation written as a four-vector relation is $p_1 + p_2 = p_3 + p_4 + p_X$ in an obvious notation. Because the four vectors $p_1 + p_2$ and $p_3 + p_4 + p_X$ are equal, so are their invariant lengths,

$$(p_1 + p_2)^2 = (p_3 + p_4 + p_X)^2.$$

Evaluate the left-hand side of this equation in the lab frame and the right-hand side in the center of momentum frame. To obtain the maximum rest mass m_X, all three final-state particles should be produced at rest in the center of momentum frame. So,

$$c^2(p_3 + p_4 + p_X)^2 = \left(m_oc^2 + m_oc^2 + m_Xc^2\right)^2.$$

The left-hand side is easily evaluated in terms of lab quantities,

$$c^2(p_1 + p_2)^2 = \left(E_1 + m_oc^2\right)^2 - c^2\overrightarrow{p}^2 = m_o^2c^4 + 2m_oc^2E_1 + E_1^2 - \overrightarrow{p}_1^2c^2.$$

But $E_1^2 - \overrightarrow{p}_1^2c^2 = m_o^2c^4$, so,

$$c^2(p_1 + p_2)^2 = 2m_o^2c^4 + 2m_oc^2E_1.$$

Combining these two results,

$$2m_o^2c^4 + 2m_oc^2E_1 = \left(2m_oc^2 + m_Xc^2\right)^2,$$

which gives,

$$m_Xc^2 = \sqrt{2m_oc^2(m_oc^2 + E_1)} - 2m_oc^2 = 219 \text{ GeV}$$

after substituting in $m_oc^2 = 0.938 \text{ GeV/c}^2$, and $E_1 = 300 \text{ GeV} + 0.938 \text{ GeV}$.

A nice aspect of this method of solution is that we did not have to calculate any velocities in intermediate steps on the way to determining m_X.

Problem 6-36

a. One photon emerges with energy E_1 at right angles to the incident line of motion. Let the other photon emerge with energy E_2 at an angle θ from the incident line of motion. The total initial energy (in MeV) is 0.511 (rest mass of positron) + 0.511 (kinetic energy of positron) + 0.511 (rest mass of electron). So we get

$$3 \cdot 0.511 = E_1 + E_2.$$

The initial momentum is due to the positron and is in the incident direction. Its value is

$$pc = \sqrt{E^2 - m^2c^4} = \sqrt{(0.511 \cdot 2)^2 - 0.511^2} = 0.511\sqrt{3}\ \text{MeV}.$$

The momentum conservation in the right-angle direction to the incident line of motion is

$$\frac{E_1}{c} = \frac{E_2}{c} \sin\theta$$

and in the incident direction

$$\frac{E_2}{c} \cos\theta = \frac{0.511\sqrt{3}}{c}.$$

Squaring and adding the last two relations ($\sin^2\theta + \cos^2\theta = 1$), we get

$$E_2^2 = 0.511^2 \cdot 3 + E_1^2.$$

Squaring the energy conservation equation, we get

$$E_2^2 = 9 \cdot 0.511^2 - 6 \cdot 0.511 \cdot E_1 + E_1^2.$$

Equating the last two equations, we get

$$E_1 = 0.511\ \text{MeV}.$$

Energy conservation gives

$$E_2 = 3 \cdot 0.511 - E_1 = 1.022\ \text{MeV}.$$

b. The angle is found from the momentum conservation relation,

$$E_1 = E_2 \sin\theta$$

$$\sin\theta = \frac{E_1}{E_2},$$

which gives $\theta = 30$ degrees.

Problem 6-38

a. The binding energy is found from the definition given in the problem:

$$\frac{1}{c^2}\Delta E = m_{Be} - 4m_p - 3m_n$$

$$= 6536 - 4\cdot 938.28 - 3\cdot 939.57 = -35.83\ \text{MeV}/c^2.$$

b. The initial energy is $E_{Be} + E_n = 6536 + 939.57 = 7475.57$ MeV. The energy is split up equally between the alpha particles, so

$$7475.57 = 2(K + mc^2) = 2(K + 3728),$$

which gives

Problem 6.39

a. In the center of momentum, all the final-state particles are produced at rest at the threshold energy, so $2E_{beam} = 4m_pc^2$ which gives $E_{beam} = 2m_pc^2 = 1876$ MeV.

b. Boost the solution to part a. from the center of momentum frame to the lab frame. Consider the invariant,

$$(p_1 + p_2)^2 = (p_3 + p_4 + p_5 + p_6)^2$$

Evaluate the left-hand side in the lab frame and the right-hand side in the center of momentum,

$$\left(E_1 + m_pc^2\right)^2 - \boldsymbol{p}_1^2c^2 = \left(4m_pc^2\right)^2$$

$$E_1^2 + 2m_pc^2E_1 + m_p^2c^4 - \boldsymbol{p}_1^2c^2 = 16m_p^2c^4$$

But $E_1^2 - \boldsymbol{p}_1^2c^2 = m_p^2c^4$, so $E_1 = 7m_pc^2 = 6566$ MeV.

So, much less beam energy is required to produce antiprotons using a collider as compared to a fixed target accelerator.

B.6 CHAPTER 8 PROBLEMS

Problem 8.1

The derivation follows the steps in Section 8.1 except now,

$$x' = \gamma(x - vt)$$

It is convenient to introduce the vector between the charge at time t and the observation point $\boldsymbol{r}$ where $\boldsymbol{E}$ is measured, $\boldsymbol{R} = \boldsymbol{r} - \boldsymbol{v}t$ It makes an angle θ with the velocity $\boldsymbol{v}$, as shown in Fig. B.8.

Then, the desired electric field reads,

$$\boldsymbol{E}(\boldsymbol{r}, t) = kq \frac{1 - v^2/c^2}{(1 - v^2 \sin^2 \theta/c^2)^{3/2}} \frac{\widehat{\boldsymbol{R}}}{R^2}$$

And the magnetic field is which points out of the page as shown in Fig. B.8,

$$\boldsymbol{B} = \frac{1}{c^2}(\boldsymbol{v} \times \boldsymbol{E})$$

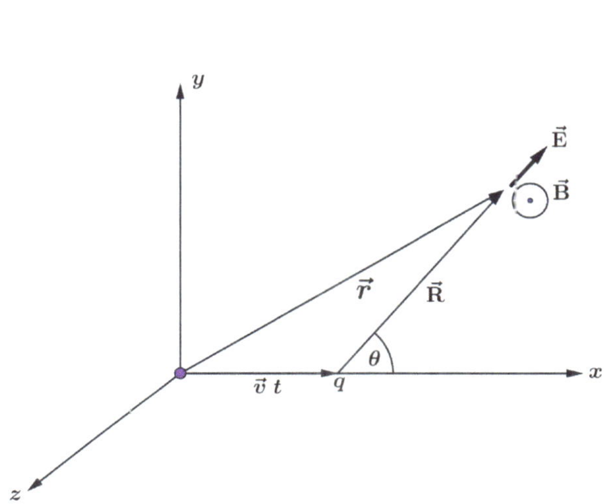

Figure B.8 Coordinate set-up for the calculation of the electric and magnetic fields generated by a moving charge in Problem 8.1.

Problem 8.4

b. You showed in part a, using Gauss's law, that $E_y \sim \sigma$. Under a boost, the x extension of the capacitor contracts by a factor of γ, so $E_y' \sim \sigma' = \gamma\sigma$, so $E_y' = \gamma E_y$.

c. In this case, the area is transverse to the direction of the boost so, $\sigma' = \sigma$ and $E_x' = E_x$.

e. The boosted solenoid is contracted by a factor of γ in the x direction. Since $l' = l/\gamma$ and n is the turns per length, $n' = \gamma n$. But by time dilation, the number of charges passing a point on the wire is dilated by a factor of γ by the boost, so $I' = I/\gamma$. Since Ampere's law gave $B_x \sim nI$ in part d, $B_x' = B_x$.

B.7 CHAPTER 9 PROBLEMS

Problem 9.3

a. $\oint \boldsymbol{E}\cdot d\boldsymbol{l} = -\frac{d}{dt}\oiint \boldsymbol{B}\cdot d\boldsymbol{a}$

The rate of change of the enclosed area is vl. So,

$$V = \oint \boldsymbol{E}\cdot d\boldsymbol{l} = RI = Bvl$$

which gives,

$$I(t) = \frac{Bl}{R}v(t)$$

b. Force $= BlI$, by the Lorentz force law, and it points to the left.

c. Equation of motion of the rod,

$$m\frac{dv}{dt} = -BlI = -\frac{B^2l^2}{R}v$$

So,

$$v(t) = v_0 \exp(-B^2l^2t/mR).$$

d. Power loss $= RI^2(t) = \frac{B^2l^2v_0^2}{R}\exp(-2B^2l^2t/mR)$.

Energy loss

$$= \int_0^{\infty} RI^2 dt = \frac{B^2l^2v_0^2}{R}\int \exp(-2B^2l^2t/mR)dt = \frac{B^2l^2v_0^2}{R}\cdot\frac{mR}{2B^2l^2} = \frac{1}{2}mv_0^2$$

Problem 9.5

There is an electric field at the rim according to Faraday's law,

$$\oint \mathbf{E}\cdot d\mathbf{l} = -\frac{d}{dt}\oiint \mathbf{B}\cdot d\mathbf{a} = -\pi a^2\frac{dB}{dt}$$

A charge element λdl on the rim experiences a force $E(\lambda dl)$ which produces a torque (torque = lever arm × force, which is the rate of change of the angular momentum, dL/dt),

$$\frac{dL}{dt} = b\left(\oint E\lambda dl\right) = -b\lambda\pi a^2\frac{dB}{dt}$$

So, the total angular momentum is

$$L = \int \frac{dL}{dt}dt = -b\lambda\pi a^2\int_0^{\infty}\frac{dB}{dt}dt = b\lambda\pi a^2 B_0$$

where the magnetic field started at B_0 at $t = 0$ and fell to zero. The angular momentum is lever arm × momentum and velocity is lever arm × angular velocity ω, so $L = bMv = bM\omega b = Mb^2\omega$, so the final angular velocity is,

$$\omega = \frac{\lambda\pi a^2 B_0}{Mb}$$

and we have assumed that the mass M is concentrated at the rim, a distance b from the axis of rotation.

B.8 CHAPTER 12 PROBLEMS

Problem 12.8

a. We will use Eq. (12.44) to calculate the Christoffel symbols from the metric,

$$\Gamma^{\mu}_{\rho\beta} = \frac{1}{2}\sum_{\alpha} g^{\mu\alpha}\left(\partial_{\rho}g_{\beta\alpha} + \partial_{\beta}g_{\alpha\rho} - \partial_{\alpha}g_{\rho\beta}\right)$$

Label the variables $1 = r$ and $2 = \theta$ and compute,

$$\Gamma^1_{22} = \frac{1}{2} g^{11}(\partial_2 g_{21} + \partial_2 g_{12} - \partial_1 g_{22}) = -\frac{1}{2}\frac{\partial}{\partial r} r^2 = -r$$

$$\Gamma^2_{12} = \frac{1}{2} g^{22}(\partial_2 g_{21} + \partial_1 g_{22} - \partial_2 g_{12}) = \frac{1}{2r^2}\frac{\partial}{\partial r} r^2 = \frac{1}{r}$$

Similar algebra shows that the other components of the Christoffel symbol vanish.

b. Substituting $\Gamma^1_{22} = -r$ and $u^1 = dr/dt$ and $u^2 = d\theta/dt$ into the first geodesic equation gives,

$$\frac{d^2 r}{dt^2} - r\left(\frac{d\theta}{dt}\right)^2 = -\frac{GM}{r^2}$$

c. The angular momentum is $L = r^2 \frac{d\theta}{dt}$ and its conservation law reads,

$$\frac{d}{dt}\left(r^2 \frac{d\theta}{dt}\right) = 2r\frac{dr}{dt}\frac{d\theta}{dt} + r^2\frac{d^2\theta}{dt^2} = 0$$

which, upon dividing through by r^2, is the second geodesic equation.

Problem 12.9

a. Given the tangent vectors $\boldsymbol{e}_r = \cos\theta \boldsymbol{i} + \sin\theta \boldsymbol{j}$ and $\boldsymbol{e}_\theta = -r\sin\theta \boldsymbol{i} + r\cos\theta \boldsymbol{j}$, calculate the derivatives,

$$\frac{\partial \boldsymbol{e}_r}{\partial r} = 0$$

$$\frac{\partial \boldsymbol{e}_r}{\partial \theta} = -\sin\theta\, \boldsymbol{i} + \cos\theta\, \boldsymbol{j} = \frac{1}{r}\, \boldsymbol{e}_\theta$$

$$\frac{\partial \boldsymbol{e}_\theta}{\partial r} = -\sin\theta \boldsymbol{i} + \cos\theta \boldsymbol{j} = \frac{1}{r}\boldsymbol{e}_\theta$$

$$\frac{\partial \boldsymbol{e}_\theta}{\partial \theta} = -r\cos\theta \boldsymbol{i} - r\sin\theta \boldsymbol{j} = -r\boldsymbol{e}_r$$

b. Write out $\partial_\beta \boldsymbol{e}_\alpha = \sum_\gamma \Gamma^\gamma_{\alpha\beta} \boldsymbol{e}_\gamma$,

$$\frac{\partial \boldsymbol{e}_r}{\partial r} = 0 = \Gamma^r_{rr}\boldsymbol{e}_r + \Gamma^\theta_{rr}\boldsymbol{e}_\theta, \quad \text{so} \quad \Gamma^r_{rr} = 0, \quad \Gamma^\theta_{rr} = 0$$

$$\frac{\partial \boldsymbol{e}_r}{\partial \theta} = \frac{1}{r}\boldsymbol{e}_\theta = \Gamma^r_{\theta r}\boldsymbol{e}_r + \Gamma^\theta_{\theta r}\boldsymbol{e}_\theta, \text{ so } \Gamma^r_{\theta r} = 0, \quad \Gamma^\theta_{\theta r} = \frac{1}{r}$$

$$\frac{\partial \boldsymbol{e}_\theta}{\partial r} = \frac{1}{r}\boldsymbol{e}_\theta = \Gamma^r_{r\theta}\boldsymbol{e}_r + \Gamma^\theta_{r\theta}\boldsymbol{e}_\theta, \text{ so } \Gamma^r_{r\theta} = 0, \quad \Gamma^\theta_{r\theta} = \frac{1}{r}$$

$$\frac{\partial \boldsymbol{e}_\theta}{\partial \theta} = -r\boldsymbol{e}_r = \Gamma^r_{\theta\theta}\boldsymbol{e}_r + \Gamma^\theta_{\theta\theta}\boldsymbol{e}_\theta, \text{ so } \Gamma^r_{\theta\theta} = -r, \quad \Gamma^\theta_{\theta\theta} = 0$$

Problem 12.11

$$\Gamma^\alpha_{\mu\nu} = \frac{1}{2}\sum_\beta g^{\alpha\beta}\left(\partial_\mu g_{\nu\beta} + \partial_\nu g_{\beta\mu} - \partial_\beta g_{\mu\nu}\right)$$

a. Differentiate $\Gamma^\alpha_{\mu\nu}$, then set $\partial_\sigma g^{\alpha\beta}(P) = 0$ and find,

$$\partial_\sigma \Gamma^\alpha_{\mu\nu} = \frac{1}{2}\sum_\beta g^{\alpha\beta}\left(\partial_\sigma\partial_\mu g_{\nu\beta} + \partial_\sigma\partial_\nu g_{\beta\mu} - \partial_\sigma\partial_\beta g_{\mu\nu}\right)$$

b. Since $\Gamma^\alpha_{\mu\nu}(P) = 0$, the expression for the Riemann tensor reduces to $R^\alpha_{\beta\mu\nu} = \partial_\mu \Gamma^\alpha_{\nu\beta} - \partial_\nu \Gamma^\alpha_{\mu\beta}$. Substituting the result of part a and noting two cancelling terms, find,

$$R^\alpha_{\beta\mu\nu} = \frac{1}{2}\sum_\sigma g^{\alpha\sigma}\left(\partial_\beta\partial_\mu g_{\sigma\nu} - \partial_\beta\partial_\nu g_{\sigma\mu} + \partial_\sigma\partial_\nu g_{\beta\mu} - \partial_\sigma\partial_\mu g_{\beta\nu}\right)$$

$$R_{\alpha\beta\mu\nu} = \sum_\lambda g_{\alpha\lambda}R^\lambda_{\beta\mu\nu} = \frac{1}{2}\left(\partial_\beta\partial_\mu g_{\alpha\nu} - \partial_\beta\partial_\nu g_{\alpha\mu} + \partial_\alpha\partial_\nu g_{\beta\mu} - \partial_\alpha\partial_\mu g_{\beta\nu}\right)$$

Problem 12.14

a. $\left(1 - \frac{2GM}{c^2 r}\right)^{-1} dr^2 = \lambda^2 d\rho^2 = \frac{r^2}{\rho^2}d\rho^2$. Set $k = \frac{2GM}{c^2}$ and integrate,

$$\int \frac{dr}{\sqrt{r}\sqrt{r-k}} = \int \frac{d\rho}{\rho}$$

$$2\log\left|\left(\sqrt{r-k} + \sqrt{r}\right)/2\right| = \log\rho$$

where the constant of integration was chosen so that $r \to \rho$ for large r. So,

$$\frac{\left(\sqrt{r-k}+\sqrt{r}\right)}{2}=\sqrt{\rho}$$

Solving for r,

$$r=\rho\left(1+\frac{k}{4\rho}\right)^2=\rho\left(1+\frac{GM}{2c^2\rho}\right)^2$$

Problem 12.15

a. We have the transformation law $\widetilde{g}'_{\mu\nu}=\widetilde{g}_{\mu\nu}(x)-\partial_\mu\varepsilon_\nu-\partial_\nu\varepsilon_\mu+O(\varepsilon^2)$ and want to compute the transformation law for $\overline{h}_{\mu\nu}=\widetilde{g}_{\mu\nu}-\frac{1}{2}g^{(0)}_{\mu\nu}h$ where $h=\sum_\alpha\widetilde{g}^\alpha_\alpha$.

Compute,

$$\overline{h}'_{\mu\nu}=\widetilde{g}'_{\mu\nu}-\frac{1}{2}g^{(0)}_{\mu\nu}h'=\widetilde{g}_{\mu\nu}(x)-\partial_\mu\varepsilon_\nu-\partial_\nu\varepsilon_\mu-\frac{1}{2}g^{(0)}_{\mu\nu}h'$$

We need,

$$\begin{aligned}h'&=\sum_\alpha\widetilde{g}'^\alpha_\alpha=\sum_{\alpha\mu}g^{(0)\alpha\mu}\widetilde{g}'_{\mu\alpha}\\&=\sum_{\alpha\mu}g^{(0)\alpha\mu}\left(\widetilde{g}_{\mu\alpha}-\partial_\mu\varepsilon_\alpha-\partial_\alpha\varepsilon_\mu\right)\\&=\sum_\alpha\left(\widetilde{g}^\alpha_\alpha-\partial^\alpha\varepsilon_\alpha-\partial^\alpha\varepsilon_\alpha\right)=h-2\sum_\alpha\partial^\alpha\varepsilon_\alpha\end{aligned}$$

Collecting these two results,

$$\overline{h}'_{\mu\nu}=\overline{h}_{\mu\nu}-\partial_\mu\varepsilon_\nu-\partial_\nu\varepsilon_\mu+g^{(0)}_{\mu\nu}\left(\sum_\alpha\partial^\alpha\varepsilon_\alpha\right)$$

Now we can compute the transformation law of the divergence of $\overline{h}_{\mu\nu}$,

$$\sum_\nu\partial^\nu\overline{h}'_{\mu\nu}=\sum_\nu\partial^\nu\overline{h}_{\mu\nu}-\sum_\nu\partial^\nu\partial_\mu\varepsilon_\nu-\sum_\nu\partial^\nu\partial_\nu\varepsilon_\mu+\sum_\nu g^{(0)}_{\mu\nu}\partial^\nu\left(\sum_\alpha\partial^\alpha\varepsilon_\alpha\right)$$

But $\partial_\mu=\sum_\nu g^{(0)}_{\mu\nu}\partial^\nu$, so the second and fourth terms on the right-hand side cancel, leaving,

$$\sum_\nu\partial_\nu\overline{h}'^{\mu\nu}=\sum_\nu\partial_\nu\overline{h}^{\mu\nu}-\sum_\nu\partial^\nu\partial_\nu\varepsilon^\mu$$

Problem 12.16

a. The same analysis required here has already been done in Problem 12.11 part b and 12.12 part a. The terms in the Riemann tensor which are quadratic in the Christoffel symbols do not contribute here because they are second order in $\widetilde{g}_{\mu\nu}$. Since the Minkowski metric in Cartesian coordinates are constants, they drop out as well leaving only terms which vary as the second derivatives of $\widetilde{g}_{\mu\nu}$. These were written out explicitly in Problem 12.12 part a.

b. Begin with the transformed Riemann tensor,

$$R'_{\alpha\beta\gamma\delta} = \frac{1}{2}\left(\partial_\beta\partial_\gamma \widetilde{g}'_{\alpha\delta} + \partial_\alpha\partial_\delta \widetilde{g}'_{\beta\gamma} - \partial_\beta\partial_\delta \widetilde{g}'_{\alpha\gamma} - \partial_\alpha\partial_\gamma \widetilde{g}'_{\beta\delta}\right)$$

where $\widetilde{g}'_{\mu\nu} = \widetilde{g}_{\mu\nu}(x) - \partial_\mu\varepsilon_\nu - \partial_\nu\varepsilon_\mu$. Substituting into the expression for $R'_{\alpha\beta\gamma\delta}$, inspect the terms that depend on $\partial_\mu\varepsilon_\nu$ and note that they all cancel(!) leaving,

$$R'_{\alpha\beta\gamma\delta} = R_{\alpha\beta\gamma\delta}$$

c. $\overline{h}_{\mu\nu} = \widetilde{g}_{\mu\nu} - \frac{1}{2}g^{(0)}_{\mu\nu}h$, so $\sum_\mu \overline{h}^\mu_\mu = \sum_\mu \widetilde{g}^\mu_\mu - \frac{1}{2}\sum_\mu g^{(0)\mu}_\mu h$ which reads

$\overline{h} = h - \frac{1}{2}(4)h = -h$ since $\sum_\mu g^{(0)\mu}_\mu = 1+1+1+1 = 4$.

So, $\widetilde{g}_{\mu\nu} = \overline{h}_{\mu\nu} + \frac{1}{2}g^{(0)}_{\mu\nu}h = \overline{h}_{\mu\nu} - \frac{1}{2}g^{(0)}_{\mu\nu}\overline{h}$.

d. To calculate the Einstein tensor to first order in $\widetilde{g}_{\mu\nu}$: we need the Ricci tensor and the Ricci scalar to first order. Using the result of Problem 12.11 part b or 12.12 part a,

$$R_{\beta\delta} = \frac{1}{2}\sum_\gamma\left(\partial_\beta\partial_\gamma \widetilde{g}^\gamma_\delta + \partial^\gamma\partial_\delta \widetilde{g}_{\beta\gamma} - \partial_\beta\partial_\delta \widetilde{g}^\gamma_\gamma - \partial^\gamma\partial_\gamma \widetilde{g}_{\beta\delta}\right)$$

$$R = \sum_\beta R^\beta_\beta = \frac{1}{2}\sum_\beta\left(\partial^\beta\partial_\gamma \widetilde{g}^\gamma_\beta + \partial^\gamma\partial_\beta \widetilde{g}^\beta_\gamma - \partial^\beta\partial_\beta \widetilde{g}^\gamma_\gamma - \partial^\gamma\partial_\gamma \widetilde{g}^\beta_\beta\right)$$

$$= \sum_{\beta\gamma}\left(\partial^\beta\partial_\gamma \widetilde{g}^\gamma_\beta - \partial^\beta\partial_\beta \widetilde{g}^\gamma_\gamma\right)$$

To first order the Einstein tensor is,

$$G_{\beta\delta} = R_{\beta\delta} - \frac{1}{2}Rg^{(0)}_{\beta\delta}$$

Our final chore is to substitute in $\widetilde{g}_{\mu\nu} = \overline{h}_{\mu\nu} - \frac{1}{2}g^{(0)}_{\mu\nu}\overline{h}$, and collect terms and find,

$$G_{\alpha\beta} = -\frac{1}{2}\left(\sum_{\mu}\partial^{\mu}\partial_{\mu}\overline{h}_{\alpha\beta} + g^{0}_{\alpha\beta}\sum_{\mu\nu}\partial^{\mu}\partial^{\nu}\overline{h}_{\mu\nu} - \sum_{\mu}\partial^{\mu}\partial_{\beta}\overline{h}_{\alpha\mu} - \sum_{\mu}\partial^{\mu}\partial_{\alpha}\overline{h}_{\beta\mu}\right)$$

Problem 12.17

We found in the text that in the Newtonian limit, $\overline{h}^{00} = 4\Phi/c^2$ and the other components of $\overline{h}^{\alpha\beta}$ are negligible, $\overline{h}^{\alpha\beta} \approx 0$.

Recall that $\overline{h}_{\mu\nu} = \widetilde{g}_{\mu\nu} - \frac{1}{2}g^{(0)}_{\mu\nu}h$ and $h = \sum_{\alpha}\widetilde{g}^{\alpha}_{\alpha}$.

So, $\overline{h}_{11} = \widetilde{g}_{11} - \frac{1}{2}g^{(0)}_{11}h \approx 0$ and similarly for 22 and 33. Since $g^{(0)}_{11} = -1$, we learn that $h = -2\widetilde{g}_{11} = -2\widetilde{g}_{22} = -2\widetilde{g}_{33}$. In addition,

$$h \equiv \widetilde{g}^{0}_{0} + \widetilde{g}^{1}_{1} + \widetilde{g}^{2}_{2} + \widetilde{g}^{3}_{3} = \widetilde{g}_{00} - \widetilde{g}_{11} - \widetilde{g}_{22} - \widetilde{g}_{33} = \widetilde{g}_{00} - 3\widetilde{g}_{11}$$

Since $h = -2\widetilde{g}_{11}$, we have $\widetilde{g}_{00} = \widetilde{g}_{11}$.

Finally, $\overline{h}_{00} = \frac{4\Phi}{c^2} = \widetilde{g}_{00} - \frac{1}{2}g^{(0)}_{00}h = -\frac{1}{2}h - \frac{1}{2}h = -h$. So,
$h = -4\Phi/c^2$, $\widetilde{g}_{00} = \widetilde{g}_{11} = \widetilde{g}_{22} = \widetilde{g}_{33} = 2\Phi/c^2$

Problem 12.19

a. $g_{00} = 1 + 2\Phi/c^2$ and $g_{ii} = -1 + 2\Phi/c^2$ and $\Gamma^{\alpha}_{\mu\nu} = \frac{1}{2}\sum_{\beta}g^{\alpha\beta}\left(\partial_{\mu}g_{\nu\beta} + \partial_{\nu}g_{\beta\mu} - \partial_{\beta}g_{\mu\nu}\right)$. So, the Christoffel symbols that dominate the geodesic equation in the Newtonian limit are,

$$\Gamma^{0}_{00} = \frac{1}{2}g^{00}(\partial_0 g_{00} + \partial_0 g_{00} - \partial_0 g_{00}) = \frac{1}{2}\partial_0 g_{00} = \frac{1}{2}\frac{\partial}{\partial ct}\left(\frac{2\Phi}{c^2}\right) = \frac{\partial\Phi}{c^2\partial ct}$$

$$\Gamma^{i}_{00} = \frac{1}{2}g^{ii}(\partial_0 g_{0i} + \partial_0 g_{i0} - \partial_i g_{00}) = -\frac{1}{2}\frac{\partial}{\partial x^i}\left(\frac{2\Phi}{c^2}\right) = \frac{\partial\Phi}{c^2\partial x^i}$$

b. For nonrelativistic motion, $dx^0/cd\tau \approx 1$ and $dx^i/cd\tau \ll 1$, so the geodesic equation becomes,

$$\frac{d^2x^0}{c^2d\tau^2} + \Gamma^{0}_{00} \approx \frac{1}{c^2}\frac{d}{dt}\left(\frac{dx^0}{d\tau} + \Phi\right) \approx 0$$

Multiply through by the mass of the particle and identify the particle's total energy, $\frac{d}{dt}(E + \Phi) \approx 0$, which expresses energy conservation in a gravitational potential.

The geodesic equation for the spatial components of the particle's position reads

$$\frac{d^2x^i}{c^2d\tau^2} + \Gamma^i_{00} \approx \frac{1}{c^2}\left(\frac{d}{dt}\frac{dx^i}{d\tau} + \frac{\partial \Phi}{\partial x^i}\right) \approx 0$$

Again, multiply through by m,

$$\frac{d}{dt}\boldsymbol{p} + m\nabla\Phi \approx 0$$

and we have Newton's second law.

Problem 12.20

a. The centripetal force on particle 2 is provided by its attraction to particle 1,

$$\frac{v_2^2}{r_2} = \frac{Gm_1}{r^2}$$

But,

$$\frac{v_2^2}{r_2} = r_2\omega^2 = \frac{m_1}{m_1 + m_2}r\omega^2$$

where we used the center of mass condition, $m_1r_1 = m_2r_2$, to calculate r_2 in terms of r and the particle masses,

$$r = r_2 + r_1 = r_2 + \frac{m_2}{m_1}r_2 = r_2\frac{m_1 + m_2}{m_1}$$

Now,

$$\frac{m_1}{m_1 + m_2}r\omega^2 = \frac{Gm_1}{r^2}$$

which can be solved for ω,

$$\omega = \sqrt{\frac{G(m_1 + m_2)}{r^3}}$$

b. Using the kinematics of part a, simplify the total energy of the system,

$$E = \frac{1}{2}m_1 v_1^2 + \frac{1}{2}m_2 v_2^2 - \frac{Gm_1 m_2}{r} = \frac{1}{2}m_1 \frac{Gm_2}{r^2} r_1 + \frac{1}{2}m_2 \frac{Gm_1}{r^2} r_2 - \frac{Gm_1 m_2}{r}$$

So,

$$E = \frac{1}{2}\frac{Gm_1 m_2}{r^2}(r_1 + r_2) - \frac{Gm_1 m_2}{r} = -\frac{Gm_1 m_2}{2r}$$

c. Differentiate the result of part b,

$$\frac{dE}{dt} = \frac{Gm_1 m_2}{2r^2}\frac{dr}{dt} = -\frac{32G^4}{5c^5 r^5} m_1^2 m_2^2 (m_1 + m_2)$$

So,

$$\frac{dr}{dt} = -\frac{64G^3 m_1 m_2}{5c^5 r^3}(m_1 + m_2)$$

Problem 12.24

a. Write down energy conservation when the particle is at distance r with velocity v and when it is at distance $r \to \infty$ where $v \to 0$,

$$\frac{1}{2}mv^2 - \frac{GmM}{r} = 0$$

Solving for the velocity,

$$v = \left(\frac{2GM}{r}\right)^{1/2} = \left(\frac{2GM}{c^2}\frac{c^2}{r}\right)^{1/2} = c\sqrt{\frac{r_{Sch}}{r}}$$

b. The acceleration of gravity is $g = \frac{GM}{r^2}$. The Schwarzschild radius is $r_{Sch} = \frac{2GM}{c^2}$. So, according to Newtonian mechanics, the acceleration of gravity at the Schwarzschild radius of a body of mass M is

$$g = \frac{GM}{4G^2M^2/c^4} = \frac{c^4}{4GM}$$

For a black hole having Earth's mass, substitute the parameters from Appendix A, giving $g \approx 0.5 \times 10^{45}$ m/s^2, which should be compared to the acceleration of gravity at the earth's surface, 9.8 m/s^2

c. The Newtonian potential energy of a particle of mass m at the Schwarzschild radius of a black hole of mass M is,

$$-\frac{GMm}{r_{Sch}} = -GMm\frac{c^2}{2GM} = -\frac{1}{2}mc^2$$

So, an amount of work (energy) $\frac{1}{2}mc^2$ is required to move the particle to large r.

APPENDIX C

Mathematics Background

C.1 HANDY APPROXIMATIONS AND EXPANSIONS

There is one approximation we use again and again in this book. In its simplest form it reads,

$$\frac{1}{1-x} \approx 1 + x + O(x^2) \tag{C.1}$$

This equation is useful when $x \ll 1$ so that the second-order correction, denoted $O(x^2)$ in Eq. (C.1), is numerically much smaller than x itself.

The simplest proof of Eq. (C.1) consists in just multiplying through by $1 - x$, noting that the product $(1 + x)\,(1 - x)$ is $1 - x^2$, which differs from 1 by terms of second order. Although Eq. (C.1) can be derived using differential calculus by invoking Taylor's theorem, all that power is not needed. Simple algebra is enough.

For example, let $x = 0.01$. Then Eq. (C.1) reads that $1/0.99$ is well approximated by 1.01, with an error of order 0.0001. Clearly the linear approximation that Eq. (C.1) gives makes it very handy.

The exact equality, an infinite expansion, behind Eq. (C.1) is

$$\frac{1}{1-x} = 1 + x + x^2 + x^3 + x^4 + \cdots \tag{C.2}$$

where $|x| < 1$ to guarantee convergence.

Another expansion we use in this book, especially when we face boosts with $v/c \ll 1$, is

$$\sqrt{1-x} \approx 1 - \frac{x}{2} + O(x^2) \tag{C.3}$$

To prove this one, just square both sides.

Using these bits of algebra, the approximations used in the book for γ when $v/c \ll 1$ follow. For example,

$$\frac{1}{\sqrt{1 - v^2/c^2}} \approx 1 + \frac{v^2}{2c^2} + O\left(\frac{v^4}{c^4}\right)$$

The general result here is the binomial expansion,

$$(1+x)^n = 1 + nx + \frac{n(n-1)}{2}x^2 + \frac{n(n-1)(n-2)}{3\cdot 2}x^3 + \cdots$$

C.2 CHAIN RULE OF DIFFERENTIATION AND COORDINATE TRANSFORMATIONS: APPLICATIONS TO COVARIANT AND CONTRAVARIANT VECTORS

Suppose $y(x) = f(g(x))$ and we want the slope of y on the x axis. We do this through the "chain rule," which reads,

$$\frac{dy}{dx} = \frac{df}{dg}\frac{dg}{dx} \tag{C.4}$$

The extension of the chain rule to functions of several variables is also used in the text. Suppose $F(s, t) = F(x(s, t), y(s, t))$. Then,

$$\frac{\partial F}{\partial s} = \frac{\partial F}{\partial x}\frac{\partial x}{\partial s} + \frac{\partial F}{\partial y}\frac{\partial y}{\partial s}$$

$$\frac{\partial F}{\partial t} = \frac{\partial F}{\partial x}\frac{\partial x}{\partial t} + \frac{\partial F}{\partial y}\frac{\partial y}{\partial t} \tag{C.5}$$

Throughout the text we consider coordinate transformations and need to transform derivatives and differentials between frames. Suppose,

$$y^1 = g^1(x^1, x^2, \ldots, x^m)$$

$$y^2 = g^2(x^1, x^2, \ldots, x^m)$$

$$y^n = g^n(x^1, x^2, \ldots, x^m) \tag{C.6}$$

Then,

$$\frac{\partial}{\partial x^i} = \sum_{j=1}^{n} \frac{\partial y^j}{\partial x^i}\frac{\partial}{\partial y^j} \tag{C.7}$$

where i ranges between 1 and m.

For differentials,

$$dy^j = \sum_{i=1}^{m} \frac{\partial g^j}{\partial x^i} dx^i \tag{C.8}$$

Suppose there is one coordinate mesh x^a in space–time and a mapping to the second one $x'^b = x'^b(x^a)$. Both meshes cover the space–time without singularities and the mapping is one-to-one and invertible. If the mapping is a linear transformation like a boost in special relativity then,

$$x'^a = \sum_b L^a_b x^b \text{ and } \frac{\partial x'^a}{\partial x^b} = L^a_b$$

as discussed in the text.

In the general case, imagine transforming from x^a to x'^b and back to x^a. We can express this sequence of mappings using the chain rule,

$$\sum_b \frac{\partial x^a}{\partial x'^b}\frac{\partial x'^b}{\partial x^c} = \frac{\partial x^a}{\partial x^c} = \delta^a_c$$

where δ^a_c is the kronecker symbol, $\delta^a_a = 1$ and $\delta^a_b = 0$ if $a \neq b$.

These partial derivatives determine the mappings of vectors and tensors between the two coordinate systems $x'^b(x^a)$ and x^a. For example, consider a curve $C(\tau)$ in space–time and its tangent field $u^a = dx^a/d\tau$. We can also calculate the tangent field using the second coordinate system $x'^b(x^a)$, $u'^b = dx'^b/d\tau$. Using the chain rule,

$$\frac{dx'^a}{d\tau} = \sum_b \frac{\partial x'^a}{\partial x^b}\frac{dx^b}{\partial \tau}$$

the transformation rule for the vector field becomes,

$$u'^a = \sum_b \frac{\partial x'^a}{\partial x^b} u^b \tag{C.9}$$

Fields with upper indices are called "contravariant" and Eq. (C.9) gives their transformation law. See the text for further discussion and examples.

Another example comes from the gradient of a scalar function,

$$F_a = \partial_a f = \frac{\partial f}{\partial x^a}$$

The chain rule gives,

$$\frac{\partial f}{\partial x^a} = \sum_b \frac{\partial f}{\partial x'^b}\frac{\partial x'^b}{\partial x^a}$$

or,

$$F_a = \sum_b \frac{\partial x'^b}{\partial x^a} F'_b \tag{C.10a}$$

which can be inverted,

$$F'_b = \sum_a \frac{\partial x^a}{\partial x'^b} F_a \tag{C.10b}$$

Fields with lower indices are called "covariant" and Eq. (C.10) gives their transformation law.

The inner product of contravariant and covariant vector fields are invariant under coordinate transformations. We have illustrated this point throughout the text. This follows from Eqs. (C.9) and (C.10),

$$\begin{aligned}\sum_a F'_a T'^a &= \sum_a \left(\sum_b \frac{\partial x^b}{\partial x'^a} F_b \right) \left(\sum_c \frac{\partial x'^a}{\partial x^c} T^c \right) = \sum_{abc} \frac{\partial x^b}{\partial x'^a} \frac{\partial x'^a}{\partial x^c} F_b T^c \\ &= \sum_{bc} \delta^b_c F_b T^c = \sum_c F_c T^c\end{aligned}$$

We showed in the text that second-rank covariant tensors such as the space—time metric transform as,

$$g'_{cd} = \sum_{ab} \frac{\partial x^a}{\partial x'^c} \frac{\partial x^b}{\partial x'^d} g_{ab}$$

and second-rank contravariant tensors transform as,

$$T'^{ab} = \sum_{cd} \frac{\partial x'^a}{\partial x^c} \frac{\partial x'^b}{\partial x^d} T^{cd}$$

Tensors that have one contravariant index and one covariant index transform as,

$$M'^a_b = \sum_{cd} \frac{\partial x'^a}{\partial x^c} \frac{\partial x^d}{\partial x'^b} M^c_d$$

The trace of such a tensor is a scalar,

$$\sum_a M'^a_a = \sum_{acd} \frac{\partial x'^a}{\partial x^c} \frac{\partial x^d}{\partial x'^a} M^c_d = \sum_{cd} \delta^d_c M^c_d = \sum_c M^c_c$$

We can invert the mapping in Eq. (C.9) by multiplying with $\frac{\partial x^c}{\partial x'^a}$ and summing over the index a,

$$\sum_a \frac{\partial x^c}{\partial x'^a} u'^a = \sum_{ab} \frac{\partial x^c}{\partial x'^a} \frac{\partial x'^a}{\partial x^b} u^b = \sum_b \delta^c_b u^b = u^c$$

So,

$$u^c = \sum_a \frac{\partial x^c}{\partial x'^a} u'^a$$

which we recognize as just a relabeling of Eq. (C.9).

C.3 MATRICES

Let us recall some features of linear systems and illustrate the general case with 2 × 2 matrices.

This is just a refresher of material you have studied before with an emphasis on indices in preparation for the mathematics of relativity in various coordinate systems.

We start with a linear transformation,

$$v_1 = a_{11} w_1 + a_{12} w_2$$

$$v_2 = a_{21} w_1 + a_{22} w_2$$

which we write in matrix notation,

$$\begin{pmatrix} v_1 \\ v_2 \end{pmatrix} = \begin{pmatrix} a_{11} & a_{12} \\ a_{21} & a_{22} \end{pmatrix} \begin{pmatrix} w_1 \\ w_2 \end{pmatrix} = \begin{pmatrix} a_{11} w_1 + a_{12} w_2 \\ a_{21} w_1 + a_{22} w_2 \end{pmatrix}$$

or writing out the indices,

$$v_i = \sum_j a_{ij} w_j$$

Note that the first index labels the row of the matrix and the second index labels the column.

We refer to $\begin{pmatrix} v_1 \\ v_2 \end{pmatrix}$ as vectors and $\begin{pmatrix} a_{11} & a_{12} \\ a_{21} & a_{22} \end{pmatrix}$ as matrices.

Recall the rule for successive transformations: if $v_i = \sum_j a_{ij} w_j$ and $w_j = \sum_k b_{jk} y_k$, then a short calculation gives $v_i = \sum_k c_{ik} y_k$ where c_{ik} is given by matrix multiplication $c_{ij} = \sum_k a_{ik} b_{kj}$. We write simply, $c = ab$ and must keep the order of the matrices since matrix multiplication is not commutative.

Some useful definitions:

a. A matrix a is symmetric if $a_{ij} = a_{ji}$.
b. A matrix a is antisymmetric if $a_{ij} = -a_{ji}$.
c. The transpose of a matrix interchanges rows and columns.
d. The identity matrix is $I_{ij} = \delta_{ij}$, where δ_{ij} is the "kronecker" symbol, $\delta_{ij} = 1$ if $i = j$ and zero otherwise.
e. If we multiply a matrix by a constant c, then every element in the matrix is multiplied by c.
f. The determinant of a matrix is,

$$\det(a) = \det\begin{pmatrix} a_{11} & a_{12} \\ a_{21} & a_{22} \end{pmatrix} = a_{11}a_{22} - a_{12}a_{12}$$

g. The inverse of a matrix a is denoted a^{-1} and is given by

$$a^{-1} = \frac{1}{\det(a)}\begin{pmatrix} a_{22} & -a_{12} \\ -a_{21} & a_{11} \end{pmatrix}$$

Then $a^{-1}a = aa^{-1} = I$. Also, $\det\left(a^{-1}\right) = 1/\det\,(a)$

h. If a and b are two matrices, then $\det\,(ab) = \det\,(a)\,\det\,(b)$

Let us illustrate some matrix manipulations that we will need in the text. Consider a Lorentz transformation, which is introduced in matrix form in Chapter 8,

$$L^{\mu}{}_{\nu} = \begin{pmatrix} \gamma & -\dfrac{v}{c}\gamma \\ -\dfrac{v}{c}\gamma & \gamma \end{pmatrix} = \gamma\begin{pmatrix} 1 & -\dfrac{v}{c} \\ -\dfrac{v}{c} & 1 \end{pmatrix}$$

Calculate,

$$\det\left(L^{\mu}{}_{\nu}\right) = \gamma^2\left(1 - \frac{v^2}{c^2}\right) = 1$$

The inverse of $L^{\mu}{}_{\nu}$ is,

$$(L^{-1})^{\mu}{}_{\nu} = \gamma\begin{pmatrix} 1 & \dfrac{v}{c} \\ \dfrac{v}{c} & 1 \end{pmatrix}$$

This is clearly correct: the inverse is a boost by $-v$ along the x axis, so it "undoes" the original boost by $+v$ along the x axis. We can check this by matrix multiplication,

$$\gamma^2 \begin{pmatrix} 1 & -\frac{v}{c} \\ -\frac{v}{c} & 1 \end{pmatrix} \begin{pmatrix} 1 & \frac{v}{c} \\ \frac{v}{c} & 1 \end{pmatrix} = \gamma^2 \begin{pmatrix} 1-\frac{v^2}{c^2} & 0 \\ 0 & 1-\frac{v^2}{c^2} \end{pmatrix} = \begin{pmatrix} 1 & 0 \\ 0 & 1 \end{pmatrix}$$

C.4 COVARIANT AND CONTRAVARIANT FOUR-VECTORS AND DUALITY

Covariant and contravariant vectors were introduced in the text. The reader may find the material of this appendix useful to motivate the discussion in Chapter 6, Section 5, for example. Let us discuss the geometry underlying covariant and contravariant vectors.

Consider a two-dimensional surface parametrized by a two-dimensional mesh, $u^i=(u^1, u^2) = (u,v)$. The surface lies in a three-dimensional Euclidean space. A point on the surface can be labeled with either the Cartesian coordinates $\boldsymbol{r}$ or the coordinates (u^1, u^2) on the mesh. For a point P on the surface, $\boldsymbol{r}$ is a function of (u^1, u^2). The tangent plane at the point P can be spanned by the basis $\boldsymbol{e}_i = \partial\boldsymbol{r}/\partial u^i$. The basis vector $\boldsymbol{e}_1 = \partial\boldsymbol{r}/\partial u$ is tangent to a curve of constant v and $\boldsymbol{e}_2 = \partial\boldsymbol{r}/\partial v$ is tangent to a curve of constant u. These basis vectors need not be orthogonal at P and, in fact, their magnitudes and the angle between them can vary as P moves throughout the mesh on the surface.

There is another natural basis tangent to the surface at point P: instead of considering vectors along curves of constant u and v, we could choose vectors in the directions where u and v change at their maximal rates, ∇u and ∇v. Call this basis $\boldsymbol{e}^i = \nabla u^i$, with upper or contravariant indices.

Let us check that the bases are "dual" and establish an inner product,

$$\boldsymbol{e}^i \cdot \boldsymbol{e}_j = \delta^i_j$$

So, we should calculate,

$$\boldsymbol{e}^i \cdot \boldsymbol{e}_j = \nabla u^i \cdot \partial\boldsymbol{r}/\partial u^j = \sum_m \frac{\partial u^i}{\partial r^m}\frac{\partial r^m}{\partial u^j} = \frac{\partial u^i}{\partial u^j} = \delta^i_j$$

where we used the chain rule from Section C.2 in the last step. Success!

This property gives the inner product. Consider vectors $\boldsymbol{A}$ and $\boldsymbol{B}$ in the tangent plane at point P. Write $\boldsymbol{A} = A^1\boldsymbol{e}_1 + A^2\boldsymbol{e}_2$ using the $\{\boldsymbol{e}_i\}$ basis and $\boldsymbol{B} = B_1\boldsymbol{e}^1 + B_2\boldsymbol{e}^2$ in the $\{\boldsymbol{e}^i\}$ basis. Then,

$$\boldsymbol{A}\cdot\boldsymbol{B} = (A^1\boldsymbol{e}_1 + A^2\boldsymbol{e}_2)\cdot(B_1\boldsymbol{e}^1 + B_2\boldsymbol{e}^2) = A^1B_1 + A^2B_2$$

where we used. $\boldsymbol{e}^i\cdot\boldsymbol{e}_j = \delta^i_j$.

Next let us check that the covariant metric tensor g_{ij} is the inverse of the contravariant metric tensor g^{ij},

$$g_{ij} = \boldsymbol{e}_i\cdot\boldsymbol{e}_j = \frac{\partial \boldsymbol{r}}{\partial u^i}\cdot\frac{\partial \boldsymbol{r}}{\partial u^j}, \quad g^{ij} = \boldsymbol{e}^i\cdot\boldsymbol{e}^j = \nabla \mathrm{u}^i\cdot\nabla u^j$$

Let us compute the product of the matrices,

$$\begin{aligned}\sum_l g^{il}g_{lj} &= \sum_l \nabla u^i\cdot\nabla u^l\frac{\partial \boldsymbol{r}}{\partial u^l}\cdot\frac{\partial \boldsymbol{r}}{\partial u^j} = \sum_{lmn}\frac{\partial u^i}{\partial r^n}\frac{\partial u^l}{\partial r^n}\frac{\partial r^m}{\partial u^l}\frac{\partial r^m}{\partial u^k}\\ &= \sum_{mn}\frac{\partial u^i}{\partial r^n}\left(\sum_l \frac{\partial u^l}{\partial r^n}\frac{\partial r^m}{\partial u^l}\right)\frac{\partial r^m}{\partial u^k} = \sum_{mn}\frac{\partial u^i}{\partial r^n}\delta^m_n\frac{\partial r^m}{\partial u^k}\\ &= \sum_n \frac{\partial u^i}{\partial r^n}\frac{\partial r^n}{\partial u^k} = \frac{\partial u^i}{\partial u^k} = \delta^i_k\end{aligned}$$

Perfect! Note that the sums over i, l and k run from 1 to 2, while those over n and m run over 1, 2, and 3 because $\boldsymbol{r}$ has three components. The crucial step here used the chain rule, $\sum_l \frac{\partial u^l}{\partial r^n}\frac{\partial r^m}{\partial u^l} = \delta^m_n$.

Finally let us check that g^{ij} "raises" indices, in particular,

$$\boldsymbol{e}^i = \sum_j g^{ij}\boldsymbol{e}_j$$

or in this case,

$$\begin{aligned}\nabla \mathrm{u}^i &= \sum_j g^{ij}\frac{\partial \boldsymbol{r}}{\partial u^j} = \sum_j \nabla \mathrm{u}^i\cdot\nabla u^j\frac{\partial \boldsymbol{r}}{\partial u^j} = \sum_{mj}\frac{\partial u^i}{\partial r^m}\frac{\partial u^j}{\partial r^m}\frac{\partial \boldsymbol{r}}{\partial u^j} = \sum_m \frac{\partial u^i}{\partial r^m}\left(\sum_j \frac{\partial \boldsymbol{r}}{\partial u^j}\frac{\partial u^j}{\partial r^m}\right)\\ &= \sum_m \frac{\partial u^i}{\partial r^m}\frac{\partial \boldsymbol{r}}{\partial r^m} = \frac{\partial u^i}{\partial r^1}(1,0,0) + \frac{\partial u^i}{\partial r^2}(0,1,0) + \frac{\partial u^i}{\partial r^3}(0,0,1)\\ &= \left(\frac{\partial u^i}{\partial r^1}, \frac{\partial u^i}{\partial r^2}, \frac{\partial u^i}{\partial r^3}\right) = \nabla \mathrm{u}^i\end{aligned}$$

which checks out.

Let us finish this appendix with a simple example: Consider a vector $\boldsymbol{A}$ in a two-dimensional *flat* space, which is spanned by nonorthogonal basis vectors $\boldsymbol{e}_1$ and $\boldsymbol{e}_2$, which have a fixed angle θ between them. We write $\boldsymbol{A}$ in terms of contravariant components A^1 and A^2,

$$\boldsymbol{A} = A^1\boldsymbol{e}_1 + A^2\boldsymbol{e}_2$$

The length of an infinitesimal vector, $d\boldsymbol{x} = du\boldsymbol{e}_1 + dv\boldsymbol{e}_2$, is,

$$d\boldsymbol{x}\cdot d\boldsymbol{x} = (du\boldsymbol{e}_1 + dv\boldsymbol{e}_2)\cdot(du\boldsymbol{e}_1 + dv\boldsymbol{e}_2) = g_{11}du^2 + 2g_{12}dudv + g_{22}dv^2$$

where we identify the metric,

$$g_{11} = \boldsymbol{e}_1\cdot\boldsymbol{e}_1 = 1 \quad g_{12} = g_{21} = \boldsymbol{e}_1\cdot\boldsymbol{e}_2 = \cos\theta \quad g_{22} = \boldsymbol{e}_2\cdot\boldsymbol{e}_2 = 1$$

and θ is the angle between the two axes $\boldsymbol{e}_1$ and $\boldsymbol{e}_2$.

The metric can be expressed as a 2×2 matrix,

$$g_{ij} = \begin{pmatrix} g_{11} & g_{12} \\ g_{12} & g_{22} \end{pmatrix} = \begin{pmatrix} 1 & \cos\theta \\ \cos\theta & 1 \end{pmatrix}$$

The inverse of the metric is written with upper (contravariant) indices,

$$g^{ij} = \begin{pmatrix} g^{11} & g^{12} \\ g^{12} & g^{22} \end{pmatrix} = \frac{1}{\sin^2\theta}\begin{pmatrix} 1 & -\cos\theta \\ -\cos\theta & 1 \end{pmatrix}$$

One can easily check that they are inverses, $\sum_j g^{ij}g_{jk} = \delta^i_k$.

The metric g_{ij} and g^{ij} is used to map between covariant and contravariant indices,

$$\boldsymbol{A} = A^1\boldsymbol{e}_1 + A^2\boldsymbol{e}_2 = A_1\boldsymbol{e}^1 + A_2\boldsymbol{e}^2$$

$$A_i = \sum_j g_{ij}A^j \quad \boldsymbol{e}^i = \sum_j g^{ij}\boldsymbol{e}_j$$

So, in the case at hand,

$$A_1 = A^1 + \cos\theta A^2 \quad A_2 = A^2 + \cos\theta A^1$$

and

$$\boldsymbol{e}^1 = \frac{1}{\sin^2\theta}(\boldsymbol{e}_1 - \cos\theta\,\boldsymbol{e}_2) \quad \boldsymbol{e}^2 = \frac{1}{\sin^2\theta}(\boldsymbol{e}_2 - \cos\theta\,\boldsymbol{e}_1)$$

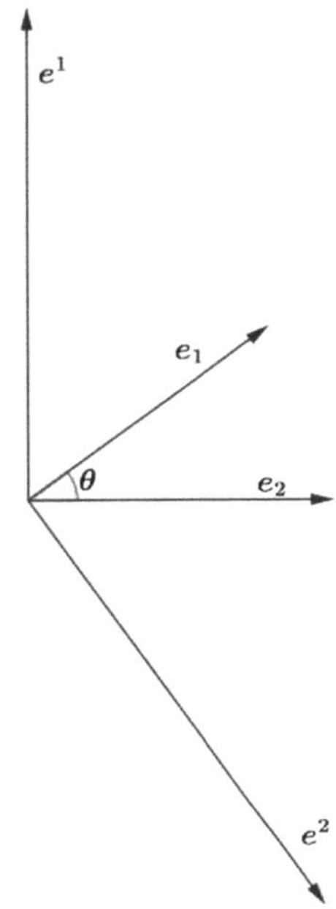

Figure C.1 The basis vectors ($\boldsymbol{e}_1$, $\boldsymbol{e}_2$) and their duals ($\boldsymbol{e}^1$, $\boldsymbol{e}^2$).

The basis vectors $(\boldsymbol{e}_1, \boldsymbol{e}_2)$ and their duals $\left(\boldsymbol{e}^1, \boldsymbol{e}^2\right)$ are plotted in Fig. C.1.

Note that

$$\boldsymbol{e}^i \cdot \boldsymbol{e}_j = \left(\sum_k g^{ik} \boldsymbol{e}_k\right) \cdot \boldsymbol{e}_j = \sum_k g^{ik} g_{kj} = \delta^i_j$$

This example demonstrates the utility of upper and lower indices in the case of nonorthogonal coordinates, which are natural to special relativity. Upper and lower indices are even more useful when using curvilinear coordinates which occur in general relativity.

APPENDIX D

Theorems and Concepts of Vector Calculus

D.1 CARTESIAN AND SPHERICAL COORDINATES

We gather here some relations that are relevant to the discussions in the text. This presentation is just a summary and reference. The student should consult textbooks or the Web for more details.

In Cartesian coordinates a vector $\boldsymbol{A}$ can be written in terms of the unit triad,

$$\boldsymbol{A} = A_x\widehat{\boldsymbol{x}} + A_y\widehat{\boldsymbol{y}} + A_z\widehat{\boldsymbol{z}}$$

where the unit triad $\widehat{\boldsymbol{x}}, \widehat{\boldsymbol{y}}, \widehat{\boldsymbol{z}}$ is shown in Fig. D.1.

In spherical coordinates,

$$\boldsymbol{A} = A_r\widehat{\boldsymbol{r}} + A_\theta\widehat{\boldsymbol{\theta}} + A_\varphi\widehat{\boldsymbol{\varphi}}$$

where the unit triad $\widehat{\boldsymbol{r}}, \widehat{\boldsymbol{\theta}}, \widehat{\varphi}$ is shown in Fig. D.2.

The relationships between the coordinate systems and their unit vectors read,

$$x = r\sin\theta\cos\varphi, \quad y = r\sin\theta\sin\varphi, \quad z = r\cos\theta$$

$$\widehat{\boldsymbol{r}} = \sin\theta\cos\varphi\widehat{\boldsymbol{x}} + \sin\theta\sin\varphi\widehat{\boldsymbol{y}} + \cos\theta\widehat{\boldsymbol{z}}$$

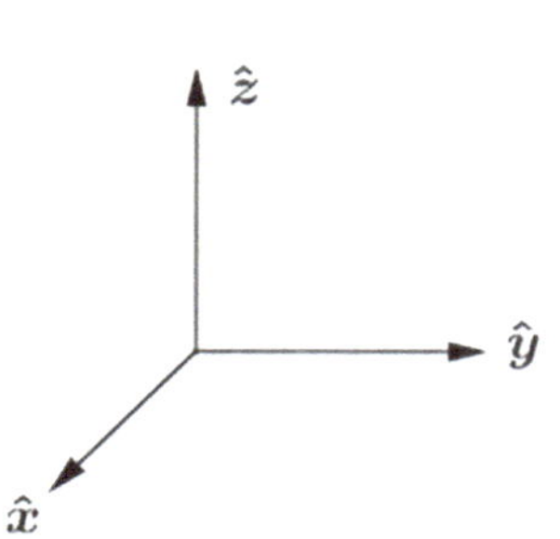

Figure D.1 Mutually orthonormal basis vectors in three dimensional Euclidean space.

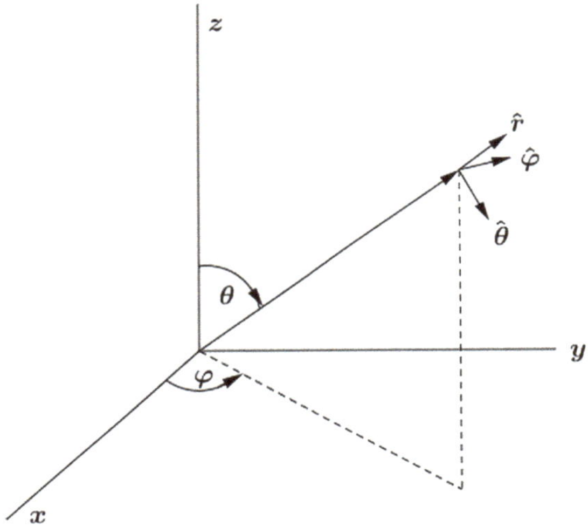

Figure D.2 Spherical coordinates and their orthonormal basis vectors in three dimensional Euclidean space.

$$\widehat{\boldsymbol{\theta}} = \cos\theta\cos\varphi\widehat{\boldsymbol{x}} + \cos\theta\sin\varphi\widehat{\boldsymbol{y}} - \sin\theta\widehat{\boldsymbol{z}}$$

$$\widehat{\boldsymbol{\varphi}} = -\sin\varphi\widehat{\boldsymbol{x}} + \cos\varphi\widehat{\boldsymbol{y}}$$

D.2 CROSS PRODUCT OF TWO VECTORS

The cross product of two vectors $\boldsymbol{A}$ and $\boldsymbol{B}$ is,

$$\mathbf{A} \times \boldsymbol{B} = \begin{vmatrix} \widehat{\boldsymbol{x}} & \widehat{\boldsymbol{y}} & \widehat{\boldsymbol{z}} \\ A_x & A_y & A_z \\ B_x & B_y & B_z \end{vmatrix}$$
$$= \widehat{\boldsymbol{x}}(A_yB_z - A_zB_y) - \widehat{\boldsymbol{y}}(A_xB_z - A_zB_x) + \widehat{\boldsymbol{z}}(A_xB_y - A_yB_x)$$

The direction of $\mathbf{A} \times \boldsymbol{B}$ is given by the "right-hand rule": orient your right hand with the fingers in the direction of $\boldsymbol{A}$ curling toward $\boldsymbol{B}$. Then your thumb points in the direction of $\mathbf{A} \times \boldsymbol{B}$. One can show that the magnitude of the cross product is the product of the magnitudes of $\boldsymbol{A}$ and $\boldsymbol{B}$ times the sin of the angle between them.

D.3 GRADIENTS, CURLS, DIVERGENCES, AND LAPLACIANS

The gradient operator is the generalization of differentiation to three dimensions. In Cartesian coordinates,

$$\nabla = \left(\frac{\partial}{\partial x}, \frac{\partial}{\partial y}, \frac{\partial}{\partial z}\right)$$

The gradient of a scalar function $f(x, y, z)$ is a vector field,

$$\nabla f = \left(\frac{\partial f}{\partial x}, \frac{\partial f}{\partial y}, \frac{\partial f}{\partial z}\right)$$

which points in the direction of the most rapid change in f.

The fundamental theorem of gradients reads,

$$\int_a^b \nabla f \cdot d\boldsymbol{l} = f(\boldsymbol{b}) - f(\boldsymbol{a})$$

where the line integral is taken on any path between points $\boldsymbol{a}$ and $\boldsymbol{b}$.

The divergence of a vector field $\boldsymbol{V}(\boldsymbol{r})$ reads,

$$\nabla \cdot \boldsymbol{V} = \frac{\partial V_x}{\partial x} + \frac{\partial V_y}{\partial y} + \frac{\partial V_z}{\partial z}$$

The curl of $\boldsymbol{V}(\boldsymbol{r})$ reads,

$$\nabla \times \boldsymbol{V} = \begin{vmatrix} \widehat{\boldsymbol{x}} & \widehat{\boldsymbol{y}} & \widehat{\boldsymbol{z}} \\ \frac{\partial}{\partial x} & \frac{\partial}{\partial y} & \frac{\partial}{\partial z} \\ V_x & V_y & V_z \end{vmatrix}$$

$$= \widehat{\boldsymbol{x}}\left(\frac{\partial V_z}{\partial y} - \frac{\partial V_y}{\partial z}\right) - \widehat{\boldsymbol{y}}\left(\frac{\partial V_z}{\partial x} - \frac{\partial V_x}{\partial z}\right) + \widehat{\boldsymbol{z}}\left(\frac{\partial V_y}{\partial x} - \frac{\partial V_x}{\partial y}\right)$$

The Laplacian operator is $\nabla \cdot \nabla = \nabla^2$,

$$\nabla^2 f = \left(\frac{\partial^2}{\partial x^2} + \frac{\partial^2}{\partial y^2} + \frac{\partial^2}{\partial z^2}\right) f$$

In applications, we need these operations in spherical coordinates,

$$\text{Gradient}: \quad \nabla f = \frac{\partial f}{\partial r}\widehat{\boldsymbol{r}} + \frac{1}{r}\frac{\partial f}{\partial \theta}\widehat{\boldsymbol{\theta}} + \frac{1}{r sin\theta}\frac{\partial f}{\partial \varphi}\widehat{\boldsymbol{\varphi}}$$

$$\text{Divergence}: \quad \nabla \cdot \boldsymbol{V} = \frac{1}{r^2}\frac{\partial}{\partial r}(r^2 V_r) + \frac{1}{r \sin\theta}\frac{\partial}{\partial \theta}(\sin\theta V_\theta) + \frac{1}{r \sin\theta}\frac{\partial}{\partial \varphi}(V_\varphi)$$

$$\text{Laplacian}: \quad \nabla^2 f = \frac{1}{r^2}\frac{\partial}{\partial r}\left(r^2\frac{\partial f}{\partial r}\right) + \frac{1}{r^2 \sin\theta}\frac{\partial}{\partial \theta}\left(\sin\theta\frac{\partial f}{\partial \theta}\right) + \frac{1}{r^2 \sin^2\theta}\frac{\partial^2 f}{\partial \varphi^2}$$

D.4 HELMHOLTZ THEOREMS FOR VECTOR FIELDS

1. It is easy to check (use Cartesian coordinates) that the curl of a vector field $\boldsymbol{V}$, which itself is a gradient, $\boldsymbol{V} = \nabla f$, vanishes identically. In fact, if $\nabla \times \boldsymbol{V} = 0$, then $\boldsymbol{V}$ can always be expressed as the gradient of some field, $\boldsymbol{V} = \nabla f$.
2. It is easy to check (use Cartesian coordinates) that the divergence of a curl of a vector field $\boldsymbol{A}$ vanishes identically, $\nabla \cdot (\nabla \times \boldsymbol{A}) = 0$. In fact, if $\nabla \cdot \boldsymbol{V} = 0$, then there is an $\boldsymbol{A}$ so that $\boldsymbol{V} = \nabla \times \boldsymbol{A}$.

D.5 INTEGRAL THEOREMS FOR VECTOR FIELDS

Gauss' theorem states that the flux of a vector field out of a closed surface is the volume integral of the divergence of the field inside the surface,

$$\oiiint \nabla \cdot \boldsymbol{V}\, d^3 r = \oiint \boldsymbol{V} \cdot d\boldsymbol{a}$$

where $d\boldsymbol{a}$ is an infinitesimal surface element pointing outward on the boundary surface.

Stoke's theorem states that the line integral of a vector field $\boldsymbol{V}$ around a closed curve equals the flux of the curl of $\boldsymbol{V}$ through any surface bounded by the closed curve,

$$\oiint (\nabla \times \boldsymbol{V}) \cdot d\boldsymbol{a} = \oint \boldsymbol{V} \cdot d\boldsymbol{l}$$

The vector elements $d\boldsymbol{a}$ and $d\boldsymbol{l}$ must be oriented similarly, (the right-hand rule, as illustrated in Fig. D.3.)

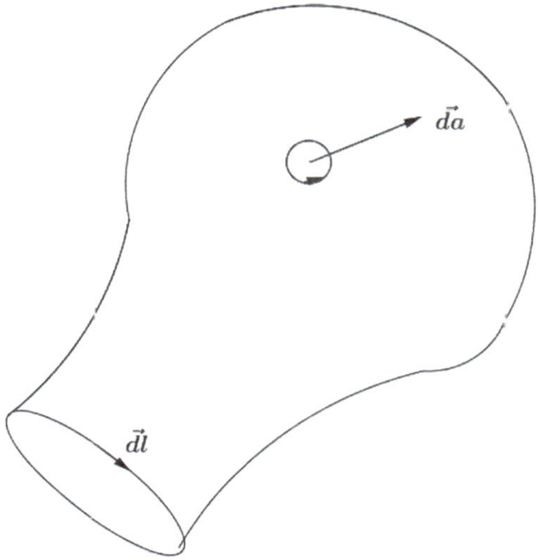

Figure D.3 The line element and surface element in Stoke's theorem.

D.6 LOCALITY AND THE DIRAC DELTA FUNCTION

We spoke about isolated point charges in the text. To incorporate these singularities ("distributions") into the mathematics, one introduces the Dirac "delta function". In one dimension, one writes $\delta(x-a)$. This function is nonzero only at $x=a$, and it has "unit weight", $\int \delta(x-a)dx = 1$ if the range of the integral includes the point a and zero otherwise. The Dirac delta function can be thought of as the limit of continuous, differentiable functions that isolate the point a with unit weight. If $f(x)$ is a continuous function then,

$$\int f(x)\delta(x-a)dx = f(a)$$

if the range of the integral includes the point a and zero otherwise.

The delta function can be generalized to three dimensions. One writes $\delta(\boldsymbol{r}-\boldsymbol{a})$, which has the properties,

$$\iiint \delta(\boldsymbol{r}-\boldsymbol{a})d^3r = 1$$

if the volume includes the point $\boldsymbol{a}$ and is zero otherwise. In addition,

$$\iiint f(\boldsymbol{r})\delta(\boldsymbol{r}-\boldsymbol{a})d^3r = f(\boldsymbol{a})$$

if the volume includes the point $\boldsymbol{a}$, and zero otherwise.

Our discussion of Gauss' law for point particles in Chapter 9.1 can be summarized in the equation,

$$\nabla\cdot\left(\frac{\widehat{\boldsymbol{r}}}{r^2}\right) = 4\pi\delta(\boldsymbol{r})$$

Using the fact that $\frac{1}{r}$ is the potential for the force $\frac{\hat{\boldsymbol{r}}}{r^2}$, $\nabla\left(\frac{1}{r}\right) = -\frac{\hat{\boldsymbol{r}}}{r^2}$, one derives,

$$\nabla^2\left(\frac{1}{r}\right) = -4\pi\delta(\boldsymbol{r})$$

which is important in electrostatics and wave equations. Taking the integral of this equation over any volume containing the origin and then applying Gauss' theorem we find,

$$\oiint \frac{q\widehat{\boldsymbol{r}}}{r^2}\cdot d\boldsymbol{a} = 4\pi q$$

which states that one can detect the presence of a point charge inside a surface by measuring the total flux of the electric field through any closed surface enclosing the charge.

APPENDIX E

Summary of Formulas

E.1 SPECIAL RELATIVITY

It is convenient to collect some of the most useful results and formulas of special relativity. All of these results were derived and discussed in the text.

E.1.1 Lorentz Transformations ("Boosts")

Consider an inertial frame S and another frame S' moving in the $+x$ direction with velocity v. We suppose that their origins coincided at $t = 0$.

The space–time measurements in S and S' are related by "Lorentz transformations,"

$$t' = \gamma(t - vx/c^2), \quad x' = \gamma(x - vt), \quad y' = y, \quad z' = z$$

where,

$$\gamma = \left(1 - v^2/c^2\right)^{-1/2}$$

If a clock is at rest in S and a (proper) time Δt passes on it, a dilated time $\Delta t' = \gamma \Delta t$ passes on clocks at rest in S'.

If a rod is at rest on the x axis in S and has a (proper) length Δl, its length when measured in S' is contracted $\Delta l' = \Delta l/\gamma$.

Clocks that are synchronized in frame S and displaced along the x axis are not synchronized in S'. At an instant t' in S', the clocks at rest in S register times, which are x dependent, $t(x) = t_c + \frac{v}{c^2}x$.

E.1.2 Addition of Velocities

If a particle has velocity $\boldsymbol{u}$ in S, then its velocity in S' is $\boldsymbol{u}'$,

$$u'_x = \frac{u_x - v}{1 - vu_x/c^2}, \quad u'_y = \frac{u_y}{\gamma(1 - vu_x/c^2)}, \quad u'_z = \frac{u_z}{\gamma(1 - vu_x/c^2)}$$

The relativistic law of the addition of velocities leaves the speed limit c invariant (the same in all inertial frames.)

E.1.3 Four-Vectors

A "four-vector" is a set of four quantities, which transform between reference frames in the same fashion as $x^{\mu} = (ct, x, y, z)$, i.e., by Lorentz transformations.

The length of a four-vector $b^{\mu} = (b_0, \boldsymbol{b})$ is defined to be $b \cdot b = b_0^2 - \boldsymbol{b}^2$. It is invariant under Lorentz transformations.

The inner product of two four-vectors $a^{\mu} = (a_0, \boldsymbol{a})$ and $b^{\mu} = (b_0, \boldsymbol{b})$ is defined to be $a \cdot b = a_0 b_0 - \boldsymbol{a} \cdot \boldsymbol{b}$. It is invariant under Lorentz transformations.

E.1.4 Energy–Momentum

The relativistic energy of a particle moving with velocity $\boldsymbol{u}$ in frame S is $E = \gamma(u)mc^2$ where m is its rest mass. The relativistic momentum of the particle is $\boldsymbol{p} = \gamma(u)m\boldsymbol{u}$. The four quantities $(E/c, \boldsymbol{p})$ form a four-vector p^{μ} so it transforms between frames S and S' in the same fashion as $x^{\mu} = (ct, \boldsymbol{x})$, by a Lorentz transformation,

$$E' = \gamma(E - vp_x), \quad p'_x = \gamma\left(p_x - \frac{v}{c^2}E\right), \quad p'_y = p_y, \quad p'_z = p_z$$

The length of the four-vector $p^{\mu} = (E/c, \boldsymbol{p})$ is $E^2/c^2 - \boldsymbol{p}^2 = m^2c^2$, so the energy–momentum relation of a freely propagating particle is,

$$E^2 = \boldsymbol{p}^2c^2 + m^2c^4$$

E.1.5 Force

Newton's second law generalizes to relativity as "force equals the time rate of change of the relativistic momentum,"

$$\boldsymbol{F} = \frac{d}{dt}\boldsymbol{p}$$

in the frame S where $\boldsymbol{p} = \gamma(u)m\boldsymbol{u}$.

The force transforms between frames S and S' as,

$$F'_x = \frac{F_x - (v/c^2)\boldsymbol{F} \cdot \boldsymbol{u}}{1 - vu_x/c^2}, \quad F'_y = \frac{F_y}{\gamma(1 - vu_x/c^2)}, \quad F'_z = \frac{F_z}{\gamma(1 - vu_x/c^2)}$$

And the power transforms as,

$$\boldsymbol{F}' \cdot \boldsymbol{u}' = \frac{\boldsymbol{F} \cdot \boldsymbol{u} - vF_x}{(1 - vu_x/c^2)}$$

where $\boldsymbol{u}$ ($\boldsymbol{u}'$) is the velocity of the particle in frame S (S').

The Lorentz force applies to a particle of charge q moving with velocity $\boldsymbol{u}$ in frame S where there is an electric field $\boldsymbol{E}$ and a magnetic field $\boldsymbol{B}$,

$$\boldsymbol{F} = q(\boldsymbol{E} + \boldsymbol{u} \times \boldsymbol{B})$$

E.1.6 Electric and Magnetic Fields

The electric and magnetic fields mix when transformed from S to S',

$$E'_x = E_x, \quad E'_y = \gamma(E_y - vB_z), \quad E'_z = \gamma(E_z + vB_y)$$

$$B'_x = B_x, \quad B'_y = \gamma\left(B_y + \frac{v}{c^2}E_z\right), \quad B'_z = \gamma\left(E_z - \frac{v}{c^2}E_y\right)$$

E.1.7 Electric Current and Current Conservation

The electric charge density ρ and current density $\boldsymbol{J}$ are the sources of electric and magnetic fields, respectively. They form a four-vector $J^\mu = (c\rho, \boldsymbol{J})$ and therefore transform under boosts such as x^μ. J^μ satisfies the "continuity equation"

$$\nabla \cdot \boldsymbol{J} + \frac{\partial \rho}{\partial t} = 0$$

which is a statement of current conservation.

E.1.8 Maxwell's Equations

The electric and magnetic fields satisfy Maxwell's equations,

$$\nabla \cdot \boldsymbol{E} = 4\pi k \rho \qquad \text{(Gauss' Law)}$$

$$\nabla \times \boldsymbol{B} = \frac{4\pi k}{c^2}\boldsymbol{J} + \frac{1}{c^2}\frac{\partial \boldsymbol{E}}{\partial t} \qquad \text{(Ampere – Maxwell Law)}$$

$$\nabla \cdot \boldsymbol{B} = 0 \qquad \text{(No Magnetic Monoples)}$$

$$\nabla \times \boldsymbol{E} = -\partial \boldsymbol{B}/\partial t \qquad \text{(Faraday's Law)}$$

E.1.9 Wave Equations Without Sources

In free space the fields satisfy the wave equation,

$$\left(\nabla^2 - \frac{\partial^2}{c^2\partial t^2}\right)\boldsymbol{E} = \left(\nabla^2 - \frac{\partial^2}{c^2\partial t^2}\right)\boldsymbol{B} = 0$$

which establishes that electromagnetic fields propagate at the speed limit c of special relativity.

E.1.10 Wave Equations With Sources

As discussed in Problem 10.4 and Appendix F, $\boldsymbol{E}$ and $\boldsymbol{B}$ can be derived from a four-vector potential, $A^\mu = (V/c, \boldsymbol{A})$,

$$\boldsymbol{B} = \nabla \times \boldsymbol{A} \quad \boldsymbol{E} = -\frac{\partial \boldsymbol{A}}{\partial t} - \nabla V$$

Maxwell's equations are equivalent to the four-vector wave equations with sources,

$$\left(\nabla^2 - \frac{1}{c^2}\frac{\partial^2}{\partial t^2}\right)V = -4\pi k\rho \quad \left(\nabla^2 - \frac{1}{c^2}\frac{\partial^2}{\partial t^2}\right)\boldsymbol{A} = -\frac{4\pi k}{c^2}\boldsymbol{J}$$

where the four-vector potential satisfies the Lorenz gauge condition,

$$\nabla\cdot\boldsymbol{A} + \frac{1}{c^2}\frac{\partial V}{\partial t} = 0$$

As discussed in Appendix F, the solutions for V and $\boldsymbol{A}$ read,

$$V(x, y, z, t) = -4\pi k\iiint \frac{1}{R}\,\rho\left(x', y', z', t - \frac{R}{c}\right)d^3r' + V_0$$

$$\boldsymbol{A}(x, y, z, t) = -\frac{4\pi k}{c^2}\iiint \frac{1}{R}\boldsymbol{J}\left(x', y', z', t - \frac{R}{c}\right)d^3r'$$

where $R^2 = (x - x')^2 + (y - y')^2 + (z - z')^2$. $t_r = t - \frac{R}{c}$ is called the "retarded time" because it accounts for the finite speed of light c.

E.1.11 The Electromagnetic Field Strength Tensor and the Lorentz Force Law

The electromagnetic field strength tensor $F^{\mu\nu}$ reads,

$$F^{\mu\nu} = \begin{pmatrix} 0 & E_x/c & E_y/c & E_z/c \\ -E_x/c & 0 & B_z & -B_y \\ -E_y/c & -B_z & 0 & B_x \\ -E_z/c & B_y & -B_x & 0 \end{pmatrix}$$

And its dual $\widetilde{F}^{\mu\nu}$ reads,

$$\widetilde{F}^{\mu\nu} = \begin{pmatrix} 0 & B_x & B_y & B_z \\ -B_x & 0 & -E_z/c & E_y/c \\ -B_y & E_z/c & 0 & -E_x/c \\ -B_z & -E_y/c & E_x/c & 0 \end{pmatrix}$$

In this language, Maxwell's equations can be written in explicitly covariant form,

$$\sum_\nu \partial_\nu F^{\mu\nu} = \frac{4\pi k}{c^2} J^\mu \quad \sum_\nu \partial_\nu \widetilde{F}^{\mu\nu} = 0$$

The Lorentz force law becomes,

$$dp^\mu / d\tau = q \sum_\nu u_\nu F^{\nu\mu}$$

We can express the electromagnetic field strength tensor using the four-vector potential,

$$A^\mu = (V/c, \boldsymbol{A})$$

Then,

$$F^{\mu\nu} = \partial^\mu A^\nu - \partial^\nu A^\mu$$

The two homogeneous Maxwell equations, $\sum_\nu \partial_\nu \widetilde{F}^{\mu\nu} = 0$, are automatically satisfied in this parametrization.

The Lorenz gauge condition reads,

$$\sum_{\mu} \partial_{\mu} A^{\mu} = 0$$

Then Maxwell's equations become the wave equations,

$$\sum_{\nu} \partial_{\nu} \partial^{\nu} A^{\mu} = \left(\frac{1}{c^2} \frac{\partial^2}{\partial t^2} - \nabla \cdot \nabla \right) A^{\mu} = \frac{4\pi k}{c^2} J^{\mu}$$

E.1.12 The Electromagnetic Stress Tensor $T^{\mu\nu}$

$$T^{\mu\nu} = \sum_{\sigma} F^{\mu\sigma} F^{\nu}_{\sigma} - \frac{1}{4} g^{\mu\nu} \left(\sum_{\lambda\sigma} F^{\lambda\sigma} F_{\lambda\sigma} \right)$$

represents the local energy—momentum density of the electromagnetic field.

$T^{\mu\nu}$ is symmetric, $T^{\mu\nu} = T^{\nu\mu}$, traceless, $\sum_{\mu} T^{\mu}_{\mu} = 0$ and conserved $\sum_{\mu} \partial_{\mu} T^{\mu\nu} = 0$ in a source-free region of space—time.

Local energy—momentum conservation in a region of space—time where there are charged particles reads,

$$\partial_{\mu} (T^{\mu\nu} + T^{\mu\nu}_{M}) = 0$$

where the energy—momentum carried by charged matter fields is $T^{\mu\nu}_{M}$. This is the statement of energy—momentum conservation that replaces Newton's third law in relativity.

This material is introduced in Problem 10.5.

E.2 GENERAL RELATIVITY

E.2.1 Gravitational Redshift

$$\frac{\Delta \nu}{\nu} \approx -\frac{\Delta V}{c^2}$$

where $\Delta \nu$ is the frequency difference and ΔV is the potential difference between the point of detection and emission of the light wave.

E.2.2 Coordinate Transformations: Covariant Vectors and Tensors

Under the transformation of general curvilinear coordinates, $x^\mu \to x'^\mu$,

$$\frac{\partial}{\partial x'^\sigma} = \sum_\mu \frac{\partial x^\mu}{\partial x'^\sigma}\frac{\partial}{\partial x^\mu} \quad \omega'_\sigma = \sum_\mu \frac{\partial x^\mu}{\partial x'^\sigma}\omega_\mu$$

E.2.3 Coordinate Transformations: Contravariant Vectors and Tensors

Under the transformation of general curvilinear coordinates, $x^\mu \to x'^\mu$,

$$dx'^\sigma = \sum_\mu \frac{\partial x'^\sigma}{\partial x^\mu} dx^\mu \quad V'^\sigma = \sum_\mu \frac{\partial x'^\sigma}{\partial x^\mu} V^\mu$$

E.2.4 Metric Tensor

Under the transformation of general curvilinear coordinates, $x^\mu \to x'^\mu$

$$g'_{\mu\nu} = \sum_{\sigma\rho} \frac{\partial x^\sigma}{\partial x'^\mu}\frac{\partial x^\rho}{\partial x'^\nu} g_{\sigma\rho}$$

$g_{\sigma\rho}$ is a symmetric second-rank tensor. Its inverse is $g^{\mu\nu}$

$$\sum_\nu g^{\mu\nu} g_{\nu\sigma} = \delta^\mu_\sigma$$

The invariant volume element is $\sqrt{|\det(g)|}\, d^4x$ and the determinant of the metric transforms as, $\det(g') = \det\left(\frac{\partial x'^\mu}{\partial x^\nu}\right)\det(g)$.

E.2.5 Christoffel Symbol

$$\Gamma^\mu_{\rho\beta} = \frac{1}{2}\sum_\alpha g^{\mu\alpha}\left(\partial_\rho g_{\beta\alpha} + \partial_\beta g_{\alpha\rho} - \partial_\alpha g_{\rho\beta}\right)$$

E.2.6 Covariant Derivatives

Applied to contravariant vectors,

$$D_\mu V^\sigma = \partial_\mu V^\sigma + \sum_\gamma \Gamma^\sigma_{\mu\gamma} V^\gamma$$

Applied to covariant vectors,

$$D_\mu W_\sigma = \partial_\mu W_\sigma - \sum_\gamma \Gamma^\gamma_{\mu\sigma} W_\gamma$$

E.2.7 Geodesic Curves

$$\frac{d^2 x^\beta}{d^2\tau} + \sum_{\alpha\gamma} \Gamma^\beta_{\gamma\alpha} \frac{dx^\alpha}{d\tau}\frac{dx^\gamma}{d\tau} = 0$$

E.2.8 Curvature and the Riemann Tensor

$$(D_\mu D_\nu - D_\nu D_\mu)A^\rho = -\sum_\sigma R^\rho_{\sigma\mu\nu} A^\sigma$$

$$R^\rho_{\sigma\mu\nu} = \partial_\mu \Gamma^\rho_{\nu\sigma} - \partial_\nu \Gamma^\rho_{\mu\sigma} + \sum_\alpha \Gamma^\rho_{\mu\alpha}\Gamma^\alpha_{\nu\sigma} - \sum_\alpha \Gamma^\rho_{\nu\alpha}\Gamma^\alpha_{\mu\sigma}$$

E.2.9 Curvature, Tidal Forces, and the Geodesic Deviation

In Newtonian mechanics, the vector between two freely falling masses satisfies the equation of motion,

$$\frac{d^2\boldsymbol{\varepsilon}}{dt^2} = -\nabla(\boldsymbol{\varepsilon}\cdot\nabla\Phi(r))$$

For a $1/r$ potential in Cartesian coordinates,

$$\frac{d^2\varepsilon^i}{dt^2} = \frac{GM}{r^5}\sum_k \varepsilon^k\left(3x^k x^i - r^2\delta^{ki}\right)$$

The relativistic generalization reads,

$$\frac{D^2\varepsilon^\alpha}{D\tau^2} = -\sum_{\beta\gamma\delta} R^\alpha_{\beta\gamma\delta} u^\beta u^\delta \varepsilon^\gamma + O(\varepsilon^2)$$

E.2.10 Symmetries of the Riemann Tensor

1. $R_{\rho\sigma\alpha\beta} = -R_{\sigma\rho\alpha\beta}$
2. $R_{\rho\sigma\alpha\beta} = -R_{\rho\sigma\beta\alpha}$
3. $R_{\rho\sigma\alpha\beta} = R_{\alpha\beta\rho\sigma}$
4. $R_{\rho\sigma\alpha\beta} + R_{\rho\alpha\beta\sigma} + R_{\rho\beta\sigma\alpha} = 0$
5. $D_\gamma R_{\rho\sigma\alpha\beta} + D_\rho R_{\sigma\gamma\alpha\beta} + D_\sigma R_{\gamma\rho\alpha\beta} = 0$ (Bianchi identity)

E.2.11 The Einstein and Ricci Tensors

The second-rank Ricci tensor is

$$R_{\alpha\beta} = \sum_{\gamma} R^{\gamma}_{\alpha\gamma\beta} \quad R_{\alpha\beta} = R_{\beta\alpha}$$

The Ricci scalar is the fully contracted version,

$$R = \sum_{\gamma} R^{\gamma}_{\gamma} = \sum_{\alpha\beta} g^{\alpha\beta} R_{\alpha\beta}$$

The Einstein tensor is defined as,

$$G_{\sigma\rho} = R_{\sigma\rho} - \frac{1}{2} R g_{\sigma\rho}$$

It satisfies the conservation law,

$$\sum_{\gamma} D^{\gamma} G_{\rho\gamma} = 0$$

The Einstein field equation reads,

$$G_{\alpha\beta} = \frac{8\pi G}{c^4} T_{\alpha\beta}$$

where $T_{\alpha\beta}$ is the conserved and symmetric energy–momentum tensor, $\sum_{\mu} D^{\mu} T_{\mu\nu} = 0$.

The Einstein field equation can be written,

$$R_{\alpha\beta} = 8\pi G\left(T_{\alpha\beta} - \frac{1}{2} T g_{\alpha\beta}\right)$$

where T is the trace of the energy–momentum tensor.

E.2.12 Linearized Gravity

Suppose the metric deviates slightly from the Minkowski metric of special relativity, $g_{\mu\nu} = g^{(0)}_{\mu\nu} + \widetilde{g}_{\mu\nu}$ with $\left|\widetilde{g}_{\mu\nu}\right| \ll 1$. Define,

$$\overline{h}_{\mu\nu} = \widetilde{g}_{\mu\nu} - \frac{1}{2} g^{(0)}_{\mu\nu} h \quad h = \sum_{\alpha} \widetilde{g}^{\alpha}_{\alpha}$$

and impose the Lorenz gauge condition,

$$\sum_{\nu} \partial_{\nu} \overline{h}^{\mu\nu} = 0$$

Then the Einstein field equation becomes a wave equation in a background Minkowski metric,

$$\Box \overline{h}_{\mu\nu} = \frac{16\pi G}{c^4} T_{\mu\nu} \quad \Box \equiv \left(\nabla^2 - \frac{1}{c^2} \frac{\partial^2}{\partial t^2} \right)$$

to first order in $\overline{h}_{\mu\nu}$.

E.2.13 Schwarzschild Metric

$$ds^2 = \left(1 - \frac{2GM}{c^2 r}\right) c^2 dt^2 - \left(1 - \frac{2GM}{c^2 r}\right)^{-1} dr^2 - r^2 \left(d\theta^2 + \sin^2 \theta d\varphi^2\right)$$

E.2.14 Slightly Curved Space–Time, Isotropic Coordinates, the Newtonian Limit

The space–time metric, when the gravitational potential Φ satisfies the weak field condition, $2\Phi/c^2 \ll 1$, can be written in the form,

$$ds^2 = \left(1 + \frac{2\Phi}{c^2}\right) c^2 dt^2 - \left(1 - \frac{2\Phi}{c^2}\right) (dx^2 + dy^2 + dz^2)$$

with corrections $O(c^{-4})$.

APPENDIX F

The Wave Equation With Sources: The Four-Vector Potential and the Lorenz Gauge

F.1 THE WAVE EQUATION AND RADIATION

There are several objectives for this appendix. First, we want to derive the wave equations with sources from Maxwell's equations. This completes the discussion of the wave equation in Chapter 9. Second, we want to analyze these equations and see how they incorporate causality, the fact that information propagates at the speed of light. This produces the generalization of Coulomb's law for time-dependent charge densities and currents, and this completes a logical issue central to the developments in the text. And third, the resulting equations are the starting point of quantitative discussions of radiation that are best discussed in a course dedicated to applications of electromagnetism, perhaps using Landau and Lifshitz [1], which was the source for this appendix.

Our first task is to simplify Maxwell's equations so they are easier to apply to concrete situations. Recall that this is a good strategy in Newtonian mechanics where in many cases forces can be derived from potentials ($\boldsymbol{F} = -\nabla V$), and this substitution leads to simpler mathematics and physical insights (potentials provide a path to energy and work).

Let us try this strategy for $\boldsymbol{E}$ and $\boldsymbol{B}$.

One of Maxwell's equations reads $\nabla \cdot \boldsymbol{B} = 0$. Invoking a Helmholtz theorem from vector calculus (see Appendix D), this means that $\boldsymbol{B}$ can be expressed as the curl of a vector field, $\boldsymbol{B} = \nabla \times \boldsymbol{A}$. Now we can use the same analysis on Faraday's law to conclude that there exists a scalar potential V so that,

$$\boldsymbol{E} = -\frac{\partial \boldsymbol{A}}{\partial t} - \nabla V$$

Next, realize that although the electric and magnetic fields are physical quantities, the potentials are introduced for mathematical convenience.

What freedom exists in choosing them? If we choose another pair of $A'$and V', they must generate the same $\boldsymbol{E}$ and $\boldsymbol{B}$, but can be otherwise free. It is easy to show that A' and V' can differ from $\boldsymbol{A}$ and V,

$$\boldsymbol{A}' = \boldsymbol{A} + \nabla G, \quad V' = V - \frac{\partial G}{\partial t}$$

This is called a "gauge" transformation. G is the gauge function. Gauge transformations leave the physical fields $\boldsymbol{E}$ and $\boldsymbol{B}$ unchanged, as the reader should verify!

Let us take stock. The introduction of the potentials *solves* two of Maxwell's equations explicitly (the divergence free nature of the magnetic field and Faraday's equation.). Good! And there is "gauge" freedom in choosing the potentials V and $\boldsymbol{A}$. But are the other two Maxwell's equations simplified in this new language?

Let us write Gauss' law and the Ampere—Maxwell equation in terms of the potentials,

$$\nabla^2 V + \frac{\partial}{\partial t}(\nabla \cdot \boldsymbol{A}) = -4\pi k\rho$$

$$\left(\nabla^2 - \frac{1}{c^2}\frac{\partial^2}{\partial t^2}\right)\boldsymbol{A} - \nabla\left(\nabla \cdot \boldsymbol{A} + \frac{1}{c^2}\frac{\partial V}{\partial t}\right) = -\frac{4\pi k}{c^2}\boldsymbol{J}$$

where we obtained the last equation by using an identity from vector calculus, $\nabla \times (\nabla \times \boldsymbol{C}) = \nabla(\nabla \cdot \boldsymbol{C}) - \nabla^2 \boldsymbol{C}$, which is true for any vector field $\boldsymbol{C}$.

These are complicated, coupled differential equations! If we could choose $\boldsymbol{A}$ and V so that $\nabla \cdot \boldsymbol{A} + \frac{1}{c^2}\frac{\partial V}{\partial t} = 0$, then they would decouple into two wave equations,

$$\left(\nabla^2 - \frac{1}{c^2}\frac{\partial^2}{\partial t^2}\right)V = -4\pi k\rho$$

$$\left(\nabla^2 - \frac{1}{c^2}\frac{\partial^2}{\partial t^2}\right)\boldsymbol{A} = -\frac{4\pi k}{c^2}\boldsymbol{J}$$

This would be great! We recognize the invariant wave operator here so that the speed limit of relativity and causality are built in! Furthermore, we read off that $\boldsymbol{A}$ and V form a four-vector because $\boldsymbol{J}$ and ρ do! In this language, boosting problems from one frame to another will be as simple as possible! The "gauge condition" $\nabla \cdot \boldsymbol{A} + \frac{1}{c^2}\frac{\partial V}{\partial t} = 0$ was invented by Lorenz (*not*

Lorentz!) and bares his name. It is very useful. Note that it is a *covariant* expression because $\boldsymbol{A}$ and V form a four-vector.

Back to the problem at hand. Can we always find a gauge transformation G so that $\nabla\cdot\boldsymbol{A}+\frac{1}{c^2}\frac{\partial V}{\partial t}=0$ is true?

It is easy to use the gauge transformation to obtain,

$$\left(\nabla\cdot\boldsymbol{A}'+\frac{1}{c^2}\frac{\partial V'}{\partial t}\right)=\left(\nabla^2-\frac{1}{c^2}\frac{\partial^2}{\partial t^2}\right)G+\left(\nabla\cdot\boldsymbol{A}+\frac{1}{c^2}\frac{\partial V}{\partial t}\right)$$

We learn that *if* G satisfies the wave equation,

$$\left(\nabla^2-\frac{1}{c^2}\frac{\partial^2}{\partial t^2}\right)G=\left(\nabla\cdot\boldsymbol{A}'+\frac{1}{c^2}\frac{\partial V'}{\partial t}\right)$$

then $\boldsymbol{A}$ and V will be in the Lorentz gauge. Luckily for most applications we only need to know that such a G exists. We seldom need it. This argument just guarantees that we can choose the Lorenz gauge even if we started our considerations with potentials that did not satisfy it.

Now let us derive expressions for the general solutions of our wave equations with sources,

$$\left(\nabla^2-\frac{1}{c^2}\frac{\partial^2}{\partial t^2}\right)V=-4\pi k\rho$$

$$\left(\nabla^2-\frac{1}{c^2}\frac{\partial^2}{\partial t^2}\right)\boldsymbol{A}=-\frac{4\pi k}{c^2}\boldsymbol{J}$$

Let us concentrate on the equation for the scalar potential V. If we can solve it for a point particle at the origin then the general solution will follow by linear superposition. This problem has spherical symmetry so we should write it in spherical polar coordinates. The Laplacian ∇^2 is written in Appendix D in spherical coordinates. For a point charge at the origin, V will depend on r, the distance from the origin where the source is, but it will *not* have any angular dependence. So, the wave equation for $r\neq 0$ becomes,

$$\frac{1}{r^2}\frac{\partial}{\partial r}\left(r^2\frac{\partial V}{\partial r}\right)-\frac{1}{c^2}\frac{\partial^2 V}{\partial t^2}=0$$

If we parametrize V in the form,

$$V(r,t) = \frac{\varphi(r,t)}{r}$$

then by substitution into the equation above we derive that φ satisfies a one-dimensional wave equation in r and t,

$$\left(\frac{\partial^2}{\partial r^2} - \frac{1}{c^2}\frac{\partial^2}{\partial t^2}\right)\varphi = 0$$

From our discussion in Section 9.2, we know that this equation is solved by running waves moving to the left or the right,

$$\varphi(r,t) = \varphi_1\left(t - \frac{r}{c}\right) + \varphi_2\left(t + \frac{r}{c}\right)$$

Only φ_1 will be physically relevant to the problem at hand, so we have for $r \neq 0$,

$$V(r,t) = \frac{\varphi_1\left(t - \frac{r}{c}\right)}{r}$$

Next, φ_1 must be determined by the source at the origin. We choose a point charge with a time-varying infinitesimal charge, $-4\pi k e(t)\delta(\boldsymbol{r})$. Here $\delta(\boldsymbol{r})$ is the three-dimensional Dirac delta function, which is discussed further in Appendix D. It has unit weight and is nonzero only at the origin, so it models a point charge. $e(t)$ is the infinitesimal time-dependent charge at the origin. Now let $r \to 0$ in the wave equation,

$$\left(\nabla^2 - \frac{1}{c^2}\frac{\partial^2}{\partial t^2}\right)V = -4\pi k\rho$$

In this limit $V \to \infty$ and its derivatives with respect to r overwhelm its time derivatives. So as the origin is approached the wave equation reduces to,

$$\nabla^2 V = -4\pi k q(t)\delta(\boldsymbol{r})$$

which is Coulomb's law with a time-dependent charge $q(t)$ at the origin. So, V behaves as $\frac{1}{r}$ near the origin, and in general is,

$$V(r,t) = -4\pi k\frac{q\left(t - \frac{r}{c}\right)}{r}$$

to match on to the functional form $\frac{\varphi_1\left(t-\frac{r}{c}\right)}{r}$ away from the origin. The argument of the charge, $q\left(t-\frac{r}{c}\right)$, incorporates the fact that information flows at the speed of light so V at time t at position r is determined by the source at the *earlier* "retarded" time $t-\frac{r}{c}$!

Now we can use linear superposition to write down the solution for a *general* time-dependent charge density,

$$V(x, y, z, t) = -4\pi k \iiint \frac{1}{R} \rho\left(x', y', z', t - \frac{R}{c}\right) d^3r' + V_0$$

where $R^2 = (x - x')^2 + (y - y')^2 + (z - z')^2$. Note that the amount of time delay varies across ρ as the integral is done.

A completely analogous result holds for $\boldsymbol{A}$ in terms of $\boldsymbol{J}$.

These are formulas one needs to calculate radiation from antennas and radiation due to the collision of charged particles. Such problems have their own special challenges!

See Landau and Lifshitz.

F.2 THE FOUR-VECTOR POTENTIAL AND MAXWELL'S EQUATIONS

In the discussion above, we made some discoveries in passing that we should discuss more. We introduced the potentials V and $\boldsymbol{A}$ to write Maxwell's equations as four-wave equations with sources. We discovered that the potentials formed a "four-vector potential" for the electromagnetic field,

$$A^\mu = (V/c, \boldsymbol{A})$$

which could be used to generate the electric and magnetic fields,

$$\boldsymbol{E} = -\nabla V - \frac{\partial \boldsymbol{A}}{\partial t}, \quad \boldsymbol{B} = \nabla \times \boldsymbol{A} \tag{F.1}$$

It is interesting to write these expressions in the language of the electromagnetic field strength tensor, $F^{\mu\nu}$, to obtain explicitly covariant expressions. The challenge is to write the field tensor $F^{\mu\nu}$ in terms of ∂^ν and A^μ and obtain the covariant form of the expressions for $\boldsymbol{E}$ and $\boldsymbol{B}$ in terms of V and $\boldsymbol{A}$.

Since $F^{\mu\nu}$ is antisymmetric, a candidate formula that has the correct transformation laws is,

$$F^{\mu\nu} = \partial^\mu A^\nu - \partial^\nu A^\mu \tag{F.2}$$

In the problem set, you will show that Eq. (F.2) produces Eq. (F.1) when you write out the Cartesian coordinates. Note that this is a conceptually satisfying discovery because it shows us that you can make the tensor $F^{\mu\nu}$ out of two four-vectors!

In the discussion in Section F.1 we chose the Lorenz gauge to find the wave equation for A^μ. The Lorenz gauge condition was,

$$\nabla \cdot \boldsymbol{A} = -\frac{1}{c^2}\frac{\partial V}{\partial t}$$

This can be written in a manifestly covariant form,

$$\sum_\mu \partial_\mu A^\mu = 0$$

which shows that it is fully relativistic, i.e., if it is chosen in one frame, it is true in all frames.

We were able to introduce V and $\boldsymbol{A}$ above because they made two of Maxwell's equations, the sourceless ones, tautologies: setting $\boldsymbol{B} = \nabla \times \boldsymbol{A}$ solved $\nabla \cdot \boldsymbol{B} = 0$, and setting $\boldsymbol{E} = -\nabla V - \partial \boldsymbol{A}/\partial t$ solved $\nabla \times \boldsymbol{E} = -\partial \boldsymbol{B}/\partial t$ (Faraday's law). In the covariant language where the four-vector A^μ generates the field strength tensor, $F^{\mu\nu} = \partial^\mu A^\nu - \partial^\nu A^\mu$, it is easy to show that this parametrization of $F^{\mu\nu}$, automatically solves the two homogeneous Maxwell equations, $\sum_\nu \partial_\nu \widetilde{F}^{\mu\nu} = 0$. See the problem set for Chapter 10.

Finally, if one writes out the remaining Maxwell equations $\sum_\nu \partial_\nu F^{\mu\nu} = \frac{4\pi k}{c^2} J^\mu$ in terms of A^μ using $F^{\mu\nu} = \partial^\mu A^\nu - \partial^\nu A^\mu$ and one works in the Lorenz gauge $\sum_\mu \partial_\mu A^\mu = 0$, the four Maxwell's equations with sources become the four wave equations,

$$\sum_\nu \partial_\nu \partial^\nu A^\mu = \left(\frac{1}{c^2}\frac{\partial^2}{\partial t^2} - \nabla \cdot \nabla\right) A^\mu = \frac{4\pi k}{c^2} J^\mu$$

Note from this wave equation that the Lorenz gauge choice implies current conservation, $\sum_\mu \partial_\mu J^\mu = 0$.

REFERENCE

[1] L.D. Landau, E.M. Lifshitz, The Classical Theory of Fields, Pergamon Press, Oxford, 1962.

APPENDIX G

Equation of Motion in Curved Space and Christoffel Symbols, Two-Dimensional Surfaces, and the Geodesic Deviation and Gravity Waves

G.1 FREE FALL AND CHRISTOFFEL SYMBOLS

Let us consider an example of a particle's equation of motion in curved space and along the way obtain the expression for the Christoffel symbols in terms of the metric and its first derivatives.

Start in N-dimensional Euclidean space with Cartesian coordinates y_k, $k = 1, 2, \ldots, N$. The distance between y_k and $y_k + dy_k$ is given by the Pythagorean theorem in N dimensions,

$$ds^2 = \sum_k dy_k dy_k \tag{G.1}$$

There is no distinction between covariant and contravariant indices here, so we will use ordinary lower indices most everywhere.

Now let us suppose that we parametrize the space with curvilinear coordinates, or we have a curved surface embedded in this space. Call the coordinates we use for this purpose x_i, $i = 1, 2, \ldots, M$. Each of the original differentials dy_k can be expressed in terms of the dx_i using the chain rule,

$$dy_k = \sum_i \frac{\partial y_k}{\partial x_i} dx_i \tag{G.2}$$

We can visualize the quantities $\partial y_k/\partial x_i$. Denote a vector in the N-dimensional Euclidean space $\boldsymbol{r}$. It has components y_k, $\boldsymbol{r} = (y_1, y_2, \ldots, y_N)$. Suppose that $\boldsymbol{r}$ denotes a point on an M-dimensional surface parametrized by the coordinates $(x_1, x_2, \ldots, x_M)$. Then the derivatives $\partial \boldsymbol{r}/\partial x_i$ at point x span the tangent space to the surface there. We shall see an example of this in the case of a sphere in the problem sets. Now substitute Eq. (G.2) into Eq. (G.1),

$$ds^2 = \sum_{ij} \left(\sum_k \frac{\partial y_k}{\partial x_i} \frac{\partial y_k}{\partial x_j} \right) dx_i dx_j$$

where we identify the metric on the surface $g_{ij}(x)$,

$$g_{ij}(x) = \sum_k \frac{\partial y_k}{\partial x_i} \frac{\partial y_k}{\partial x_j} \tag{G.3}$$

To gain familiarity with this formalism, Problem (12.7) asks you to work out these formulas for metrics and tangent spaces in the case of a two-dimensional sphere.

Next, consider the derivatives of the metric,

$$\frac{\partial g_{ij}}{\partial x_l} = \sum_k \frac{\partial^2 y_k}{\partial x_i \partial x_l} \frac{\partial y_k}{\partial x_j} + \sum_k \frac{\partial y_k}{\partial x_i} \frac{\partial^2 y_k}{\partial x_j \partial x_l} \tag{G.4}$$

In preparation for the calculation of Christoffel symbols, we can consider Eq. (G.4) for three arrangements of the indices i, j, and l and verify the identity,

$$\frac{\partial g_{ij}}{\partial x_l} + \frac{\partial g_{jl}}{\partial x_i} - \frac{\partial g_{li}}{\partial x_j} = 2\sum_k \frac{\partial^2 y_k}{\partial x_i \partial x_l} \frac{\partial y_k}{\partial x_j} \tag{G.5}$$

Now consider a curve $C(\tau)$ on the surface which is parametrized by a scalar quantity τ, which varies from 0 to 1 as the curve moves from its beginning to its end,

$$C_k(\tau) = y_k(x(\tau))$$

Think of $C_k(\tau)$ as the curve swept out by a particle as time τ proceeds from 0 to 1. Then the velocity of the particle is, using the chain rule,

$$\frac{dC_k}{d\tau} = \sum_i \frac{\partial y_k}{\partial x_i} \frac{dx_i}{d\tau} \tag{G.6}$$

and the acceleration is,

$$\frac{d^2 C_k}{d\tau^2} = \sum_{ij} \frac{\partial^2 y_k}{\partial x_i \partial x_j} \frac{dx_i}{d\tau} \frac{dx_j}{d\tau} + \sum_i \frac{\partial y_k}{\partial x_i} \frac{d^2 x_i}{d\tau^2} \tag{G.7}$$

The particle is constrained to the M-dimensional subspace, so there is a force of constraint which acts normal to the subspace. This means that the projection of the particle's acceleration onto the tangent space of the M-dimensional subspace must vanish in a force-free space–time. We learned above that the tangent space at the point of the particle is spanned by the M vectors, $\partial y_k/\partial x_i$, $i = 1, 2,\ldots, M$. So, the inner product of $\frac{d^2C_k}{d\tau^2}$ and $\partial y_k/\partial x_i$ must vanish for each i, $i = 1, 2,\ldots, M$,

$$\sum_k \frac{d^2C_k}{d\tau^2}\frac{\partial y_k}{\partial x_i} = 0, \quad i = 1, 2, \ldots, M \tag{G.8}$$

Next, substitute Eq. (G.7) into (G.8) and find,

$$\sum_{ki} \frac{\partial y_k}{\partial x_i}\frac{\partial y_k}{\partial x_l}\frac{d^2x_i}{d\tau^2} + \sum_{kij}\frac{\partial^2 y_k}{\partial x_i \partial x_j}\frac{\partial y_k}{\partial x_l}\frac{dx_i}{d\tau}\frac{dx_j}{d\tau} = 0 \tag{G.9}$$

We can identify the metric g_{il} in the first term. Then we use Eq. (G.5) to simplify the second term and write it in terms of derivatives of the metric. Finally, we can eliminate g_{il} in the first term by multiplying through with g^{kl}, summing over l, and use the fact that g^{kl} is the inverse of g_{il}, $\sum_l g^{kl}g_{li} = \delta_i^k$. Eq. (G.9) then becomes,

$$\frac{d^2x_k}{d\tau^2} + \sum_{ij}\Gamma_{ij}^k \frac{dx_i}{d\tau}\frac{dx_j}{d\tau} = 0 \tag{G.10}$$

where we have an explicit formula for the Christoffel symbols,

$$\Gamma_{ij}^k = \frac{1}{2}\sum_l g^{kl}\left(\frac{\partial g_{il}}{\partial x_j} + \frac{\partial g_{jl}}{\partial x_i} - \frac{\partial g_{ij}}{\partial x_l}\right)$$

Compare Eq. (G.10) with the discussion in Chapter 12 and we see the geodesic equation with explicit equations for the Christoffel symbols expressed in terms of the metric and its derivatives. We also learn from this derivation that the velocity-dependent "forces" $\sum_{ij}\Gamma_{ij}^k \frac{dx_i}{d\tau}\frac{dx_j}{d\tau}$ appear because of the use of curvilinear coordinates. In the case of general relativity, the curvilinear coordinates are a necessity because space–time is warped (curved) by sources of energy–momentum. However, sometimes we use curvilinear coordinates in ordinary problems in three-dimensional

Euclidean space if the symmetries of the problem suggest it. For example, centripetal and Coriolis forces experienced on a rotating earth are of this sort. These points are pursued further in the problem set for Chapter 12.

G.2 RIEMANN TENSOR AND TWO-DIMENSIONAL SURFACES

It is natural to ask how the Riemann tensor analysis approach to curved spaces is related to traditional differential geometry introduced in Section 12.1 of Chapter 12.

If we specialize to two-dimensional surfaces so that the indices on the Riemann tensor are each either 1 or 2, we can enumerate the 16 R_{ijkl} and use the symmetries discussed in the text to conclude that only one component is independent. Choose that nonzero component to be R_{1212}.

We learned that two-dimensional surfaces are characterized by one scalar curvature quantity, the intrinsic Gaussian curvature K. So, the Ricci scalar and K must be identical, up to a numerical factor in two dimensions.

The Ricci scalar is the fully contracted Riemann tensor,

$$R = \sum_{iklm} g^{il} g^{km} R_{iklm}$$

If we collect all the nonzero terms on the right-hand side of this equation and use the symmetry properties of R_{iklm}, we find,

$$\begin{aligned} R &= g^{11}g^{22}R_{1212} + g^{22}g^{11}R_{2121} + g^{21}g^{12}R_{2112} + g^{12}g^{21}R_{1221} \\ &= 2(g^{11}g^{22} - g^{12}g^{12})R_{1212} \end{aligned}$$

On the right-hand side, we recognize the determinant of g^{ij}, which is the inverse of the metric g_{ij}, $\det(g^{ij}) = 1/\det(g_{ij}) \equiv 1/\det(g)$. We learn that,

$$R = 2\frac{R_{1212}}{\det(g)}$$

But we had in Section 12.1 of Chapter 12,

$$K = \frac{\det(b)}{\det(g)}$$

So, R_{1212} is essentially the determinant of the *second* fundamental form. If we work an example, the surface of a sphere as done in the problem sets, we clarify the conventions and establish,

$$R = 2K$$

So,

$$R_{1212} = \det(b)$$

Note that this observation provides an alternative proof to Gauss's Theorema Egregium sketched in Problem 12.3 because it establishes that the Gaussian curvature can be computed just from the metric and its derivatives: $K = R_{1212}/\det(g)$ and the definition and formulas for the Riemann tensor only involve intrinsic properties of the surface, the metric, and its derivatives.

If we consider an orthogonal mesh $(u \cdot v)$ on the surface, then $g_{12} = 0$ and $ds^2 = g_{11}du^2 + g_{22}dv^2 \equiv Edu^2 + Gdv^2$, in the notation of Section 12.1. For such a "simple" metric it is easy to calculate the Christoffel symbols in terms of the derivatives of E and G from Eq. (12.44) and the non-vanishing component of the Riemann tensor, R_{1212} from Eq. (12.46). Then straightforward algebra produces an explicit formula for the Gaussian curvature $K = R_{1212}/\det(g)$, in terms of the metric E and G and their derivatives,

$$K = -\frac{1}{\sqrt{EG}}\left[\frac{\partial}{\partial u}\left(\frac{1}{\sqrt{E}}\frac{\partial\sqrt{G}}{\partial u}\right) + \frac{\partial}{\partial v}\left(\frac{1}{\sqrt{G}}\frac{\partial\sqrt{E}}{\partial v}\right)\right]$$

in agreement with the result of Problem 12.3, but obtained with much, much less work!

G.3 TIDAL FORCES AND GRAVITY WAVES

In the text we argued that gravity waves effect the LIGO detector by squeezing space in one direction and elongating it in the perpendicular direction. This typically tiny effect is measured using interferometry between freely floating mirrors. The pattern of squeezing and elongation originated in the oscillating quadrupole moment of the radiating source, the merging of two black holes in the 2015 historic event. We have also seen the squeezing and elongation of space in tidal forces. Let us relate the two.

We had the equation for the deviation between two freely falling test masses in terms of the Riemann tensor, Eq. (12.84),

$$\frac{D^2\varepsilon^\alpha}{D\tau^2} = -\sum_{\beta\gamma\delta} R^\alpha_{\beta\gamma\delta} u^\beta u^\delta \varepsilon^\gamma + O(\varepsilon^2) \tag{12.84}$$

In the LIGO application, the space–time metric fluctuates slightly by the amount $\widetilde{g}_{\mu\nu}$ from the Minkowski metric $g^{(0)}_{\mu\nu}$. Let us show that the deviation between the mirrors of the LIGO detector ε^α is proportional to the fluctuation in the metric itself $\widetilde{g}_{\mu\nu}$, so ε^α provides a measure of the amplitude, direction, and frequency of the passing gravitational wave. The mirrors of the LIGO detector move slightly from rest in the frame of the device. So, in Eq. (12.84) we put $u^\alpha \approx (c, 0, 0, 0)$. In addition, the four vector ε^α between the mirrors at the ends of each arm of the detector is transverse to the z axis, the propagation direction of the wave, $\varepsilon^\alpha = (0, \varepsilon, 0, 0)$ or $(0, 0, \varepsilon, 0)$. Then Eq. (12.84) becomes,

$$\frac{\partial^2\varepsilon^\alpha}{c^2\partial t^2} = -\varepsilon R^\alpha_{0x0} \tag{G.11}$$

We know the Riemann tensor for linearized gravity from Problem 12.16, part a,

$$R_{\alpha\beta\gamma\delta} = \frac{1}{2}\left(\partial_\beta\partial_\gamma\widetilde{g}_{\alpha\delta} + \partial_\alpha\partial_\delta\widetilde{g}_{\beta\gamma} - \partial_\beta\partial_\delta\widetilde{g}_{\alpha\gamma} - \partial_\alpha\partial_\gamma\widetilde{g}_{\beta\delta}\right)$$

So, we need R_{x0x0} and R_{y0x0} in this application,

$$R_{x0x0} = \frac{1}{2}\left(\partial_0\partial_x\widetilde{g}_{x0} + \partial_x\partial_0\widetilde{g}_{0x} - \partial_0\partial_0\widetilde{g}_{xx} - \partial_x\partial_x\widetilde{g}_{00}\right) = -\frac{1}{2}\frac{\partial^2}{c^2\partial t^2}\widetilde{g}_{xx} \tag{G.12a}$$

And,

$$R_{y0x0} = \frac{1}{2}\left(\partial_0\partial_x\widetilde{g}_{y0} + \partial_y\partial_0\widetilde{g}_{0x} - \partial_0\partial_0\widetilde{g}_{xy} - \partial_y\partial_x\widetilde{g}_{00}\right) = -\frac{1}{2}\frac{\partial^2}{c^2\partial t^2}\widetilde{g}_{xy} \tag{G.12b}$$

Collecting we have,

$$\frac{\partial^2\varepsilon^x}{\partial t^2} = \frac{1}{2}\varepsilon^x_0\frac{\partial^2}{\partial t^2}\widetilde{g}_{xx}, \qquad \frac{\partial^2\varepsilon^y}{\partial t^2} = \frac{1}{2}\varepsilon^x_0\frac{\partial^2}{\partial t^2}\widetilde{g}_{xy} \tag{G.13a}$$

where ε^x_0 is the initial value of ε^x.

Similarly, if the mirrors were initially displaced in the y-direction,

$$\frac{\partial^2 \varepsilon^x}{\partial t^2} = \frac{1}{2}\varepsilon_0^y \frac{\partial^2}{\partial t^2}\tilde{g}_{xy}, \qquad \frac{\partial^2 \varepsilon^y}{\partial t^2} = -\frac{1}{2}\varepsilon_0^y \frac{\partial^2}{\partial t^2}\tilde{g}_{xx} \tag{G.13b}$$

where ε_0^y is the initial value of ε^y. We can write Eq. (G.13) in matrix form,

$$\frac{\partial^2}{\partial t^2}\begin{pmatrix} \varepsilon^x \\ \varepsilon^y \end{pmatrix} = \frac{1}{2}\frac{\partial^2}{\partial t^2}\begin{pmatrix} \tilde{g}_{xx} & \tilde{g}_{xy} \\ \tilde{g}_{xy} & -\tilde{g}_{xx} \end{pmatrix}\begin{pmatrix} \varepsilon_0^x \\ \varepsilon_0^y \end{pmatrix} \tag{G.14}$$

Since $\tilde{g}_{ij}$ has sinusoidal dependence, $\frac{\partial^2}{\partial t^2}\tilde{g}_{ij} = -\omega^2\tilde{g}_{ij}$, Eq. (G.14) can be solved,

$$\begin{pmatrix} \varepsilon^x(t) \\ \varepsilon^y(t) \end{pmatrix} = \begin{pmatrix} 1 + \frac{1}{2}\tilde{g}_{xx}(t) & \frac{1}{2}\tilde{g}_{xy}(t) \\ \frac{1}{2}\tilde{g}_{xy} & 1 - \frac{1}{2}\tilde{g}_{xx}(t) \end{pmatrix}\begin{pmatrix} \varepsilon_0^x \\ \varepsilon_0^y \end{pmatrix} \tag{G.15}$$

where we incorporated the initial values $\varepsilon^x(0) = \varepsilon_0^x$ and $\varepsilon^y(0) = \varepsilon_0^y$.

In Eq. (G.15) we see the tidal forces at work as discussed and illustrated in the text. This equation gives an alternative but identical description of the LIGO experiment as presented in the text.

Let us contrast this result to the gravitational wave's effect on a single, isolated particle described through the geodesic equation of motion,

$$\frac{d^2 x^\mu}{d\tau^2} + \sum_{\nu,\sigma}\Gamma^\mu_{\nu\sigma}\frac{dx^\nu}{d\tau}\frac{dx^\sigma}{d\tau} = 0 \tag{12.25}$$

We are interested in the x coordinate ($\mu = 1$) of one of LIGO's mirrors and are searching for linear effects $O(\tilde{g})$. Since $\Gamma^\mu_{\nu\sigma}$ and $dx^i/d\tau$ ($i = 1, 2, 3$) are each at most $O(\tilde{g})$, only the indices $\rho = \sigma = 0$ can contribute to the geodesic Eq. (12.25) at this order of accuracy. But Γ^1_{00} (and Γ^2_{00}) vanishes to first order,

$$\Gamma^1_{00} = \frac{1}{2}g^{11}\left(\partial_0\tilde{g}_{10} + \partial_0\tilde{g}_{10} - \partial_1\tilde{g}_{00}\right) = 0$$

So, $d^2x^1/d\tau^2 \approx 0$ and the mirror does not accelerate relative to the freely falling coordinate mesh. This agrees with our earlier argument in the text based on the equivalence principle. In this class of reference frames, the mirror and a bit of "dust" labeling the coordinate mesh at the position of the mirror are affected *identically* by the gravitational wave and there is no relative acceleration. On the other hand, the geodesic deviation ε^α provides an excellent method of detecting gravitational waves. This result relied on the fact that the geodesic deviation equation is proportional to the curvature tensor $R_{\alpha\beta\gamma\delta}$. In that case we saw that ε^α tracks $\tilde{g}_{\rho\sigma}$ precisely.

How can it be that ε^α is successful here while x^μ is not? The point is that ε^α depends on the positions of *two* separate masses and thus is sensitive to the curvature of space–time. Perhaps an analogy is in order. Imagine a single buoy floating in the Boston Harbor. In many situations, the buoy does not travel at all with respect to the wave it is floating on. Now consider a second buoy some distance away and attached to the first buoy with a very, very light bungee cord. It also floats on the water and does not move with respect to the water locally. Now a long wavelength, gradual wave passes through the harbor. As more water enters or recedes from the harbor, the distance between the buoys change as can be measured using the bungee cord, but the buoys themselves do not move relative to the water they are locally floating on. The distance between the buoys changes because the amount of water between them, carried in or out by the wave, changes.

We have already noted that electromagnetic waves can be observed by their effect on isolated point charges. The Lorentz force law in curved space–time reads,

$$\frac{d^2x^\mu}{d\tau^2} + \sum_{\nu,\sigma} \Gamma^\mu_{\nu\sigma} u^\nu u^\sigma = \frac{q}{m} \sum_\lambda u_\lambda F^{\lambda\mu} \tag{12.27b}$$

The force appears explicitly on the right-hand side of the equation of motion. The charge q is accelerated relative to the charge-neutral coordinate mesh by the electromagnetic field $F^{\lambda\mu}$, which might be describing a passing electromagnetic wave. No subtlety here.

Finally, when we discussed the Newtonian limit of the geodesic equation in the vicinity of a static mass M we again found Newton's second law for gravity, which predicts acceleration relative to the coordinate mesh. This works out because the presence of the static mass M at a certain space–time point breaks translation symmetry and the motion of a small test mass m can have a nonzero acceleration relative to the fixed source M of space–time's curvature. Again, no subtlety here.

INDEX

'*Note:* Page numbers followed by "f" indicate figures.'

FSC
www.fsc.org
MIX
Papier | Fördert gute Waldnutzung
FSC® C083411

Kolibri 360 GmbH
Ferdinand-Jühlke-Straße 7
99095 Erfurt, Deutschland
produktsicherheit@kolibri360.de